# Electronics V

**D C Green**

MTech CEng MIEE

Longman
Scientific &
Technical

**Longman Scientific & Technical**,
Longman Group UK Limited,
Longman House, Burnt Mill, Harlow,
Essex, CM20 2JE, England
*and Associated Companies throughout the world.*

First published 1993

GAL® , ispGAL® , E$^2$CMOS® , ULtraMOS®  are all Registered
Trademarks and Generic Array Logic™, pDS™, pLSI™, ispLSI™ are
all trademarks of Lattice Semiconductor Corporation.

ISBN 0 582 089611

**British Library Cataloguing in Publication Data**
A CIP record for this book is available from the British Library

Set by 4 in Compugraphic 10 pt Times

Printed in Malaysia by PA

# Contents

# Preface

The Business and Technician Education Council (BTEC) has produced a bank of material for students studying electronics at the Higher Certificate/Diploma levels 4 and 5. A large part of the contents of this bank has already been covered in my two previous books, *Electronics IV* and *Digital Electronic Technology*. This book has been written to cover the remainder of the BTEC bank.

The emphasis of this book is on the design of circuits rather than the analysis and on the choice of device for a particular application. Expressions for various quantities have not been derived here if the derivation has already been done in one of the previous books, instead the result is quoted without proof. However, all expressions that are met for the first time in this book have been derived. An electronic system, analogue or digital, is assumed to consist of the interconnection of a number of subsystem blocks and the emphasis has been placed upon the design of such blocks. Usually, there are several ways in which circuits may be connected together to form a complete system and the method selected in a particular case is determined by the skill, experience and personal preferences of the designer.

Traditionally, digital systems have been designed using standard 'off-the-shelf' SSI/MSI/LSI packages, but increasingly today programmable logic devices are employed. The former approach is covered in Chapter 6, while an introduction to the latter is given by Chapter 8.

Throughout the book it has been assumed that the reader already possesses a knowledge of both analogue and digital electronics of the standard reached in the first year of a BTEC Higher Certificate/Diploma course. Many worked examples are given in the text to illustrate the principles and techniques that have been discussed and a number of examples are given at the end of the book. Answers to the numerical exercises are also given.

I wish to express my thanks to GEC Plessey Semiconductors, to Lattice Semiconductor Corporation, to Motorola Ltd, and to Texas Instruments for providing me with data on their products, and for their permission to use, and reproduce in this book, their copyright material.

D. C. G.

# 1 Analogue and Digital Devices

Most modern electronic circuits consist of one, or more, integrated circuits (ICs) that are connected together to provide a wanted circuit function. A *linear*, or analogue, IC is strictly one whose output signal has a linear relationship with its input signal, but the term is also used to include all other ICs that are not digital types. Thus, linear ICs include amplifiers, analogue multiplexers, analogue-to-digital converters, and digital-to-analogue converters, timers and voltage regulators. Usually, a number of passive components, capacitors and resistors, and, less often, inductors, must be connected to the appropriate package pins. Discrete transistors are mainly employed to provide specialized circuit functions, to increase the power output of a linear IC, and to construct 'one-off' circuits.

A *digital* IC is one whose output signal(s) can only be at either the logical 1, or the logical 0, voltage level and must bear a logical relationship to the input signal(s). Digital ICs come in various levels of complexity and they are categorized by the number of gates, or equivalent circuits, that they contain. Those ICs with fewer than 10 gates are known as *small-scale integration* (SSI) devices; those with between 10 and 100 gates are called *medium-scale integration* (MSI) devices; ICs with between 100 and 5000 gates are known as *large-scale integration* (LSI) circuits; lastly, *very large-scale integration* (VLSI) devices contain the equivalent of more than 5000 gates. Such a wide variety of MSI/LSI/VLSI digital ICs are available that SSI devices are rarely used for other than simple interconnection purposes.

System design using MSI or LSI ICs consists of the correct interconnection of such chips and consequently a good knowledge of their functions is necessary. Usually when LSI devices are employed there are some interfacing and/or decoding tasks that can best be carried out using standard MSI ICs. Digital system design using VLSI devices can also use any available standard ICs but also often involves the use of ICs known as application specific circuits (ASICs). These devices include full- and semi-custom circuits as well as programmable circuits such as programmable logic arrays (PLAs), programmable array logic (PAL), etc. Nowadays, ASICs are also available to perform both analogue circuit and mixed analogue/digital functions.

**Logic Technologies**

The main logic technologies that still find application in modern electronic circuitry are *transistor—transistor logic* (TTL), *complementary—symmetry metal oxide semiconductor* (CMOS) and *emitter-coupled logic* (ECL). The logic technology TTL has several subclasses or families; namely, standard TTL (74 series), low-power Schottky (LS), advanced low-power Schottky (ALS), advanced Schottky (AS) and FAST (F). The 74 series are rarely employed nowadays. The technology of CMOS includes the standard 4000 series, high-speed CMOS (HCMOS) and advanced CMOS; there are two subgroups of each; the HC family devices are HCMOS with CMOS compatible inputs and HCT devices are HCMOS with TTL compatible inputs, while advanced CMOS devices may have CMOS compatible inputs (AC) or TTL compatible inputs (ACT). The AC and ACT devices combine the high speed of the AS and ALS TTL families with the low power dissipation of CMOS. Finally, FACT is an advanced family of CMOS devices that has been introduced by both Motorola and National Semiconductor. In many respects it has a similar performance to HCMOS but it is both faster and able to sink/source a higher output current.

The choice of the logic family for a particular application is usually made after a careful consideration of the following criteria: (*a*) switching speed; (*b*) power dissipation; (*c*) cost; (*d*) availability; (*e*) noise immunity. The main characteristics of the logic families are listed in Table 1.1 in which typical figures are quoted.

The main advantage offered by CMOS technology is its low static power dissipation. The standard 4000 series CMOS devices are of low speed, but advanced CMOS logic (ACL) combines the low power dissipation of CMOS technology with the high speed of bipolar

**Table 1.1**

| Family | Propagation delay (ns) | Static power dissipation (mW) | Noise immunity (V) | Supply voltage (V) | Maximum clock frequency (MHz) | Speed— power product (pJ) | Minimum output current (mA) | Maximum input current (mA) |
|---|---|---|---|---|---|---|---|---|
| 74 | 10 | 10 | 0.4 | 5 | 35 | 100 | 16 | 1.6 |
| LS | 10 | 2 | 0.3 | 5 | 40 | 20 | 8 | 0.4 |
| AS | 1.5 | 8 | 0.3 | 5 | 200 | 13 | 20 | 0.5 |
| ALS | 4 | 1 | 0.4 | 5 | 70 | 5 | 8 | 0.1 |
| 4000 | 30 | 0.001 | 1.5 | 3–15 | 12 | 11 | 1.6 | 0.001 |
| 74HC | 8 | $25 \times 10^{-7}$ | 0.2 | 2–6 | 50 | 1.4 | 4 | 0.001 |
| 74HCT | 8 | $25 \times 10^{-7}$ | 0.25 | 4.5–5.5 | 50 | 0.02 | 4 | 0.001 |
| AC | 4.8 | $25 \times 10^{-7}$ | 0.25 | 2–6 | 160 | 0.01 | 24 | 0.001 |
| ACT | 4.8 | $25 \times 10^{-7}$ | 1.25 | 5 | 160 | 0.01 | 24 | 0.001 |
| ECL | | | | | | | | |
| 10k | 2 | 34 | | −5.2 | 125 | 68 | | |
| 100k | 0.95 | 57 | | −4.2 to −4.8 | 400 | 54 | | |
| 100k300 | 0.9 | 29 | | −4.2 to −5.7 | 400 | 26 | | |
| ECLinPOS | 0.5 | 30 | | −4.2 to −4.8 | 700 | 15 | | |

technology. The 4000 series is no longer used for new designs. All CMOS/HCMOS devices dissipate negligible power when they are not switching from one state to the other. When a device is switched its power dissipation increases in direct proportion with increase in the clock frequency, and at some frequency it will become greater than the power dissipation of the corresponding LS device. For example, the $25 \times 10^{7}$ mW static power dissipation of a 74HC device rises to about 0.17 mW at 100 kHz. However, system level power savings are still obtained at most clock frequencies.

Low power dissipation is a very important consideration in a digital system since it saves power, reduces costs, simplifies the power supplies — perhaps allowing a battery to be employed — and increases reliability because less heat is generated. Also, a cooler PCB (printed circuit board) allows the component packing density to be greater with consequent reductions in both equipment size and weight.

In the TTL logic family the AS and ALS devices are optimized for high speed and low power dissipation respectively and FAST devices have a performance that is a compromise between AS and ALS. The devices are pin-for-pin compatible with the corresponding circuits in the 74 and LS series.

It is an advantage if the power supply voltage to a digital IC is of low voltage since it then allows a lithium battery to be used to provide a back-up supply. Also, many memory chips use voltages which are below 5 V. The LS devices operate with a supply voltage $V_{CC}$ of 5 V $\pm$ 5%, HCT devices with $V_{DD} = 4.5-5.5$ V, and HC devices with $V_{DD} = 2-6$ V. The older 4000 series devices can operate from a wide range of supply voltages such as $3-15$ V. Devices in the ECL family suffer from the disadvantage of requiring a negative supply voltage. The noise immunity of an IC is numerically equal to a percentage of its power supply voltage. For LS devices this is:

(a) low level, 8% of $V_{CC} = (8/100) \times 5 = 0.4$ V;
(b) high level, 14% of $V_{CC} = (14/100) \times 5 = 0.7$ V, giving a noise margin of $0.7 - 0.4 = 0.3$ V.

For HC devices the corresponding figures are (for $V_{DD} = 5$ V):

(a) low level, 18% of $V_{DD} = (18/100) \times 5 = 0.9$ V;
(b) high level, 28% of $V_{DD} = (28/100) \times 5 = 1.4$ V, giving a noise margin of $1.4 - 0.9 = 0.5$ V.

The 4000 series CMOS devices are relatively slow to switch but the speed of the HCMOS versions is comparable with that of the LS family, but is slower than both the AS and the ALS families. The HC devices are pin-for-pin compatible with CMOS 4000 devices and HCT ICs are compatible with TTL devices. This arrangement allows HCMOS ICs to be direct replacements for CMOS/LS chips. Advanced CMOS logic (ACL) devices have their power supply voltage and earth pins at the centres of each side of the package, e.g. pin 5 for earth and pin 19 for $V_{DD}$ for a 24-pin package. This change from the normal layout reduces package self-inductance and hence minimizes the noise voltages that are induced when a device switches. It is also claimed that the layout of PCB boards is simplified.

For very fast applications, such as minicomputers, work stations, radar, etc. the ECL logic family is generally employed. This technology has always been associated with high power dissipation, a narrow supply voltage range, and difficult design, and hence its use has been restricted. A new ECL family, known as the F100K 300 series, has been introduced by National Semiconductor; it offers between 30 and 50% less power dissipation than standard ECL and an expanded supply voltage range (see Table 1.1). Motorola's ECLinPOS is a 500 ps, 800 MHz logic family that is compatible with 10K/100K ECL devices that combines a very high speed with lower power dissipation than the standard ECL family.

Each of the logic families contains a wide variety of different digital circuits ranging from simple gates, through flip-flops, counters and registers, to decoders, multiplexers, etc.

## BiCMOS

BiCMOS is another logic family that provides devices, such as latches, registers, line drivers and line transceivers, that are designed for use in three-state bus interface applications. When one IC is to drive another device that is some distance away a current must be provided and this is a requirement that a CMOS device is unable to deliver. A BiCMOS device consists of a CMOS gate whose output is connected to a pair of bipolar transistors. When the CMOS gate switches so do the bipolar transistors and they are able to provide the current that must be delivered to the bus to drive the next IC. At the other end of the interconnection the signal is converted back to CMOS voltage levels. Thus BiCMOS combines the best features of bipolar and CMOS technology and is able to provide an output current of up to 48/64 mA.

Many SSI/MSI circuits, from simple gates, through flip-flops and counters, to decoders and adders are available in most, if not all, of the logic families. A representative sample is shown by Table 1.2.

N-channel mos (nmos) devices were often used in LSI and VLSI

**Table 1.2**

| Circuit | 74 | LS | ALS | F | AS | HC | HCT | AC | ACT |
|---|---|---|---|---|---|---|---|---|---|
| 74 dual D flip-flop | X | X | X | X | X | X | X | X | X |
| 109 dual J−K flip-flop | X | X | X | X | X | X | — | X | X |
| 138 3-to-8 line decoder | — | X | X | X | X | X | X | X | X |
| 151 8-to-1 line data selector | X | X | X | X | X | X | — | X | X |
| 190 synchronous 4-bit decade counter | X | X | X | X | — | X | — | X | X |
| 283 4-bit full-adder | X | X | — | X | — | X | — | — | — |

chips, because they have the highest component density of any technology, but they are being replaced by HCMOS devices.

**Data Sheets**

The manufacturers of transistors and ICs publish information in the form of *data sheets* to assist the circuit designer to choose the most suitable device for a particular application. The data sheet for an IC provides all the necessary technical information about the device in the form of tables, graphs and diagrams. Before the information given can be understood and made use of, it is necessary for the user to understand the operation of the circuit and the meanings of the various terms that are employed. The technical specification of an IC, obtained from its data sheet, ought not to be the only factor which is considered in the selection of a particular device. Also of importance are such factors as its cost, its ready availability, and whether or not it is *second-sourced*. Many ICs, although originally introduced by one manufacturer, are now produced by more than one firm and they are then said to be second-sourced. Second-sourcing implies pin compatibility although the parameters of the device may not be identical. Second-sourcing is advantageous to the user since it means that future supplies of the IC are better assured. The prime example of second-sourcing in the linear field is the 741 op-amp; most, if not all, semiconductor manufacturers offer a version of this device.

The first step in the selection of an IC is to define the performance that will be expected from the chosen device. This will involve calculations of the maximum values of current, power and voltage that the device may have to handle, the maximum frequency and voltage of the input and output signals, etc. A list should be made of these values and this information can then be consulted as short-form data are studied. Short-form data give figures for the most important parameters of various ICs that perform the same function and it allows a short list of possible devices to be drawn up. For example, short-form data for some commonly used op-amps are given by Tables 1.3–1.8.

The main features of the different devices can then be compared and some possible choices put on to a short list. The full data sheets of the short-listed ICs should then be obtained so that a final choice can be made.

**Linear Integrated Circuits**

The data sheet for a linear IC will always start with a brief description of the device and its particular features, e.g. bandwidth, common-mode rejection ratio, open-loop gain, as well as both the pin connections of the device and its schematic diagram. The next section of the data sheet gives the absolute maximum ratings which must not be exceeded, even for a short period of time, or the IC may be damaged, particularly if more than one is simultaneously exceeded. These ratings include the maximum supply voltage, and the maximum internal power dissipation as well as the ambient temperature range over which the device may operate. The three most common temperature ranges are: (i) commercial, $0-70$ °C, (ii) industrial, $-25$ °C to 85 °C; (iii) military, $-55$ °C to 125 °C. A section headed 'electrical characteristics' then follows and this presents most of the information that is required for a circuit design. Quoted at the top of these data is the supply voltage and the ambient temperature at which the measurements were made. This section is usually followed by the 'operating characteristics' that give the a.c. performance of the device and are measured at the same voltage and temperature. Lastly, a number of graphs may be given to illustrate how various parameters of the IC vary with the supply voltage, frequency and ambient temperature. The curves show the behaviour of a typical device but since there is always a spread in the values of parameters they are not guaranteed.

Most linear ICs employ bipolar techniques. Some op-amps have field-effect transistor (FET) input circuits to provide them with a very high input impedance and low input offset and bias currents. These op-amps are known as BiFETs if the input FETs are junction types, and as BiMOS if the input circuit uses MOSFETs. Also, CMOS op-amps are available; these offer the usual CMOS advantages but suffer from the disadvantage that their input offset voltage drifts with time, with change in temperature, and/or with supply voltage variations. This problem is overcome by a more recent technology known as LinCMOS, which includes a whole family of devices that are able to operate from a single-polarity, low-voltage power supply with minimal power dissipation. The LinCMOS family includes op-amps, voltage comparators, timers and both analogue-to-digital and digital-to-analogue converters. Advanced LinCMOS technology is now also available and this allows both analogue and digital circuitry to be combined within a single silicon chip.

## Op-amps

An example of an op-amp's data sheet is given in Fig. 1.1; it is for the Texas Instruments uA 741C op-amp. The various terms that are employed in the data sheet should already be understood but they are outlined here.

### Supply Voltage

The majority of op-amps require a dual power supply, such as $\pm$ 15 V, but some types have been designed to operate from a single-polarity supply, e.g the LM 124. Op-amps that require a dual supply can be operated from a single-polarity supply by using the arrangements shown in Figs 1.2(a) and (b). In the inverting amplifier circuit of Fig. 1.2(a) the non-inverting terminal of the op-amp is held

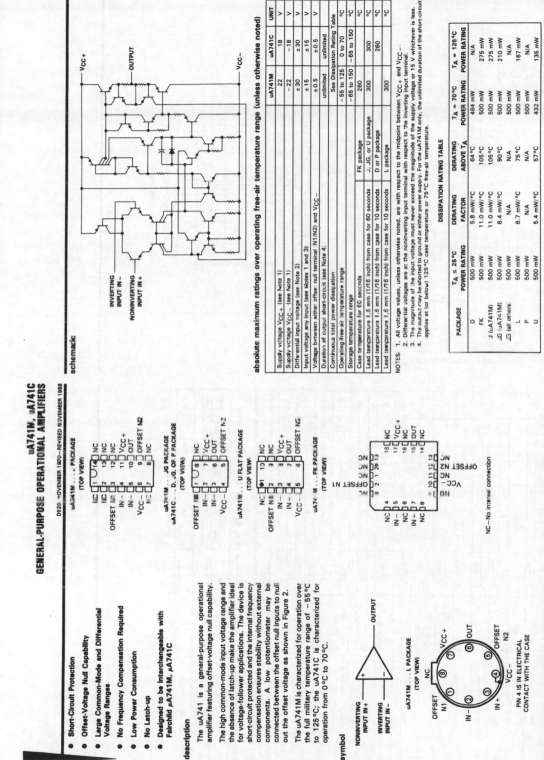

**Fig. 1.1** Data sheet for uA 741C op-amp (*courtesy of Texas Instruments*)

## TYPICAL CHARACTERISTICS

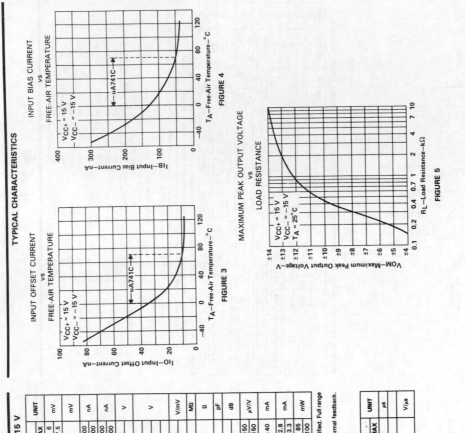

**FIGURE 3** — INPUT OFFSET CURRENT vs FREE-AIR TEMPERATURE
($V_{CC+} = 15$ V, $V_{CC-} = -15$ V; $I_{IO}$—Input Offset Current—nA vs $T_A$—Free-Air Temperature—°C; uA741C)

**FIGURE 4** — INPUT BIAS CURRENT vs FREE-AIR TEMPERATURE
($V_{CC+} = 15$ V, $V_{CC-} = -15$ V; $I_{IB}$—Input Bias Current—nA vs $T_A$—Free-Air Temperature—°C; uA741C)

**FIGURE 5** — MAXIMUM PEAK OUTPUT VOLTAGE vs LOAD RESISTANCE
($V_{CC+} = 15$ V, $V_{CC-} = -15$ V, $T_A = 25$°C; $V_{OM}$—Maximum Peak Output Voltage—V vs $R_L$—Load Resistance—kΩ)

### electrical characteristics at specified free-air temperature, $V_{CC+} = 15$ V, $V_{CC-} = -15$ V

| PARAMETER | TEST CONDITIONS | | uA741M MIN | TYP | MAX | uA741C MIN | TYP | MAX | UNIT |
|---|---|---|---|---|---|---|---|---|---|
| $V_{IO}$ Input offset voltage | $V_O = 0$ | 25°C | | 1 | 5 | | 1 | 6 | mV |
| | $V_O = 0$ | Full range | | | 6 | | | 7.5 | mV |
| $\Delta V_{IO(adj)}$ Offset voltage adjust range | $V_O = 0$ | 25°C | | ±15 | | | ±15 | | mV |
| $I_{IO}$ Input offset current | $V_O = 0$ | 25°C | | 20 | 200 | | 20 | 200 | nA |
| | $V_O = 0$ | Full range | | | 500 | | | 300 | nA |
| $I_{IB}$ Input bias current | $V_O = 0$ | 25°C | | 80 | 500 | | 80 | 500 | nA |
| | $V_O = 0$ | Full range | | | 1500 | | | 800 | nA |
| $V_{ICR}$ Common-mode input voltage range | | 25°C | ±12 | ±13 | | ±12 | ±13 | | V |
| | | Full range | ±12 | | | ±12 | | | V |
| $V_{OM}$ Maximum peak output voltage swing | $R_L = 10$ kΩ | 25°C | ±12 | ±14 | | ±12 | ±14 | | V |
| | $R_L = 10$ kΩ | Full range | ±12 | | | ±12 | | | |
| | $R_L = 2$ kΩ | 25°C | ±10 | ±13 | | ±10 | ±13 | | V |
| | $R_L \geq 2$ kΩ | Full range | ±10 | | | ±10 | | | |
| $A_{VD}$ Large-signal differential voltage amplification | $R_L \geq 2$ kΩ, $V_O = \pm10$ V | 25°C | 50 | 200 | | 20 | 200 | | V/mV |
| | | Full range | 25 | | | 15 | | | |
| $r_i$ Input resistance | | 25°C | 0.3 | 2 | | 0.3 | 2 | | MΩ |
| $r_o$ Output resistance | $V_O = 0$, See Note 5 | 25°C | | 75 | | | 75 | | Ω |
| $C_i$ Input capacitance | | 25°C | | 1.4 | | | 1.4 | | pF |
| CMRR Common-mode rejection ratio | $V_{IC} = V_{ICR}$ min | 25°C | 70 | 90 | | 70 | 90 | | dB |
| $k_{SVS}$ Supply voltage sensitivity ($\Delta V_{IO}/\Delta V_{CC}$) | $V_{CC} = \pm9$ V to ±15 V | 25°C | | 30 | 150 | | 30 | 150 | µV/V |
| | | Full range | | | 150 | | | 150 | |
| $I_{OS}$ Short-circuit output current | | 25°C | | ±25 | ±40 | | ±25 | ±40 | mA |
| $I_{CC}$ Supply current | No load, $V_O = 0$ | 25°C | | 1.7 | 2.8 | | 1.7 | 2.8 | mA |
| | No load, $V_O = 0$ | Full range | | | 3.3 | | | 3.3 | |
| $P_D$ Total power dissipation | No load, $V_O = 0$ | 25°C | | 50 | 85 | | 50 | 85 | mW |
| | | Full range | | | 100 | | | 100 | |

† All characteristics are measured under open-loop conditions with zero common-mode input voltage unless otherwise specified. Full range for uA741M is −55°C to 125°C and for uA741C is 0°C to 70°C.

NOTE 5: This typical value applies only at frequencies above a few hundred hertz because of the effects of drift and thermal feedback.

### operating characteristics, $V_{CC+} = 15$ V, $V_{CC-} = -15$ V, $T_A = 25$°C

| PARAMETER | TEST CONDITIONS | uA741M MIN | TYP | MAX | uA741C MIN | TYP | MAX | UNIT |
|---|---|---|---|---|---|---|---|---|
| $t_r$ Rise time | $V_I = 20$ mV, $R_L = 2$ kΩ, $C_L = 100$ pF, See Figure 1 | | 0.3 | | | 0.3 | | µs |
| Overshoot factor | | | 5% | | | 5% | | |
| SR Slew rate at unity gain | $V_I = 10$ V, $R_L = 2$ kΩ, $C_L = 100$ pF, See Figure 1 | | 0.5 | | | 0.5 | | V/µs |

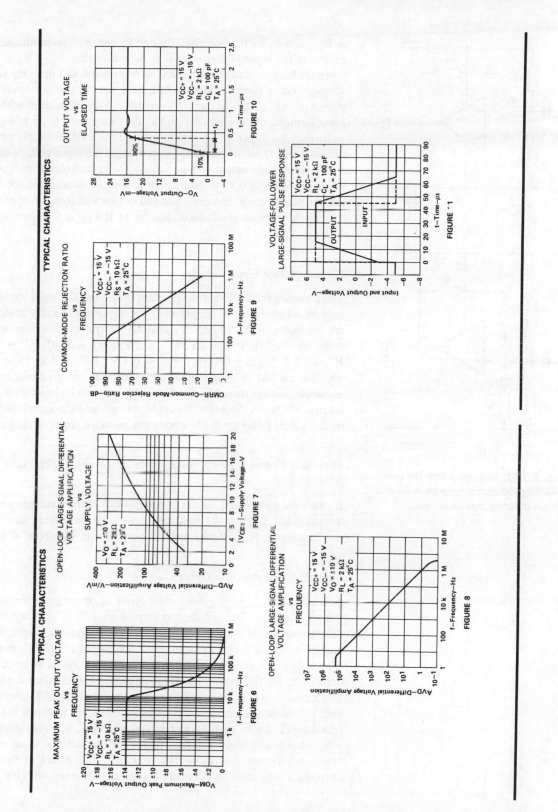

TYPICAL CHARACTERISTICS

MAXIMUM PEAK OUTPUT VOLTAGE
vs
FREQUENCY

FIGURE 6

OPEN-LOOP LARGE-SIGNAL DIFFERENTIAL
VOLTAGE AMPLIFICATION
vs
SUPPLY VOLTAGE

FIGURE 7

OPEN-LOOP LARGE-SIGNAL DIFFERENTIAL
VOLTAGE AMPLIFICATION
vs
FREQUENCY

FIGURE 8

TYPICAL CHARACTERISTICS

COMMON-MODE REJECTION RATIO
vs
FREQUENCY

FIGURE 9

OUTPUT VOLTAGE
vs
ELAPSED TIME

FIGURE 10

VOLTAGE-FOLLOWER
LARGE-SIGNAL PULSE RESPONSE

FIGURE 11

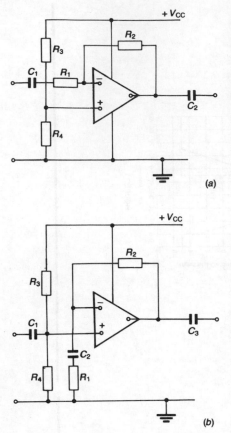

**Fig.** 1.2 (*a*) Inverting and (*b*) non-inverting op-amps operated from a single-polarity power supply

at $V_{CC}/2$ volts by the potential divider $R_3$ and $R_4$. To minimize the effects of the input bias current it is necessary that $R_3 = R_2 = R_4$, when the bias currents flowing into each terminal will drop the same voltage and so these voltages will cancel out. In the non-inverting circuit of Fig. 1.2(*b*), $R_3$ and $R_4$ are effectively in parallel with the input terminal of the amplifier and so they may need to be of higher value than $R_2$. When an op-amp is selected for a single-supply circuit two parameters are of particular importance; the input voltage range and the maximum output voltage swing. The input voltage must often be taken down to 0 V to ensure that low-level signals are not lost, and the nearer the maximum input and output voltages can approach $V_{CC}$ the better (this is of more importance if $V_{CC}$ is 5 V than if it is 15 V).

*Common-mode Input Voltage*

The common-mode input voltage states how close the input terminals may be taken to the ± supply voltages without adversely affecting the operation of the device. Typical figures for three different op-amps are: (i) 741, $V^+ - 2$ V, $V^- + 2$ V; (ii) CA 3140, $V^+ - 2$ V, $V^- - 0.5$ V; (iii) LF 351, $V^+ + 0.1$ V, $V^- + 2$ V. Note that whereas the 741 op-amp's input terminals may not go closer to the ± supply voltage than 2 V, the CA 3140 can have its input terminals taken 0.5 V below the negative supply voltage, and the input terminals of the LF 351 can go 0.1 V above the positive supply voltage.

*Maximum Common-mode Voltage and Maximum Differential Input Voltage*

If either the differential, or the common-mode, input voltages exceed the value specified by the data sheet an excessive current may flow and damage the IC. The maximum values are typically about ± 30 V.

*Input Bias Current and Offset Current*

The input circuits of the three main types of op-amp are shown by Figs 1.3(*a*)–(*c*). The bipolar op-amp, Fig. 1.3(*a*), uses a bipolar differential amplifier input stage and requires input bias currents $I_B^+$ and $I_B^-$ to forward bias the transistors. Neither the BiFET op-amp, Fig. 1.3(*b*), nor the BiMOS op-amp, Fig. 1.3(*c*), needs an input bias current. If the two input bipolar transistors were identical the input bias currents would be equal to one another and would cancel out but, in practice, there is always some difference between the two transistors and so a net bias current always flows. The data sheet for a bipolar op-amp gives a typical figure for the average value of the two bias currents, i.e. $I_B = (I_B^+ + I_B^-)/2$. The effect of the bias currents is overcome by connecting a resistor in series with the non-

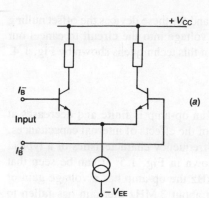

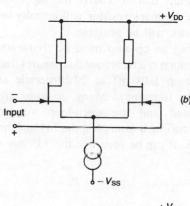

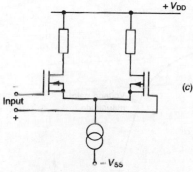

**Fig. 1.3** Input circuit of op-amps: (a) bipolar, (b) BiFET, (c) BiCMOS

inverting terminal of the device. This resistance should be of such a value that both terminals have the same driving resistance and hence equal voltages are dropped across them by the bias currents. The bias current effect is also reduced by keeping the resistance values as low as possible and, of course, by selecting an op-amp with a sufficiently small input bias current.

The difference between the two input bias currents is known as the *input offset current* $I_{OS}$ and it is usually less than $0.25I_B$. The 741 op-amp has, at normal room temperature, an input bias current $I_B$ of 80 nA typically and 500 nA maximum, and an input offset current $I_{OS}$ of 3 nA typically, and 30 nA maximum. Some op-amps are manufactured to have low values of input bias and offset currents; for example, the LM 308 typically has $I_B = 1.5$ nA and $I_{OS} = 0.2$ nA and the LM 11 typically has $I_B = 25$ pA and $I_{OS} = 0.5$ pA. The BiFET op-amps have very low values of input bias and offset currents since the gate currents of the input transistors are very small. The LF 355 has $I_B = 30$ pA and $I_{OS} = 3$ pA and the TL 070 has $I_B = 30$ pA and $I_{OS} = 5$ pA. The BiMOS op-amps have even lower values for these parameters; for example, the CA 3130 has $I_B = 2$ pA and $I_{OS} = 0.1$ pA typically, and the CA 3080 has $I_B = 15$ pA and $I_{OS} = 5$ pA typically.

The average drift of the input bias current, or of the input offset current, is the ratio of a change in that current to the change in the ambient temperature that caused it, expressed in nA/°C.

### Input Offset Voltage

The output voltage of an op-amp may not be zero even when the input voltage is zero and there are no input bias, or offset, current effects. The unwanted voltage appears at the output terminals because the base-emitter (or gate-source) voltages of the input transistors are not equal to one another. It is convenient to suppose that the offset voltage is the result of an *input offset voltage* $V_{OS}$ that has been applied in series with either of the input terminals of the op-amp. Data sheets give typical and maximum values at room temperature for the input offset voltage. For the 741 these values are $V_{OS} = 2$ mV typical and 6 mV maximum. Like the input bias and offset currents, $V_{OS}$ is temperature dependent and some data sheets may quote drift figures for $\Delta V_{OS}/\Delta t$ in μV/°C. Generally FET input op-amps have worse input offset voltages than bipolar types and also worse drift.

The combined effects of the input offset current and the input offset voltage can be reduced by the use of *offset nulling*. This, essentially, consists of balancing the input stage of the op-amp by means of an external potentiometer. Many op-amps, including the 741, are provided with two nulling terminals to which the potentiometer can be connected. Details are always given in the data sheet for the device (see Fig. 1.1 for the 741 arrangement). Some other op-amps are not

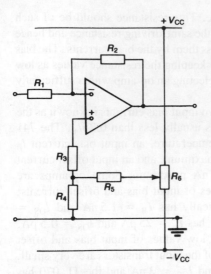

**Fig. 1.4** Offset nulling

provided with nulling terminals and for these devices the offset nulling is achieved by injecting a d.c. voltage into the circuit to cancel out the offset voltage. An example of this technique is shown by Fig. 1.4.

*Frequency Response*

The open-loop voltage gain of an op-amp is finite and decreases at the higher frequencies because of the effects of internal capacitances. The gain/frequency and phase/frequency characteristics of a typical uncompensated op-amp are shown in Fig. 1.5. It can be seen that at frequencies up to about 40 kHz the op-amp has a voltage gain of 100 dB ($10^6$) $\angle 0°$, but that at about 3 MHz the gain has fallen to 90 db (31 623) $\angle 180°$. This means that at 3 MHz the supposed negative feedback applied via the feedback resistor will actually be positive feedback and so the circuit will be unstable.

To prevent this from happening an op-amp must be *frequency compensated*. Frequency compensation is a technique that ensures that the open-loop gain of the op-amp falls off at 20 dB/decade at frequencies above a certain low frequency. Many op-amps are internally frequency compensated and the open-loop voltage gain/frequency characteristics of two such examples, the 741 and the RC 4136, are shown in Fig. 1.6. It can be seen that the 741 has a

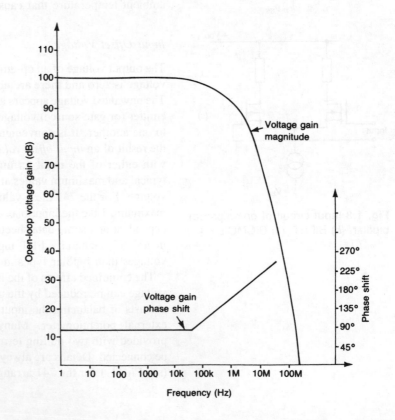

**Fig. 1.5** Gain–frequency and phase–frequency characteristics of an uncompensated op-amp

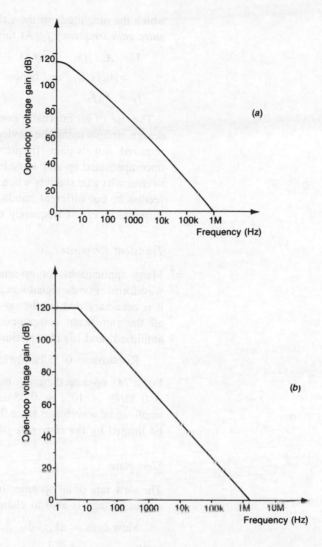

**Fig. 1.6** Gain–frequency characteristic of (a) 741 and (b) RC 4136 op-amps

large gain at 0 Hz but that at about 5 Hz it starts to roll off at $-20$ dB/decade and at 1 MHz the gain has fallen to unity, or 0 dB. At any frequency below 1 MHz the product of the gain and the frequency, known as the *gain–bandwidth product*, always has the same value, namely $1 \times 10^6$. Most general-purpose op-amps have a gain–bandwidth product somewhere in the region 500 kHz to 10 MHz but some more expensive devices may have gain–bandwidth products of many hundreds of megahertz. The gain–frequency characteristic of an op-amp can be written in the form

$$A_v = A_0/[1 + j(f/f_3)] \tag{1.1}$$

where $A_0$ is the gain at 0 Hz, and $f_3$ the frequency at which the gain has fallen by 3 dB from its maximum value $A_0$. The frequency at which the magnitude of the gain has fallen to unity is known as the *unity gain frequency* $f_t$. At this frequency

$$1 = A_0/\sqrt{(1 + f_t^2/f_3^2)}$$

$$\simeq A_0 f_3/f_t, \quad \text{and hence}$$

$$f_t = A_0 f_3 \tag{1.2}$$

The use of an internally compensated op-amp makes it easier to design circuits using the device and fewer external components are required, but its gain—frequency characteristic will be limited. An uncompensated op-amp must have some external components added to ensure its gain stability when used in a circuit that employs negative feedback, but different bandwidths can be obtained by the use of different levels of frequency compensation.

*Transient Response*

Many applications for op-amps involve rectangular and/or pulse waveforms. For the signal waveform to be amplified without distortion it is necessary that (*a*) the op-amp's bandwidth is wide enough for all the significant components contained in the waveform to be amplified, and (*b*) the risetime of the op-amp is fast enough.

$$\text{Risetime} = 0.35/\text{bandwidth} \tag{1.3}$$

For a 741 op-amp the unity bandwidth is 1 MHz and so its risetime is $0.35/(1 \times 10^6) = 0.35\,\mu\text{s}$. This figure remains true only for small-signal waveforms since the risetime of larger-voltage pulses will be limited by the *slew rate* of the op-amp.

*Slew Rate*

The slew rate of an op-amp, in V/$\mu$s, is the fastest rate at which its output voltage is able to change. Thus

$$\text{Slew rate} = dV_{\text{out}}/dt_{(\text{max})} \tag{1.4}$$

When operated in the inverting mode the slew rates in the positive and negative directions are equal to one another. But when the op-amp is operated in the non-inverting mode the two slew rates are not equal to one another. Some data sheets specify the slew rate for each direction of voltage change while others quote the average value. A 741, for example, has a slew rate of 0.5 V/$\mu$s and so it will take $10/0.05 = 20\,\mu\text{s}$ to complete a 10 V output voltage swing.

When the op-amp drives a capacitive load the maximum slew rate attainable is determined by the current available to charge the load capacitance. Then slew rate $= dV_{\text{out}}/dt = I/C$.

*Example 1.1*

A 741 op-amp is to be used as an amplifier of square waves whose amplitude varies between 0 and 10 V. Calculate the maximum closed-loop gain that may be obtained.

*Solution*

Required risetime $= (9 - 1)/(0.5 \times 10^6) = 16\,\mu s$. Therefore,

$$16 \times 10^{-6} = 0.35/f \quad \text{and} \quad f = 0.35/(16 \times 10^{-6}) = 21.875\,\text{kHz}.$$

Thus,

$$1 \times 10^6 = A_{v(F)} \times 21\,875 \quad \text{and} \quad A_{v(F)} = 45.7. \qquad (Ans.)$$

### Full-power Bandwidth

The full-power bandwidth of an op-amp refers to the effect of slew rate upon the performance of an op-amp when it is handling a sinusoidal signal.

$$v_{out} = V_{out}\sin 2\pi ft$$

$$\mathrm{d}v_{out}/\mathrm{d}t = 2\pi fV_{out}\cos 2\pi ft \quad \text{and} \quad \mathrm{d}v_{out(max)} = 2\pi fV_{out}$$

To avoid distortion of the output signal waveform caused by slew rate limitation it is necessary that $fV_{out} <$ (slew rate)$/2\pi$.

This means that there is a trade-off between the frequency and the amplitude of the output signal voltage. For the 741 op-amp with $V_{CC} = 15\,\text{V}$, the maximum output voltage without waveform distortion at different frequencies is shown by Table 1.3.

The *full-power bandwidth* is the highest frequency at which the maximum undistorted output voltage can be obtained. Thus

$$\text{Full-power bandwidth} = (\text{slew rate})/2\pi V_{out(max)} \qquad (1.5)$$

A lower supply voltage permits a higher supply current — and hence a wider full-power bandwidth — without exceeding the rated power dissipation of the op-amp. When driving a capacitive load $C$, the full-power bandwidth is

$$\text{Full-power bandwidth} = I/(2\pi CV_{out})$$

### Power Supply Rejection Ratio

Because the op-amp's circuitry is not perfectly symmetrical any change in the power supply voltage will produce a change in the output voltage. The *power supply rejection ratio* (PSRR), in $\mu$V/V, of an op-amp gives the change in the offset voltage caused by a change in the supply voltage, i.e.

$$\text{PSRR} = \Delta V_{OS}/\Delta V_{CC} \qquad (1.6)$$

Typically, the 741 has a PSRR of $30\,\mu$V/V. If the regulation of the power supply is good the effect of the PSRR on the performance of a circuit is usually small. If there should be some ripple on the

**Table 1.3**

| Frequency (kHz) | 1 | 2 | 5 | 10 | 25 | 50 | 100 | 500 | 1000 |
|---|---|---|---|---|---|---|---|---|---|
| Maximum output voltage (V) | $V_{CC}$ | $V_{CC}$ | $V_{CC}$ | 7.96 | 3.18 | 1.59 | 0.8 | 0.16 | 0.08 |

power supply line it will produce a noise output voltage. If, for example, the supply to a 741, connected to have a closed-loop gain of $-50$, has 0.08 V ripple then $V_{OS} = 0.08 \times 30 = 2.4\,\mu V$ and so the output ripple voltage will be $50 \times 2.4 = 120\,\mu V$. The PSRR of an op-amp is not a constant quantity but instead it varies with frequency in a similar way to the open-loop voltage gain.

### Common-mode Rejection Ratio

An op-amp should only produce an output voltage that is proportional to the *difference* between the voltage applied to its inverting and non-inverting terminals. However, when both the input voltages are changed by the same amount $\Delta V_{CM}$, so that their difference remains constant, the operating points of the two input transistors will be altered causing a change in the input offset voltage to occur. This means that an op-amp will also respond to a change in the common-mode input voltage. This is, of course, an undesirable effect and the ability of the op-amp to reject the common-mode voltage is expressed by its *common-mode rejection ratio* (CMRR).

$$CMRR = \Delta V_{OS}/\Delta V_{CM} \qquad (1.7)$$

in $\mu V/V$ or, more commonly, in decibels. Alternatively, the CMRR can be expressed as

$$CMRR = 20\,\log_{10}\,[\text{difference-mode gain}]/ \\ [\text{common-mode gain}]\,\text{db} \qquad (1.8)$$

and this is typically about 100 dB.

### Input and Output Impedances

An op-amp has two input impedances. One, the differential input impedance $Z_D$, is the impedance between the inverting and the non-inverting terminals. The other impedance, known as the common-mode input impedance $Z_C$, is between either terminal and earth. The two impedances are shown by Fig. 1.7. At low frequencies the impedances are purely resistive and of constant value. At higher frequencies, however, the impedances become increasingly capacitive and their magnitude falls. Conversely, the output impedance, which is purely resistive at low frequencies, becomes increasingly inductive at higher frequencies and its magnitude increases with frequency. Most data sheets specify the d.c. resistive part of these impedances and often, since it is very large, omit mention of the common-mode impedance. When a data sheet just quotes 'input resistance' it is always the differential-mode impedance that is referred to. Sometimes the input capacitance of the op-amp is also given.

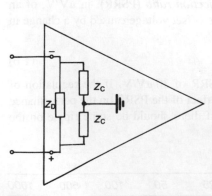

**Fig. 1.7** Input impedances of an op-amp

### Output Voltage Swing

The output voltage swing is the maximum variation in the output voltage that can be obtained. It is limited by the positive and negative saturation voltages $V_{O(SAT)}^+$ and $V_{O(SAT)}^-$. It is usually specified for a 2000 Ω load and is 1 or 2 volts less than the $\pm V_{CC}$ supply voltage. However, some op-amps are available whose output voltage may be varied to within a few millivolts of the supply voltage value.

### Output Short-circuit Current

To prevent an op-amp suffering damage caused by excessive current flow from its output terminal protective circuitry is generally included. This circuitry limits the maximum output current to a safe value. Typically, the output short-circuit current is in the milliampere range, (e.g. for a 741 op-amp 25 mA is typical), but *power op-amps* have the capability to supply much larger currents (e.g. the LM 12 can supply up to 10 A).

## Types of Op-amp

Most op-amps can be placed into one of eight categories. These are (i) general purpose; (ii) high frequency, high slew rate; (iii) high voltage, high power; (iv) low input offset voltage; (v) programmable; (vi) low noise; (vii) low voltage, low power; (viii) operational transconductance amplifier (OTA).

### General Purpose

General purpose op-amps are the cheapest and most readily available but they do not have their characteristics optimized in any way. Consequently they have medium values of input offset current and voltage, input resistance, bandwidth and slew rate. The most common example of this category is, of course, the well-known 741; alternative versions of this are the 741S which has a higher slew rate (20 V/$\mu$s), the 747 which is the dual version and the 748. The 748 is similar to the 741 but it is not internally frequency compensated.

Table 1.4 lists some popular general-purpose op-amps. Typical figures are given.

**Table 1.4**

| Device | Open-loop gain (db) | $V_{OS}$ (mV) | $I_B$ (nA) | GBP (MHz) | CMRR (dB) | Power dissipation (mW) | Slew rate (V/$\mu$s) |
|--------|------|------|------|------|------|------|------|
| LM 741 | 106 | 1 | 80 | 1 | 90 | 500 | 0.5 |
| LM 301A | 88 | 2 | 70 | 2 | 90 | 500 | |
| LM 324 | 100 | 3 | 20 | 1 | 80 | 625 | |
| TL 084 | 106 | 3 | 0.2 | 3 | 80 | 680 | 13 |

**Table 1.5**

| Device | Open-loop gain (dB) | $V_{OS}$ (mV) | $I_B$ (nA) | GBP (MHz) | Power dissipation (mW) | Slew rate (V/$\mu$s) |
|---|---|---|---|---|---|---|
| LM 318 | 106 | 4 | 150 | 15 | 500 | 50 |
| LT 1220 | 94 | 1 | 300 | 45 | 250 | |
| NE 5539 | 52 | 5 | 5000 | 700 | 550 | 30 |
| OP 37 | 123 | 0.03 | 15 | 40 | 500 | 17 |
| OP-42 | 108 | | 130 | | 500 | 50 |

*High Frequency, High Slew Rate*

An op-amp of this type has been designed to have a wide bandwidth and a high slew rate. Wide bandwidth means in excess of 5 MHz. Table 1.5 lists some high-speed op-amps.

*High Voltage, High Power*

Most op-amps are small-signal devices. Their maximum supply voltage is about ± 18 V and their maximum output current is about 10 mA. If a higher voltage and/or current output than these figures is required it can always be obtained by using discrete transistors to amplify the output signal as shown by Fig. 1.8. The maximum load current is now $h_{FE}$ times the op-amp's maximum output current. Alternatively, an op-amp with a greater output current/voltage capability can be used; two examples are (*a*) the LH 0021 can output up to 1 A and dissipate up to 3.5 W, and (*b*) the LH 004C is operated from ± 45 V power supplies and can deliver up to 38 V peak voltage to a load but its maximum power dissipation is only 400 mW. These

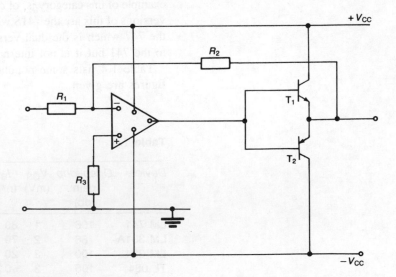

**Fig. 1.8** High output power op-amp circuit

op-amps are often FET input types in order to minimize bias current problems, since it will be necessary to employ high-value resistors in the feedback network. High output current op-amps may be able to supply currents as high as 10 A and they will usually incorporate thermal sensing and shut-down circuitry to protect the device from overheating. A heat sink will be required.

### Low Input Offset Voltage

Op-amps designed to have a low input offset voltage are also known as *precision op-amps* and they also have low values of input offset current and temperature coefficient. Examples of such devices are given in Table 1.6.

### Programmable

A programmable op-amp has one terminal that can be used to set one, or more, of the following parameters to a required value: the gain—bandwidth product, the slew rate, the current taken from the supply, the input offset current and the input noise. The designer of a circuit can make use of this terminal to optimize the noise performance of the op-amp for a given source resistance, to increase the bandwidth at the expense of the noise performance, or whatever other trade-off a particular design might require. The data sheet for a programmable op-amp gives details of how each of the various parameters varies with the current that flows into the control terminal. Some programmable op-amps are listed in Table 1.7.

**Table 1.6**

| Device | Open-loop gain (dB) | $V_{OS}$ (mV) | $I_B$ (nA) | GBP (MHz) | CMRR (dB) | Power dissipation (mW) | Slew rate (V/$\mu$s) |
|---|---|---|---|---|---|---|---|
| LM 308 | 108 | 2 | 1.5 | | 80 | 500 | |
| OP 07C | 122 | 0.061 | 1.8 | 0.6 | 120 | 500 | 0.3 |
| OP 37C | 123 | 0.03 | 15 | 45 | 120 | 500 | 17 |
| LF 411 | 98 | 2 | 2 | 3 | 100 | 500 | 13 |

**Table 1.7**

| Device | $V_{OS}$ (mV) | $I_B$ (nA) | GBP (MHz) | Noise (nV/$\sqrt{kHz}$) | CMRR (dB) |
|---|---|---|---|---|---|
| LM 346 | 6 | 250 | 0.5 | 28 | 70 |
| TCA 3002 | 5 | 500 | 2.5 | 25 | 70 |
| LMC 660 | 2.5 | $10^{-1}$ | | | 83 |

### Low Noise

Op-amps of this type have been designed to optimize their noise performance and one should be employed in the input stage of low-noise designs such as preamplifiers, and instrumentation amplifiers. Table 1.8 lists three low-noise op-amps.

### Low Voltage, Low Power

Battery-operated equipment often demands the use of op-amps that are able to operate from a low supply voltage. A low-voltage, low-power op-amp will be able to operate from a power supply voltage as low as 1 V with a supply current of perhaps only 10 $\mu$A. Examples of such devices are the TL 060 family of BiFET devices; these have supply currents that vary from about 160 to 200 $\mu$A as the supply voltage is increased from $\pm$ 1.5 V up to $\pm$ 18 V. Other low-voltage and/or low-power devices are listed in Table 1.9.

### Operational Transconductance Amplifier

The 'gain' of an OTA is the ratio (output current)/(input voltage), known as the transconductance $g_m$. The voltage gain of the device is equal to $g_m R_L$, where $R_L$ is the load resistance. The gain, the input resistance, the noise level and the input offset current and voltage, can all be varied by means of a control current. The CA 3080, for example, has a control pin, labelled as *amplifier bias input*, and its gain is directly proportional to the magnitude of the control current over the range 0.1 $\mu$A to 1 mA. Another example of an OTA is the LM 13600.

**Table 1.8**

| Device | Open-loop gain (dB) | $V_{OS}$ (mV) | $I_B$ (nA) | GBP (MHz) | Noise (nV/√kHz) | CMRR (dB) | Power dissipation (mW) | Slew rate (V/$\mu$s) |
|---|---|---|---|---|---|---|---|---|
| LM 833 | 110 | 5 | 500 | 15 | 4.5 | | 500 | 7 |
| NE 5532 | 100 | 0.5 | 200 | 10 | 8 | 100 | 800 | 9 |
| NE 5534 | 100 | 0.5 | 500 | 10 | 4 | 100 | 900 | 13 |

**Table 1.9**

| Device | Open-loop gain (db) | $V_{OS}$ (mV) | $I_B$ (nA) | CMRR (dB) | Power dissipation (mW) |
|---|---|---|---|---|---|
| LM 358 | 100 | 3 | 20 | 80 | 725 |
| LM 386 | 76 | 2 | 250 | | 660 |
| MC 33171 | 114 | 2 | 20 | | 500 |

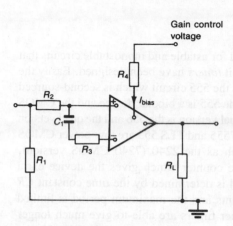

**Fig. 1.9** OTA voltage amplifier

When an OTA is used as a voltage amplifier it does not require a feedback resistor since the voltage gain is controlled by the control current, and a typical OTA voltage amplifier circuit is given by Fig. 1.9.

### Choice of an Op-amp

For most applications there will be some op-amp specifications that are more important than others, e.g. high input impedance, low input offset current and offset voltage, single-supply operation, a.c. performance, and programmable power level. The required signal level, the maximum frequency and the impedance of the signal source should all be known. The gain accuracy of an amplifier is determined by the limits of the offset current and voltage and the gain error (see p. 96). If the frequency and/or the output voltage is high the bandwidth and the slew rate of the op-amp will be of importance. A fast settling time may be important for a data conversion application. If the op-amp is required to drive a capacitive load then a high output current will be needed. In all cases the loop gain, i.e. the excess of the open-loop gain over the closed-loop gain must be large enough at the highest frequency to be amplified. If, for example, 2% accuracy for a closed-loop gain of 20 is wanted at a frequency of 10 kHz the open-loop gain must be at least equal to 1000 at 10 kHz. For a high open-loop gain an uncompensated op-amp is usually best. In general, a superior performance comes from bipolar devices but high input impedance, low power dissipation and low-voltage operation are provided by FET devices. The LinCMOS devices combine the best features of both technologies.

General-purpose op-amps are usually the cheapest types and one should be used in applications where the closed-loop gain needed is low, the wanted speed is moderate, and impedances are small.

A low-drift op-amp should be selected for those applications where the utmost accuracy must be maintained over a range of temperatures. If a low input bias current device is required a FET input type ought to be chosen. The possible applications include circuits that must have either a large value of feedback resistor or a large source resistance, i.e. circuits that have a long time constant such as integrators, buffer amplifiers and current sources. For applications such as signal conditioning, the measurement of light, and data acquisition, low noise may be of prime importance. A low-noise type of op-amp should then be selected. Wideband op-amps are those that have a bandwidth that is in excess of 5 MHz. Such op-amps will also have a high slew rate and a short settling time. This type is used for such applications as pulse and video amplifiers and multiplexer circuits. High-voltage devices are designed to provide a large output voltage swing and to operate over a wide range of power supply voltages. They usually have a FET input circuit to minimize bias current errors since high values of resistors are necessary.

## Timers

There is such a frequent need for astable and monostable circuits that a number of integrated circuit *timers* have been designed. Easily the most commonly employed is the 555 circuit which is second-sourced by several manufacturers. The 555 is a bipolar device and the CMOS equivalent is the 7555. The dual version is the 556 and the quad version is the 558/9, while the ICM 7555 and TLS 555 are low-power CMOS versions. Other timers, such as the 2240 (7240 CMOS version), incorporate a programmable counter which gives the device extra flexibility. The timed period is determined by the time constant $CR$ of external timing components and the maximum period is limited to a few seconds. Some other timers are able to give much longer timing periods, e.g. the 2240 up to 40 minutes and the ZN 1034 up to 1 day.

## Analogue-to-digital and Digital-to-analogue Converters

The function of an analogue-to-digital converter (ADC) is to convert an input analogue signal into the corresponding digital word. The quantization procedure employed, shown by Fig. 1.10, introduces inherent errors equal to one-half the least significant bit (LSB). For correct operation the ADC should have its input signal held constant during the time that the conversion is taking place and to achieve this an ADC is normally preceded by a *sample-and-hold* amplifier. A digital-to-analogue converter (DAC) is a circuit that accepts an input digital word and converts it into the equivalent analogue signal. The analogue output signal is represented by a number of discrete voltages as shown by Fig. 1.11. The smallest increment of the output voltage is contributed by the LSB of the input digital word and it is equal to the full-scale voltage $V_{FS}$ times $2^{-n}$, where $n$ is the number of bits used.

$$\text{The maximum output voltage} = V_{FS}[(2^n - 1)/2^n] \qquad (1.9)$$

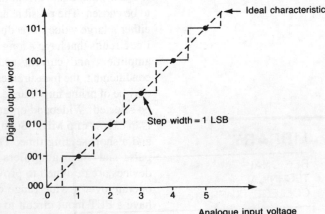

**Fig. 1.10** ADC quantization procedure

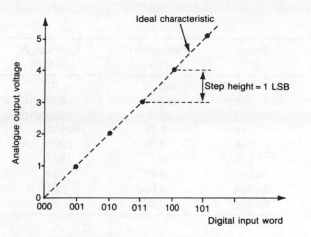

**Fig. 1.11** DAC output voltage

The ideal transfer function of a 3-bit DAC is understood to be the straight line that joins the point 0, 0 and 8, FS. Most DAC circuits use one of two different techniques, either the *weighted resistor* or $R - 2R$ method. A multiplying DAC is one that has at least two inputs, one at least of which is digital, and that generates an analogue output voltage directly proportional to the product of the two input signals. The data sheets of either kind of DAC follow similar lines to the op-amp data sheets, commencing with the main features of the device along with possible applications, pinouts, etc. Next come the absolute maximum ratings, the electrical characteristics and lastly operating instructions. A number of terms are employed in the data sheets of both ADC and DAC devices which the user must be familiar with if the information is to be understood.

## Digital to Analogue Converters

### Full-scale Output

This is the maximum current, or voltage, that can be obtained from the output of the DAC. A DAC produces its FS output when the digital input word is all 1s. It is known as the full-scale range (FSR).

### Resolution

The resolution of a DAC is the smallest change in the input voltage that will cause a change in the output voltage and it is determined by the number $n$ of bits that are used in the input digital word. The resolution is equal to $1/2^n$. If, for example, $n = 8$ there are $2^8 = 256$ possible output voltages so that the smallest possible change in the output voltage is $1/256$ times, or $0.39\%$ of, the FSR of the DAC. Table 1.10 gives the values for other bit numbers.

**Table 1.10**

| Number of bits (n) | States $2^n$ | Resolution (%) | LSB weight in ppm | Bit weight for 10 V FSR | Dynamic range (dB) |
|---|---|---|---|---|---|
| 0 | 1 | 100 | $10^6$ | 10 V | 0 |
| 1 | 2 | 50 | $500 \times 10^3$ | 5 V | 6.02 |
| 2 | 4 | 25 | $250 \times 10^3$ | 2.5 V | 12.04 |
| 4 | 16 | 6.25 | 62500 | 0.625 V | 24.08 |
| 8 | 256 | 0.39 | 3906 | 39 mV | 48.16 |
| 10 | 1 024 | 0.1 | 977 | 9.8 mV | 60.21 |
| 12 | 4 096 | 0.02 | 244 | 2.4 mV | 72.25 |
| 14 | 16 384 | 0.006 | 61 | 610 $\mu$V | 84.29 |
| 16 | 65 536 | 0.002 | 15 | 152 $\mu$V | 96.33 |

*Conversion Time*

The conversion time of a DAC is the time that elapses from the 'start conversion' command being given to the circuit and the analogue signal appearing at the output terminals.

*Absolute Accuracy*

The absolute accuracy is the largest difference between the actual analogue output voltage and the ideal output voltage when a given digital word is applied to the input terminals of the circuit. It is usually expressed in fractions of 1 LSB. The absolute accuracy should never be worse than 0.5 LSB. The errors in a DAC may be static or dynamic. Static errors include *offset error* and *gain error* which have the effect of moving the transfer characteristic of the DAC up/down, or rotating it about the origin, so that when the input digital word is zero there is a small output analogue voltage.

*Dynamic Range*

The *dynamic range* of a DAC is the ratio of the FSR to the smallest change in the input voltage to which the circuit can respond.

$$\text{Dynamic range} = 20 \log_{10} (2^n) \tag{1.10}$$

*Non-linearity*

Non-linearity in a DAC is the difference between the actual analogue output voltage and a straight line drawn between 0 V and the FS output for any input digital word. It is expressed as a percentage of the FS output.

### Differential Non-linearity

The ideal DAC has a step size equal to the LSB and the *differential non-linearity* is the maximum deviation of the actual step size from 1 LSB. It is quoted in fractions of the LSB. If, for example, the actual step size is 1.5 LSB the differential non-linearity is 0.5 LSB.

### Relative Accuracy

The relative accuracy is the largest difference between the transfer function of the circuit and a straight line drawn between the output voltage points 0 V and FSR.

### Least Significant Bit

This is the digital input line that has the smallest effect upon the output analogue voltage. It is a measure of the change in analogue output voltage when the digital word is incremented. One LSB = $(1/2^n)(\text{FSR})$.

### Monotonicity

A DAC is monotonic if an increase in the input digital word never results in a decrease in the output analogue signal, or vice versa. If the step size should, at any point in the transfer characteristic, become greater than 1 LSB the DAC will then become non-monotonic.

### Zero Offset

The zero offset is the difference between 0 V and the actual analogue output voltage when the input digital word indicates zero voltage.

### Settling Time

The settling time is the time taken for the analogue output voltage to settle to within a specified band (usually ± LSB/2), about its final value when the input digital word suddenly changes.

## Analogue-to-digital Converters

There are several different kinds of ADC in use, namely (i) the flash converter; (ii) the single-, and dual-slope integrating converters; (iii) the successive approximation converter; (iv) the ramp converter; (v) the tracking converter, but the same terms are used to describe the operation of each type. Most IC ADCs are of either the successive approximation or the dual-slope integrating types; the former are the more commonly employed because of their greater speed of operation even though the integrating converter is cheaper for a given resolution. The output of an ADC may employ either a parallel or a serial interface. A serial interface is physically smaller and easier to electrically isolate, using a transformer or an opto-coupler, since there is only the one line to deal with.

### Absolute Accuracy

The *absolute accuracy* of an ADC is the greatest deviation of its transfer characteristic from the ideal characteristic.

### Conversion Range

This is the range of analogue voltages that the circuit is able to convert into the corresponding digital word.

### Conversion Time

This is the time that elapses from the receipt of a 'start conversion' command to the digital word appearing at the ADC's output. Conversion time is one way of specifying the speed of an ADC. The alternative is to quote the number of conversions (or samples) per second. The two methods are closely related, but the conversion rate is not always equal to the reciprocal of the conversion time because some ADCs require a short time interval between successive conversions.

### Relative Accuracy

The *relative accuracy*, or integral linearity, of an ADC is a measure of the largest deviation between its transfer function and a straight line drawn from the point 0 V to the FSR. It is usually expressed as a percentage of the FSR. Integral non-linearity in an ADC is illustrated by Fig. 1.12.

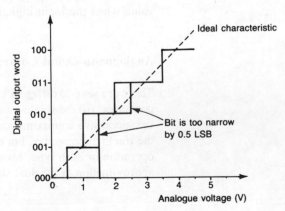

**Fig. 1.12** Integral non-linearity in an ADC

## Resolution

The *resolution* of an ADC is the smallest change in the analogue input voltage that can produce a change in the digital output signal. It is defined as the FS input analogue voltage divided by the total number of quantization steps employed, i.e.

$$\text{resolution} = FSR/2^n \qquad (1.11)$$

where $n$ is the number of bits used.

## Differential Linearity

This has the same meaning as for a DAC except that the input and output signals are reversed.

## Monotonicity

This also has the same meaning as that applied to a DAC.

## Aperture Error

The samples taken of the analogue input voltage are a series of square waves whose periodic time $T$ is equal to the reciprocal of the sampling frequency. Each sample has a width $\tau$ that is much smaller than $T$. The frequency spectrum of a square wave has an envelope whose amplitude is a sinc $x$ function (p. 301). This effect produces an *aperture time error* given by

$$\text{Aperture time error} = [1 - (\sin \pi f \tau)/(\pi f \tau)] \times 100\% \qquad (1.12)$$

## Digital Codes

A converter may employ any one of a number of different codes for its data conversion. The natural binary code may be used or any one of the (*a*) offset binary, (*b*) complementary binary, (*c*) 1s complement, (*d*) 2s complement, (*e*) binary coded decimal or (*f*) sign magnitude codes. Offset binary code is the same as natural binary except that the code 00000000 is equal to one-half of the analogue scale. The code represents analogue values from $-FSR$ to $+FSR$ with analogue zero being indicated by the code 1000000. With the complementary binary code all 1s become 0s and all 0s become 1s. The 1s complement code is one in which the positive and negative codes of the same magnitude always sum to all 1s. The 2s complement code is such that positive and negative codes of the same magnitude always sum to all 0s plus a carry; it is obtained by complementing the MSB (most significant bit) of the offset binary code. Binary coded decimal (BCD) represents decimal numbers using the 8421 code and only the numbers 0 through to 9 are possible. Lastly, the sign magnitude code uses the MSB to indicate the polarity of the signal, MSB = 1 indicates a

**Table 1.11**

| Scale | Offset binary | 1s complement | 2s complement | Sign magnitude |
|---|---|---|---|---|
| $+FSR - LSB$ | 11111111 | 01111111 | 01111111 | 11111111 |
| $+\frac{3}{4}FSR$ | 11100000 | 01100000 | 01100000 | 11100000 |
| $+\frac{1}{2}FSR$ | 11000000 | 01000000 | 01000000 | 11000000 |
| $+\frac{1}{4}FSR$ | 10100000 | 00100000 | 00100000 | 10100000 |
| $+0$ | 00000000 | 00000000 | 00000000 | 00000000 |
| $-0$ | — | 11111111 | — | 00000000 |
| $-\frac{1}{4}FSR$ | 01100000 | 11011111 | 11100000 | 00100000 |
| $-\frac{1}{2}FSR$ | 01000000 | 10111111 | 11000000 | 01000000 |
| $-\frac{3}{4}FSR$ | 00100000 | 10011111 | 10100000 | 01100000 |
| $-FSR + LSB$ | 00000001 | 10000000 | 10000001 | 01111111 |
| $-FSR$ | 00000000 | — | 10000000 | — |

**Table 1.12**

| ADC | Type | Resolution (bits) | Conversion time (ns) | Linearity error (LSB) | Power dissipation (mW) |
|---|---|---|---|---|---|
| TL 507 | SA | 7 | 1000 | ± 1/2 | 25 |
| ADC 302 | F | 8 | 20 | ± 1/2 | 550 |
| ZN 439 | SA | 8 | 5000 | ± 1/2 | 150 |
| ZN 433 | T | 10 | 1000 | ± 1/2 | 500 |

**Table 1.13**

| DAC | Resolution (bits) | Settling time (ns) | Non-linearity (% FS) | FSR | Power dissipation (mW) | Differential non-linearity (%FS) |
|---|---|---|---|---|---|---|
| ZN 558 | 8 | 800 | 0.5 | 2.55 V | 100 | 0.5 |
| ZN 425 | 8 | 10000 | 0.5 | 2.55 V | | 0.5 |
| DAC 08 | 8 | 150 | 0.2 | 2 mA | 50 | 0.2 |
| DAC 10 | 10 | 85 | 0.05 | 4 mA | 85 | 0.05 |

positive signal and MSB = 0 indicates a negative signal. Either natural binary or BCD can be used to indicate the magnitude of the signal.

The offset binary, 1s and 2s complement, and sign magnitude codes are shown by Table 1.11.

Short-form data for ADCs are employed to pick possible devices for use in a particular application. Table 1.12 gives details of some typical ADCs and Table 1.13 gives details of some typical DACs.

The data sheet for the ZN 439 ADC is given in Fig. 1.13 and that for the ZN 558 DAC is given in Fig. 1.14.

# GEC PLESSEY
SEMICONDUCTORS

## ZN439

### 8-BIT MICROPROCESSOR COMPATIBLE A-D CONVERTER

The ZN439 is an 8-bit successive approximation A-D converter, designed to be easily interfaced to microprocessors. All active circuitry is contained on-chip including clock generator trimmable 2.5V bandgap reference, control logic and double buffered latches with three-state outputs.

These features give extra flexibility in use, with just three inputs to control all ADC operations and double buffered output latches which will allow data to be read at any time irrespective of the status of the converter.

**FEATURES**

- Choice of Linearity: ¼ LSB - ZN439-9, ½ LSB - ZN439-8, 1 LSB - ZN439-7
- 5 microseconds Conversion Time
- Microprocessor, TTL and CMOS Compatible
- On-Chip Clock
- Trimmable Bandgap Reference
- Versatile Microprocessor Interfacing with Double Buffered Output Latch
- Equally Suitable for Stand-Alone Applications
- ROM Type Operation
- Commercial or Military Temperature Ranges

Pin connections - top view

ZN439E (DP22)
ZN439J (DC22)

3012-1.0

**ORDERING INFORMATION**

| Device type | Linearity error(LSB) | Operating temperature | Package |
|---|---|---|---|
| ZN439J9 | ¼ | -55°C to +125°C | DC22 |
| ZN439E8 | ½ | 0°C to +70°C | DP22 |
| ZN439J8 | ½ | -55°C to +125°C | DC22 |
| ZN439J7 | 1 | -55°C to +125°C | DC22 |

**ELECTRICAL CHARACTERISTICS** (at $V_{CC}$ = 5V, $T_{amb}$ = 25°C and $f_{CLK}$ = 1.6MHz unless otherwise specified).

| Parameter | $T_{amb}$ = +25°C Min | Typ. | Max. | Over specified Temp. range Min. | Max. | Units | Conditions |
|---|---|---|---|---|---|---|---|
| **ZN439-9** | | | | | | | |
| Linearity error | — | — | ±0.25 | — | ±0.25 | LSB | |
| Differential linearity error | — | — | ±0.5 | — | ±0.5 | LSB | |
| **ZN439-8** | | | | | | | |
| Linearity error | — | — | ±0.5 | — | ±0.5 | LSB | |
| Differential linearity error | — | — | ±0.75 | — | ±0.75 | LSB | |
| **ZN439-7** | | | | | | | |
| Linearity error | — | — | ±1 | — | ±1 | LSB | |
| Differential linearity error | — | — | ±1 | — | ±1 | LSB | |
| **ALL TYPES** | | | | | | | |
| Zero transition (00000000→00000001) | 7 | — | — | — | — | mV | ZN439E |
| | 7 | — | — | — | — | mV | ZN439J |
| Full-scale transition (11111110→11111111) | 2.550 | — | — | — | — | V | ZN439E |
| | 2.550 | — | — | — | — | V | ZN439J |
| Linearity temperature coefficient | | | ±3 typ. | | | ppm/°C | Ext. Ref. |
| Differential linearity temperature coefficient | | | ±6 typ. | | | ppm/°C | |
| Gain temperature coefficient | | | ±10 typ | | | ppm/°C | |
| Offset temperature coefficient | | | ±7 typ. | | | ppm/°C | |
| Resolution | 8 | — | — | — | — | Bits | |
| Conversion time | 5 | — | — | — | — | μs | |
| Supply reject on | 4.5 | 0.2 | — | 4.5 | — | %/V | |
| Supply voltage | | 5.0 | 5.5 | | 5.5 | V | |
| Supply current | | 30 | 45 | | 45 | mA | |
| Power consumption | | 150 | 225 | | 225 | mW | |
| Reference input range | 1.5 | | 3.0 | | 3.0 | V | |
| Ladder output impedance | | 2.7 | | | | kΩ | |
| **COMPARATOR** | | | | | | | |
| Input current | | 1.0 | | | | μA | } Outputs in high impedance state |
| Input resistance | 25 | 100 | | 25 | | kΩ | $V_{in}$ = +3V, $R_{ext}$ = 82K |
| Tail current | | | 150 | | 150 | μA | $R_{ext}$ = 82K, V-- = -5V |
| Negative supply | 3 | 5 | -30 | -3 | -30 | V | |
| Input voltage | 0.5 | | +3.5 | -0.5 | +3.5 | V | |

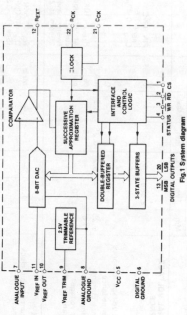

Fig.1 System diagram

**Fig. 1.13** Data sheet for the ZN 439 ADC (*courtesy of GEC Plessey Semiconductors*)

## ELECTRICAL CHARACTERISTICS (Cont.)

| Parameter | T_amb = +25°C Min. | Typ. | Max. | Over specified Temp. range Min. | Max. | Units | Conditions |
|---|---|---|---|---|---|---|---|
| High level I/P current $I_{IH}$ | – | 120 | – | – | – | μA | $V_{CC} = +5.5V$, $V_{IN} = +2.4V$ |
| Low level I/P current $I_{IL}$ | – | –370 | – | – | – | μA | $V_{CC} = +5.5V$, $V_{IN} = +0.4V$ |
| **DATA AND STATUS OUTPUTS** | | | | | | | |
| High level output voltage $V_{OH}$ | 2.4 | – | – | 2.4 | – | V | $I_{OH\ MAX}$ |
| Low level output voltage $V_{OL}$ | – | – | 0.4 | – | 0.4 | V | $I_{OL\ MAX}$ |
| High level output current $I_{OH}$ | – | – | –800 | – | – | μA | $V_{OUT} = 0.4V$ |
| Low level output current $I_{OL}$ | – | – | 2 | – | – | mA | $V_{OUT} = 2.4V$ |
| Three-state disable output leakage current (Data output only) | – | – | 2.0 | – | – | μA | |
| | – | – | 2.0 | – | – | μA | |
| Enable/disable Delay times $T_{E1}$ | 90 | 120 | 160 | – | – | ns | |
| $T_{EO}$ | 60 | 100 | 120 | – | – | ns | |
| $T_{D1}$ | 80 | 120 | 160 | – | – | ns | |
| $T_{DO}$ | 60 | 80 | 110 | – | – | ns | |
| Write pulse width | 150 | – | – | – | – | ns | |
| WR input to status O/P high | – | 280 | 350 | – | – | ns | |
| Read pulse width | 160 | – | – | – | – | ns | |
| Read input high to status output high | – | 240 | 400 | – | – | ns | |

## ABSOLUTE MAXIMUM RATINGS

| | |
|---|---|
| Supply voltage $V_{CC}$ | +7V |
| Maximum voltage, logic e and $V_{REF}$ inputs, $A_{IN}$ | $V_{CC}$, –0.5V |
| Operating temperature range | 0°C to +70°C (ZN439E) <br> –55°C to +125°C (ZN439J) |
| Storage temperature range | –55°C to +125°C |

## GENERAL CIRCUIT OPERATION

The ZN439 utilises the successive approximation technique to produce an 8-bit parallel digital output. Upon receipt of a negative going pulse on the WR input the status output goes high, and the DAC input is set to the MSB. The resulting analogue output is compared with the unknown analogue input signal by means of the comparator. If the analogue input is larger, the MSB is left in circuit and if not the MSB is removed. On the second clock pulse this sequence is repeated for the next most significant bit and so on until all the 8 bits have been compared. On the 8th negative clock edge the status goes low indicating that the conversion is complete.

The double-buffered register means the outputs can be enabled at any time, irrespective of the conversion status, and valid data will always be presented to the data bus. **Therefore the RD signal can be completely asynchronous with respect to the status.** Data can be read by taking RD low, thus enabling the three-state outputs. RD cannot be tied low as this will prevent the converter from updating it's outputs at the end of a conversion.

## ELECTRICAL CHARACTERISTICS (Cont.)

| Parameter | T_amb = +25°C Min. | Typ. | Max. | Over specified Temp. range Min. | Max. | Units | Conditions |
|---|---|---|---|---|---|---|---|
| **INTERNAL VOLTAGE REFERENCE** | | | | | | | |
| Output voltage | – | 2.588 | – | – | – | V | PIN 9 NC $R_{REF} = 1.6K$ $C_{REF} = 0.47\mu F$ |
| Output voltage tolerance | – | – | ±3 | – | – | % | |
| Slope impedance | – | 0.75 | – | – | – | Ω | |
| Reference current | 0.25 | – | 5.2 | 0.25 | 5.2 | mA | $R_{TRIM} = 10K$ At 5mA operating current (worst case) 25ppm at 2.0mA |
| Trim range | ±5 | – | – | ±5 | – | % | |
| Output voltage temperature coefficient | – | 70 | – | – | – | ppm/°C | |
| **CLOCK** | | | | | | | |
| Maximum on-chip clock frequency | – | 1.6 | – | – | – | MHz | $R_{ck} = 1.5K\Omega$ $C_{ck} = 100pF$ (See Fig. 13) |
| Clock frequency tempco | – | –0.1 | – | – | – | %/°C | |
| Clock capacitor | 100 | – | – | – | – | pF | |
| Clock resistor | 1.0 | – | – | – | – | kΩ | |
| Maximum external clock frequency | 2 | – | – | 2 | – | MHz | |
| Clock pulse width | 250 | – | – | – | – | ns | |
| High level I/P voltage $V_{IH}$ | 3.5 | – | – | 3.5 | – | V | |
| Low level I/P voltage $V_{IL}$ | – | – | 0.8 | – | 0.8 | V | |
| High level I/P current $I_{IH}$ | – | 1 | – | – | – | μA | $V_{CC} = 5.5V$ $V_{IN} = 4V$ |
| Low level I/P current $I_{IL}$ | – | 10 | – | – | – | nA | $V_{CC} = 5.5V$ $V_{IN} = 0.8V$ |
| Supply rejection | – | 3.5 | – | – | – | %/V | Int. clock Freq. |
| **LOGIC WR + CS INPUTS** | | | | | | | |
| High level I/P voltage $V_{IH}$ | 2 | – | – | 2 | – | V | $V_{CC} = +5.5V$ |
| Low level I/P voltage $V_{IL}$ | – | – | 0.8 | – | 0.8 | V | $V_{CC} = +5.5V$ |
| High level I/P current $I_{IH}$ | – | 40 | – | – | – | μA | $V_{CC} = +5.5V$ $V_{IN} = +2.4V$ |
| High level I/P current $I_{IH}$ | – | 20 | – | – | – | μA | |
| Low level I/P current $I_{IL}$ | – | –50 | – | – | – | μA | $V_{CC} = +5.5V$ $V_{IN} = +0.4V$ |
| **LOGIC RD INPUT** | | | | | | | |
| High level I/P voltage $V_{IH}$ | 2 | – | – | 2 | – | V | $V_{CC} = +5.5V$ |
| Low level I/P voltage $V_{IL}$ | – | – | 0.8 | – | 0.8 | V | $V_{CC} = +5.5V$ |
| High level I/P current $I_{IH}$ | – | 220 | – | – | – | μA | $V_{CC} = +5.5V$ $V_{IN} = +5.5V$ |

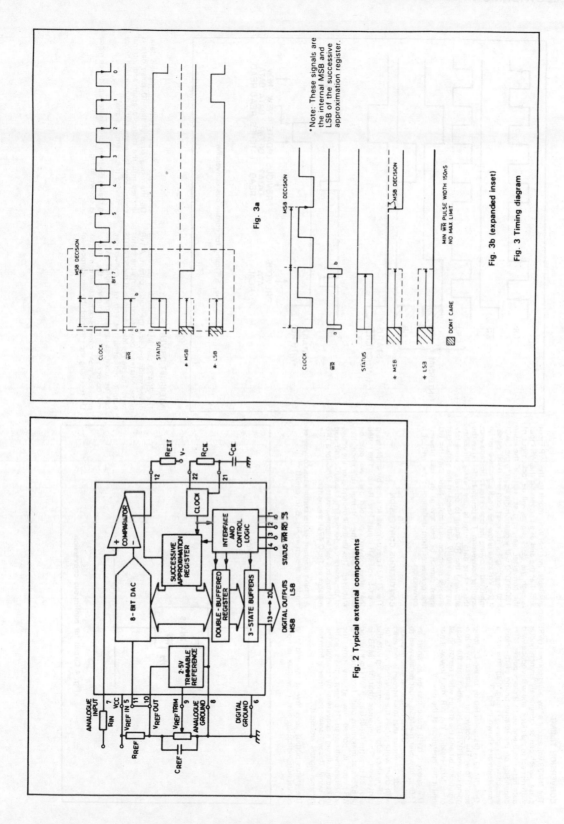

Fig. 3a

* Note: These signals are
the internal MSB and
LSB of the successive
approximation register.

Fig. 3b (expanded inset)

Fig. 3 Timing diagram

Fig. 2 Typical external components

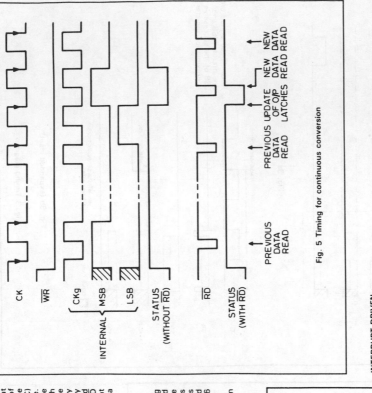

Fig. 5 Timing for continuous conversion

## CONVERSION TIMING

The ZN439 will accept a low going convert (WR) pulse, which can be completely asynchronous with respect to the clock, and will produce valid data between 8 and up to 9 clock pulses later depending on the relative timing of the clock and convert signals. Timing diagrams for a conversion are shown in Fig. 3.

The ZN439 is first selected by taking CS (chip select) low. The converter is cleared by a low going convert (WR) pulse, which sets the most significant bit and the status while resetting all other bits. Holding the WR input low will not inhibit the operation of the device.

The convert (WR) pulse can be as short as 150ns; however the MSB must be allowed to settle for at least 625ns before the MSB decision is made. To ensure that this criterion is met even with short write pulses the converter waits for a falling clock edge before commencing with the conversion. This ensures that the MSB is allowed to settle for at least a full clock period or 625ns at maximum clock frequency. If the WR input is pulsed low at any time the conversion will restart. The input signals can be locked out during a conversion by removing the CS signal. This will isolate the converter from the external signals around it.

The status output goes low at the end of a conversion indicating that new data is now

available. Internal logic monitors the WR input and if at the end of a conversion the WR input is high the clock signal will be locked out of the converter leaving it set up (i.e. the code 10000000 will appear on the input to the DAC) and waiting for its next convert (WR) pulse. If the WR input is low the clock signal will not be inhibited allowing the converter to procede with another conversion. The double buffering on the three-state data outputs gives extra flexibility allowing the RD input to operate completely asynchronously with respect to the status and always produce valid data. Note that the RD input cannot be tied low as this will prevent the converter from updating at the end of a conversion.

## CONTINUOUS CONVERSION

The ZN439 can be made to cycle by simply tying the CS and WR inputs low. It should be noted that after power up, valid data will only be available after the internal reference has stabilised. This time will depend upon the values of the reference decoupling capacitor and load resistor, but will be approximately 2mS for a 1K6 resistor and a 0.47μF capacitor.

A timing diagram for the continuous conversion mode is shown in Fig. 5 (overleaf).

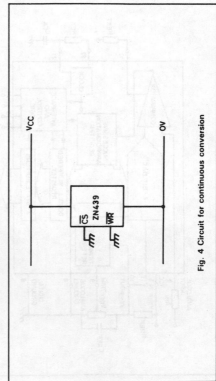

Fig. 4 Circuit for continuous conversion

## INTERRUPT DRIVEN

The ZN439 can also be used in an interrupt driven mode by using the status output. A WR pulse initiates a conversion sending the status high. The high to low transition of the STATUS output, indicating the end of a conversion, can be used as an interrupt signal by the microprocessor i.e. informing the micro-processor that a conversion has been completed. On receiving the interrupt the microprocessor

sends out an RD pulse to take in the new data. On the rising edge of the RD pulse data is latched into the microprocessor and internal control logic forces the status output high hence removing the interrupt signal.

A timing diagram for the interrupt driven mode is shown in Fig. 6.

By tying the $\overline{WR}$ and $\overline{CS}$ inputs low the device can be made to cycle. Also if the status output is connected via an inverter to the $\overline{RD}$ input the device can be updated at the end of each conversion and the output buffers enabled without the need for extra external control signals.

A timing diagram for stand alone operation is shown in Fig. 8.

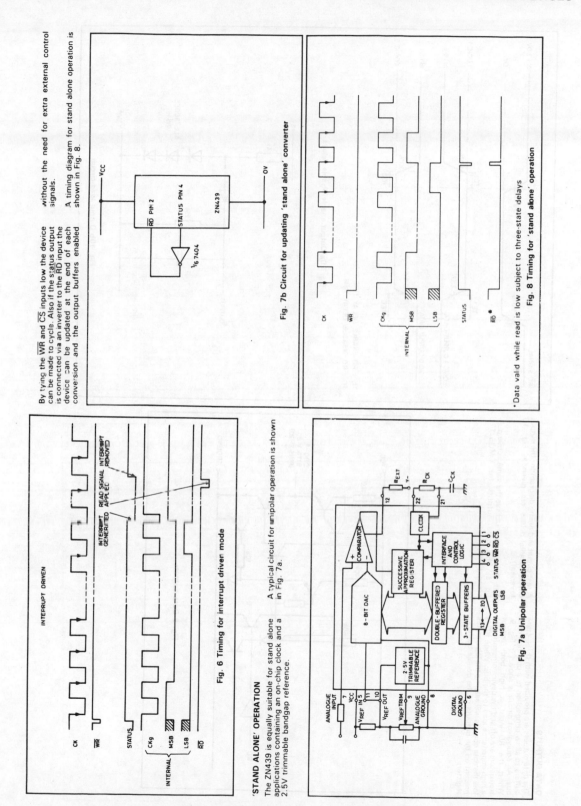

Fig. 6 Timing for interrupt driven mode

Fig. 7b Circuit for updating 'stand alone' converter

Fig. 7a Unipolar operation

Fig. 8 Timing for 'stand alone' operation

* Data valid while read is low subject to three-state delays

**'STAND ALONE' OPERATION**

The ZN439 is equally suitable for stand alone applications containing an on-chip clock and a 2.5V trimmable bandgap reference.

A typical circuit for unipolar operation is shown in Fig. 7a.

## DATA OUTPUTS

The data outputs are provided with three-state buffers to allow connection to a common data bus. An equivalent circuit is shown in Fig. 9. Whilst the RD input is high both output transistors are off and the device presents only a high impedance load to the bus. When RD is low the data outputs will assume the logic states present on the outputs of the double buffered register.

A test circuit and timing diagram for the output enable/disable delays are given in Fig. 10 (overleaf).

The status output utilises the same active pull-up as the data outputs for CMOS/TTL compatibility.

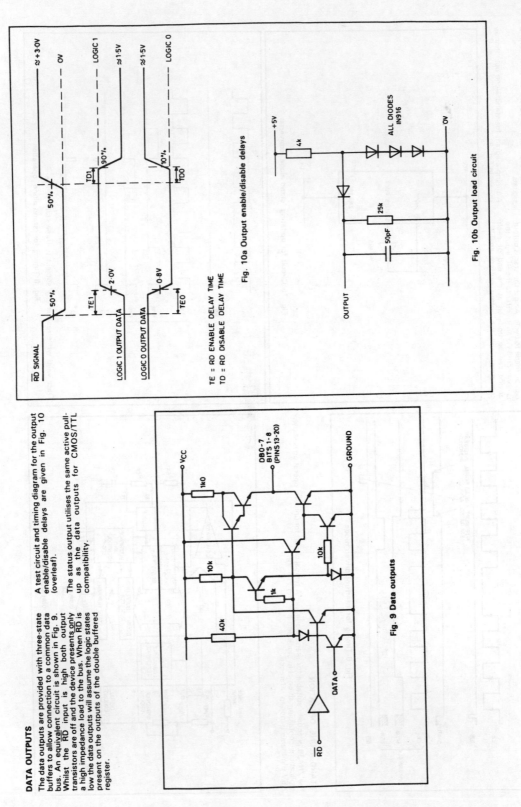

Fig. 10a Output enable/disable delays

TE = RD ENABLE DELAY TIME
TD = RD DISABLE DELAY TIME

Fig. 10b Output load circuit

Fig. 9 Data outputs

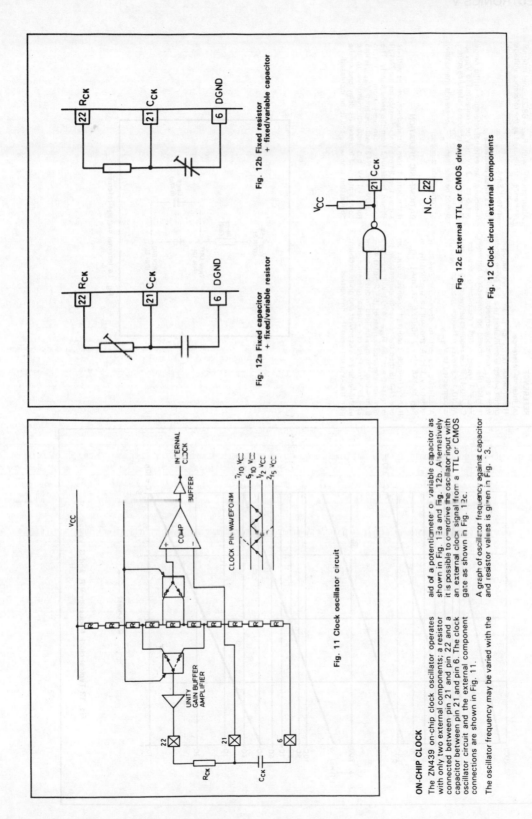

Fig. 12a Fixed capacitor + fixed/variable resistor

Fig. 12b Fixed resistor + fixed/variable capacitor

Fig. 12c External TTL or CMOS drive

Fig. 12 Clock circuit external components

Fig. 11 Clock oscillator circuit

## ON-CHIP CLOCK

The ZN439 on-chip clock oscillator operates with only two external components; a resistor connected between pin 21 and pin 22 and a capacitor connected between pin 21 and pin 6. The clock oscillator circuit and the external component connections are shown in Fig. 11.

The oscillator frequency may be varied with the aid of a potentiometer or variable capacitor as shown in Fig. 12a and Fig. 12b. Alternatively it is possible to overdrive the oscillator input with an external clock signal from a TTL or CMOS gate as shown in Fig. 12c.

A graph of oscillator frequency against capacitor and resistor values is given in Fig. 13.

## ANALOGUE CIRCUITS

### REFERENCE

#### (a) Internal reference

The internal reference is an active bandgap circuit which is equivalent to a 2.5 Zener diode with a very low slope impedance (Fig. 14). A resistor ($R_{REF}$) should be connected between $V_{CC}$ and $V_{REF\ OUT}$, and a decoupling capacitor, $C_{REF}$ (0.47$\mu$F), is required between $V_{REF\ OUT}$ and AGND. For internal reference operation $V_{REF\ OUT}$ is connected to $V_{REF\ IN}$.

A suitable current to drive one ZN439 is nominally 1.5mA and will be supplied by an $R_{REF}$ of 1K6 [(5 – 2.56)/1K6 = 1.5mA].

If the reference is required to drive more than one ZN439 then the reference current can be increased e.g. an $R_{REF}$ = 470$\Omega$ will supply a nominal reference current of (5 – 2.56)/0.47 = 5.2mA and this may be used to drive up to four ZN439's from just one internal reference. This useful feature saves power and gives excellent gain tracking between the converters.

Alternatively with $R_{REF}$ = 680$\Omega$, the internal reference can be used as the reference voltage for other external circuits and can source or sink up to 1.5mA.

#### (b) External reference

If required an external reference in the range + 1.5 to + 3.0V may be connected to $V_{REF\ IN}$. The slope resistance of such a reference source should be less than $\dfrac{2.5\Omega}{n}$, where n is the number of converters supplied.

### RATIOMETRIC OPERATION

If the output from a transducer varies with its supply then an external reference for the ZN439 should be derived from the same supply. The external reference can vary from + 1.5 to + 3.0V. The ZN439 will operate if $V_{REF\ IN}$ is less than + 1.5V but reduced overdrive to the comparator will increase its delay and so the conversion time will need to be increased.

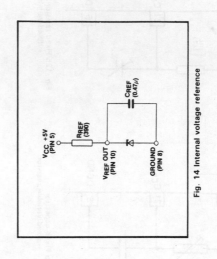

Fig. 14 Internal voltage reference

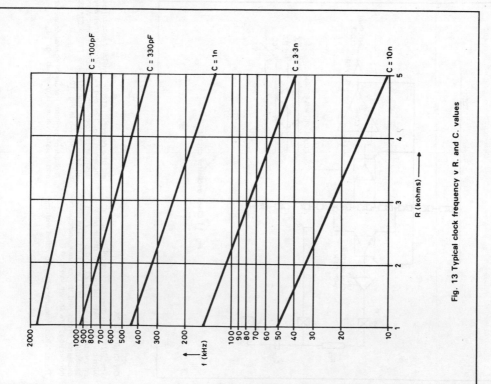

Fig. 13 Typical clock frequency v R. and C. values

## COMPARATOR

The ZN439 contains a fast comparator, the equivalent input circuit of which is shown in Fig. 15. A negative supply voltage is required to supply the tail current of the comparator.

However as this is only 25 to 150μA and need not be well stabilised it can be supplied by a simple diode pump circuit driven from the $R_{CK}$ pin (pin 22).

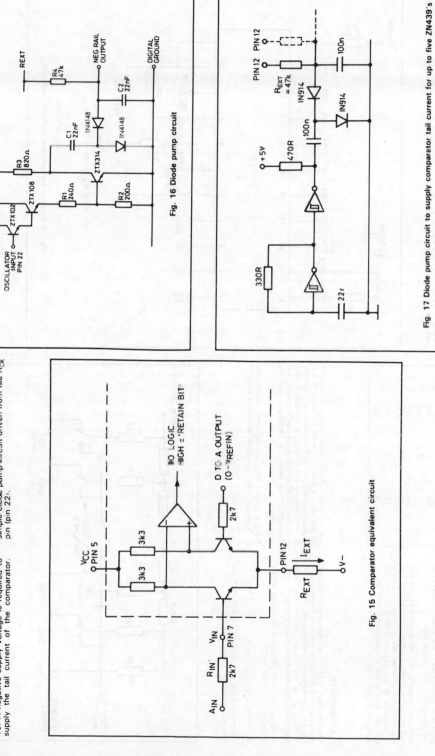

Fig. 15 Comparator equivalent circuit

Fig. 16 Diode pump circuit

Fig. 17 Diode pump circuit to supply comparator tail current for up to five ZN439's

attenuator on the comparator input, whilst for smaller input ranges the signal must be amplified to a suitable level.

Bipolar input ranges are accommodated by off-setting the analogue input ranges so that the comparator always sees a positive input voltage.

## ANALOGUE INPUT RANGES

The basic connection of the ZN439 shown in Fig. 19 has an analogue input range 0 to $V_{REF\,IN}$ which, in some applications, may be made available from previous signal conditioning/scaling circuits. Input voltage ranges greater than this are accommodated by providing an

A suitable circuit is shown in Fig. 16. This circuit can be used in any converter operation mode. The diode pump circuit shown in Fig. 16 is driven by the on-chip clock (pin 22) and applies a voltage of about $-3V$ to R4, thus providing the tail current for the comparator.

Where several ZN439's are used in a system the self-oscillating diode pump circuit of Fig. 17 is recommended. Alternatively, if a negative supply is available in the system then this may be utilised. A list of suitable resistors for different supply voltages is given in Table 1.

## D-A CONVERTER

The converter is of the voltage switching type and uses an R-2R ladder network as shown in Fig. 18. Each element is connected to either 0V or $V_{REF\,IN}$ by transistor voltage switches specially designed for low offset voltage (1mV).

A binary weighted voltage is produced at the output of the R-2R ladder:

$$\text{D-A output} = \frac{n}{256}\,(V_{REF\,IN} - V_{OS}) + V_{OS}$$

where n is the digital input to the D-A from the successive approximation register.

$V_{OS}$ is a small offset voltage that is produced by the device supply current flowing in the package lead resistance. This offset will normally be removed by the setting up procedure and since the offset temperature coefficient is low (7ppm/°C) the effect on accuracy will be negligible.

The D-A output range can be considered to be $0 - V_{REF\,IN}$ through an output resistance R(2k7).

Table 1

| V_ (volts) | $R_{EXT}$ (kΩ) |
|---|---|
| 3 | 47 |
| 5 | 82 |
| 10 | 150 |
| 12 | 180 |
| 15 | 220 |
| 20 | 330 |
| 25 | 390 |
| 30 | 470 |

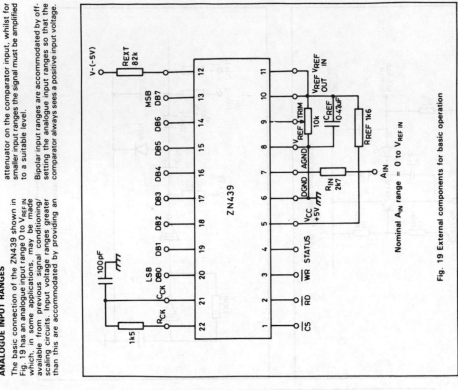

Fig. 19 External components for basic operation

Nominal $A_{IN}$ range = 0 to $V_{REF\,IN}$

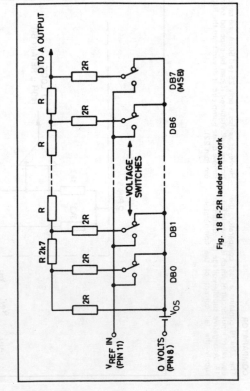

Fig. 18 R-2R ladder network

## UNIPOLAR OPERATION

The general connection for unipolar operation is shown in Fig. 20.

The values of $R_1$ and $R_2$ are chosen so that $V_{IN} = V_{REF\ IN}$ when the analogue input ($A_{IN}$) is at full-scale.

The resulting full-scale range is given by: $A_{IN}FS = \left(1 + \frac{R_1}{R_2}\right) \cdot V_{REF\ IN} = G \cdot V_{REF\ IN}$.

To match the ladder resistance $R_1/R_2$ ($R_{IN}$) = 2.7k.

The required nominal values of $R_1$ and $R_2$ are given by $R_1 = 2.7Gk$, $R_2 = \frac{2.7Gk}{G-1}$.

Using these relationships a table of nominal values of $R_1$ and $R_2$ can be constructed for $V_{REF\ IN} = 2.5V$.

| Input range | G | $R_1$ | $R_2$ |
|---|---|---|---|
| +5V | 2 | 5.4k | 5.4k |
| +10V | 4 | 10.8k | 3.6k |

## GAIN ADJUSTMENT

Due to tolerances in $R_1$ and $R_2$, tolerances in $V_{REF}$ and the gain (full-scale) error of the DAC, some adjustment should be incorporated into $R_1$ to calibrate the full-scale of the converter. When used with the internal reference and 2% resistors a preset capable of adjusting $R_1$ by at least ±5% of its nominal value is suggested.

## ZERO ADJUSTMENT

Zero adjustment must be provided to set the zero transition to the value of +½LSB. This is achieved by applying an adjustable positive offset to tie the comparator input via P2 and R3. The values shown are suitable for all input ranges greater than 1½ times $V_{REF\ IN}$.

Practical circuits values for +5 and +10V input ranges are given in Fig. 21 which incorporates both zero and gain adjustments.

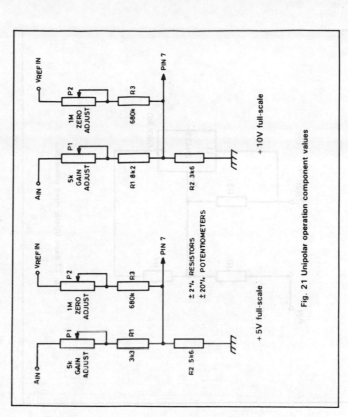

Fig. 21 Unipolar operation component values

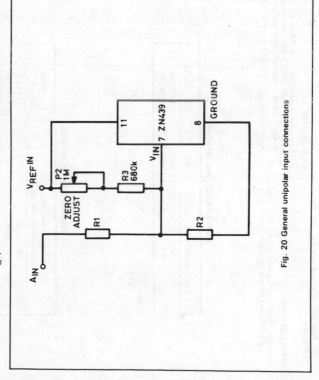

Fig. 20 General unipolar input connections

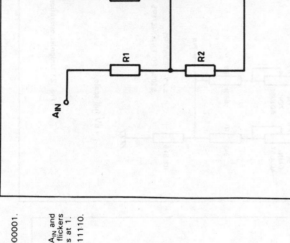

**Fig. 22 Basic bipolar input connection**

## UNIPOLAR ADJUSTMENT PROCEDURE

(i) Apply continuous $\overline{WR}$ pulses at intervals long enough to allow a complete conversion or hold $\overline{WR}$ low and monitor the digital outputs.

### OFFSET SETTING

(ii) Apply ½LSB to $A_{IN}$ and adjust zero until DBO (LSB) just flickers between 0 and 1 with all other bits at 0.
i.e. for transition 00000000 to 00000001.

### GAIN SETTING

(iii) Apply full-scale minus 1 ½LSB to $A_{IN}$ and adjust gain until DBO (LSB) just flickers between 0 and 1 with all other bits at 1.
i.e. for transition 11111111 to 11111110.

### UNIPOLAR SETTING-UP POINTS

| Input range, + FS | ½LSB | FS - 1½LSB |
|---|---|---|
| + 5V | 9.8mV | 4.9707V |
| + 10V | 19.5mV | 9.9414V |

 $1LSB = \dfrac{FS}{256}$

### UNIPOLAR LOGIC CODING

| Analogue input ($A_{IN}$) (Nominal code centre value) | Output code (Binary) |
|---|---|
| FS - 1LSB | 11111111 |
| FS - 2LSB | 11111110 |
| ¾FS | 11000000 |
| ½ FS + 1LSB | 10000001 |
| ½ FS | 10000000 |
| ½ FS - 1LSB | 01111111 |
| ¼ FS | 01000000 |
| 1LSB | 00000001 |
| 0 | 00000000 |

## BIPOLAR OPERATION

For bipolar operation the input to the ZN439 is offset by half full-scale by connecting a resistor $R_3$ between $V_{REF\ IN}$ and $V_{IN}$ (Fig. 22).

When $A_{IN} = -FS$, $V_{IN}$ needs to be equal to zero.

When $A_{IN} = +FS$, $V_{IN}$ needs to be equal to $V_{REF\ IN}$.

If the full-scale range is $\pm G. V_{REF\ IN}$ then $R_1 = (G - 1). R_2$ and $R_1 = G. R_3$ fulfil the required conditions.

To match the ladder resistance, $R_1/R_2/R_3$ ($= R_{IN}$) $= 2.7k$.

Thus the nominal values of $R_1$, $R_2$, $R_3$ are given by $R_1 = 5.4Gk$, $R_2 = 5.4G/(G - 1)k$, $R_3 = 5.4k$.

A bipolar range of $\pm V_{REF\ IN}$ (which corresponds to the basic unipolar range 0 to $V_{REF\ IN}$) results if $R_1 = R_3 = 5.4k$ and $R_2 = \infty$.

Assuming the $V_{REF\ IN} = 2.5V$ the nominal values of resistors for ±5 and ±10V input ranges are given in the following table.

## BIPOLAR ADJUSTMENT PROCEDURE

(i) Apply continuous $\overline{WR}$ pulses at intervals long enough to allow a complete conversion or hold $\overline{WR}$ low and monitor the digital outputs.

### OFFSET SETTING

(ii) Apply $-(FS - \frac{1}{2}LSB)$ to $A_{IN}$ and adjust offset until the DB0 (LSB) output just flickers between 0 and 1 with all other bits at 0.
i.e. for transition 00000000 to 00000001.

### GAIN SETTING

iii) Apply $+(FS - 1\frac{1}{2}LSB)$ to $A_{IN}$ and adjust gain until DB0 (LSB) just flickers between 0 and 1 with all other bits at 1.
i.e. for transition 11111111 to 11111110.

### BIPOLAR SETTING-UP POINTS

| Input range, $\pm FS$ | $-(FS - \frac{1}{2}LSB)$ | $+(FS - 1\frac{1}{2}LSB)$ |
|---|---|---|
| $\pm 5V$ | $-4.9805V$ | $+4.9414V$ |
| $\pm 10V$ | $-9.9609V$ | $+9.8828V$ |

$$1LSB = \frac{2FS}{256}$$

### BIPOLAR LOGIC CODING

| Analogue input ($A_{IN}$) (Nominal code centre value) | Digital output code<br>MSB        LSB |
|---|---|
| $+(FS - 1LSB)$ | 11111111 |
| $+(FS - 2LSB)$ | 11111110 |
| $+\frac{1}{2}FS$ | 11000000 |
| $+1LSB$ | 10000001 |
| 0 | 10000000 |
| $-1LSB$ | 01111111 |
| $-\frac{1}{2}FS$ | 01000000 |
| $-(FS - 1LSB)$ | 00000001 |
| $-FS$ | 00000000 |

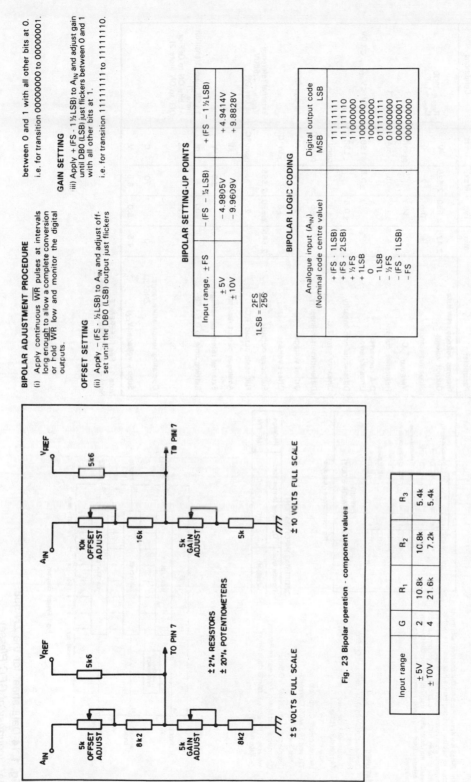

±2% RESISTORS
±20% POTENTIOMETERS

Fig. 23 Bipolar operation - component values

| Input range | G | $R_1$ | $R_2$ | $R_3$ |
|---|---|---|---|---|
| $\pm 5V$ | 2 | 10.8k | 10.8k | 5.4k |
| $\pm 10V$ | 4 | 21.6k | 7.2k | 5.4k |

Minus full-scale (offset) is set by adjusting $R_1$ about its nominal value relative to $R_3$. Plus full-scale (gain) is set by adjusting $R_2$ relative to $R_1$. Practical circuit realisations are given in Fig. 23.

**GEC PLESSEY**
S E M I C O N D U C T O R S

# ZN558D
## 8-BIT LATCHED INPUT D-A CONVERTER

The ZN558 is a monolithic 8-bit D-A converter with input latches to facilitate updating from a data bus. The latch is transparent when enable is LOW and the data is held when enable is taken HIGH. The ZN558 also contains a 2.5V reference the use of which is pin optional to retain flexibility. An external fixed or varying reference may therefore be substituted.

### FEATURES
- Contains DAC with Data Latch and On-Chip Reference
- Guaranteed Monotonic over the Full Operating Temperature Range
- Single +5V Supply
- Microprocessor Compatible
- TTL and 5V CMOS Compatible
- 800ns Settling Time
- Complementary to ZN447 A-D Series
- Commercial and Military Temperature Ranges
- Available in Miniature Plastic Surface Mount Package (MP16)

**Pin connections - top view**

ZN558D
(MP16W - WIDE BODY)

|  | |
|---|---|
| 1 BIT 6 | 16 ANALOGUE OUTPUT |
| 2 BIT 7 | 15 VREF IN |
| 3 BIT 5 | 14 VREF OUT |
| 4 BIT 4 | 13 ANALOGUE GROUND |
| 5 BIT 3 | 12 DIGITAL GROUND |
| 6 BIT 2 | 11 +Vcc (+5V) |
| 7 BIT 1 | 10 ENABLE |
| 8 (MSB) BIT 8 | 9 NC |

**ORDERING INFORMATION**

| Device type | Operating temperature | Package |
|---|---|---|
| ZN558D | 0°C to +70°C | MP16W |

## ABSOLUTE MAXIMUM RATINGS

| | |
|---|---|
| Supply voltage $V_{CC}$ | +7.0V |
| Max. voltage, logic and $V_{REF}$ input | +$V_{CC}$ |
| Operating temperature range | 0°C to +70°C |
| Storage temperature range | -55°C to +125°C |
| Analogue ground to digital ground | ±200mV |

## ELECTRICAL CHARACTERISTICS ($V_{CC}$ = + 5V, $T_{amb}$ = 25°C unless otherwise specified).

| Parameter | Min. | Typ. | Max. | Units | Conditions |
|---|---|---|---|---|---|
| **Internal voltage reference** | | | | | |
| Output voltage | 2.475 | 2.550 | 2.625 | V | $R_{REF}$ = 390Ω |
| Slope resistance | | 0.5 | 2 | Ω | $C_{REF}$ = 1μF |
| $V_{REF\ OUT}$ T.C. | | 50 | | ppm/°C | Note 1 |
| Reference current | 4 | | 15 | mA | |
| **D-A converter** | | | | | |
| Linearity error | | | ±0.5 | LSB | |
| Differential non-linearity | | ±0.5 | | LSB | 2.0V ≤ $V_{REF\ IN}$ ≤ 3.0V |
| Linearity error T.C. | | ±3 | | ppm/°C | |
| Differential non-linearity T.C. | | ±6 | | ppm/°C | |
| Offset voltage | | 2 | 5 | mV | All bits OFF |
| Offset voltage | | ±6 | | μV/°C | |
| Full scale output | 2.545 | 2.550 | 2.555 | | External reference $V_{REF\ IN}$ = 2.560V, all bits ON |
| Full scale output T.C. | | 2 | | ppm/°C | |
| Analogue output resistance | 0 | 4 | | kΩ | |
| External reference voltage | | | 3.0 | V | |
| Settling time to 0.5 LSB | | 800 | | ns | 1 LSB major transition (note 2) |
| | | 1.25 | | μs | All bits ON to OFF or OFF to ON (note 2) |
| Operating temperature range: ZN558D | 0 | | 70 | C | |
| Supply voltage ($V_{CC}$) | 4.5 | 5.0 | 5.5 | V | |

Note 1   See REFERENCE.
Note 2   $R_L$ = 10MΩ, $C_L$ = 10pF.

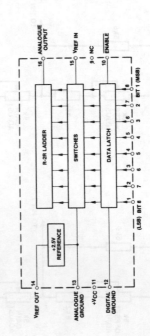

**Fig.1 System diagram**

VREF OUT 14 · ANALOGUE GROUND 13 · +Vcc 11 · DIGITAL GROUND 12 · ANALOGUE OUTPUT 16 · VREF IN 15 · NC 9 · ENABLE 10 · BIT 1 (MSB) · BIT 8 (LSB) · R-2R LADDER · SWITCHES · DATA LATCH · +2.5V REFERENCE

**Fig. 1.14** Data sheet for the ZN 558 DAC (*courtesy of GEC Plessey Semiconductors*)

## ELECTRICAL CHARACTERISTICS (Cont.)

| Parameter | Min. | Typ. | Max. | Units | Conditions |
|---|---|---|---|---|---|
| Supply current | | 20 | 30 | mA | Note 3 |
| Power consumption | | 100 | | mW | |
| **Logic** (over specified operating temperature range) | | | | | |
| High level input voltage | 2.0 | | | V | |
| Low level input voltage | | | 0.8 | V | |
| High level input current | | | 60 | µA | $V_{IN} = 5.5V$, $V_{CC} = $ Max. |
| | | | 20 | µA | $V_{IN} = 2.4V$, $V_{CC} = $ Max. |
| Low level input current | | | −5 | µA | $V_{IN} = 0.4V$, $V_{CC} = $ Max. |
| Input clamp diode voltage | | −1.5 | | V | $I_{IN} = -8mA$ |
| Enable pulse width | 100 | | | ns | Note 4 |
| Data set-up time | 150 | | | ns | Note 5 |
| Data hold time | 10 | | | ns | |

Note 3   All inputs HIGH ($V_{IH} = 3.5V$).
Note 4   Set up time before enable goes high.
Note 5   Hold time after enable goes high.

### D-A CONVERTER

The converter is of the voltage switching type and uses an R-2R ladder network as shown in Fig. 2. Each 2R element is connected to 0V or $V_{REF\ IN}$ by transistor voltage switches specially

designed for low offset voltage (<1mV). A binary weighted voltage is produced at the output of the R-2R ladder.

Analog output $= \dfrac{n}{256}(V_{REF\ IN} - V_{OS}) + V_{OS}$

where n is the digital input to the D-A from the data latch.

$V_{OS}$ is a small offset voltage produced by the D-A switch currents flowing through the package lead resistance. The value of $V_{OS}$ is typically 1mV. This offset will normally be removed by the setting up procedure (see APPLICATIONS section) and because the offset temperature coefficient is low ($\pm 6µV/°C$) the effect on accuracy is negligible.

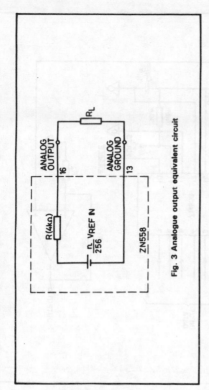

**Fig. 3 Analogue output equivalent circuit**

Fig. 3 shows an equivalent circuit of the output (ignoring $V_{OS}$). The output resistance R has a temperature coefficient of +0.2% per °C.

The gain drift due to this is $\dfrac{0.2R}{R + R_L}$ % per °C

$R_L$ should be chosen to be as large as possible to make the gain drift small. As an example if $R_L = 400k\Omega$ then the gain drift due to the T.C. of R for a 100°C change in ambient temperature will be less than 0.2%. Alternatively the ZN558 can be buffered by an amplifier (see Operating Notes).

### REFERENCE

**(a) Internal reference**

The internal reference is an active band gap circuit which is equivalent to a 2.5V Zener diode with a very low slope impedance (Fig. 4). A resistor ($R_{REF}$) should be connected between $+V_{CC}$ (pin 11) and pin 14. The recommended value of 390$\Omega$ will supply a nominal reference current of (5.0-2.5)/0.39

= 6.4mA.   A   stabilising/decoupling capacitor $C_{REF} = 1µF$ is required between pins 14 and 13 for internal reference option, $V_{REF\ OUT}$ (pin 14) being connected to $V_{REF\ IN}$ (pin 15).

Up to five ZN558's may be driven from one internal reference (there is no need to reduce $R_{REF}$). This useful feature saves power and gives excellent gain tracking between the converters.

**(b) External reference**

If required an external reference voltage may be connected to $V_{REF\ IN}$. The slope resistance of such a reference source should be less than $\dfrac{2.5\Omega}{n}$, where n is the number of converters supplied.

$V_{REF\ IN}$ can be varied from 0 to +3V for ratiometric operation. The ZN558 is guaranteed monotonic for $V_{REF\ IN}$ above 2V.

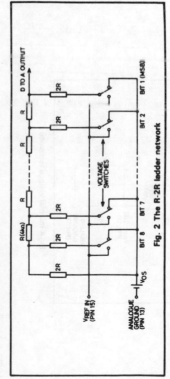

**Fig. 2 The R-2R ladder network**

## OPERATING NOTES

### (1) Unipolar D-A converter

The nominal output range of the ZN558 is 0 to $V_{REF\,IN}$ through a 4kΩ resistance. Other output ranges can readily be obtained by using an external amplifier.

The general scheme (Fig. 6) is suitable for amplifiers with input bias currents less than 1.5 μA.

The resulting full scale range is given by

$$V_{OUT}\,FS = \left(1 + \frac{R1}{R2}\right)V_{REF\,IN} = G.V_{REF\,IN}$$

The impedance at the inverting input is R1//R2 and for low drift with temperature this parallel combination should be equal to the ladder resistance (4kΩ). The required nominal values of R1 and R2 are given by R1 = 4GkΩ and $R_2 = 4G/(G-1)$kΩ.

Using these relationships a table of nominal resistance values for $R_1$ and $R_2$ can be constructed for $V_{REF\,IN} = 2.5V$.

| Output range | G | $R_1$ | $R_2$ |
|---|---|---|---|
| +5V | 2 | 8kΩ | 8kΩ |
| +10V | 4 | 16kΩ | 5.33kΩ |

+10V output ranges are given in Fig. 7. Settling time for a major transistion is 1.5μs typical.

For gain setting $R_1$ is adjusted about its nominal value. Practical circuit realisations (including amplifier stabilising components) for +5 and

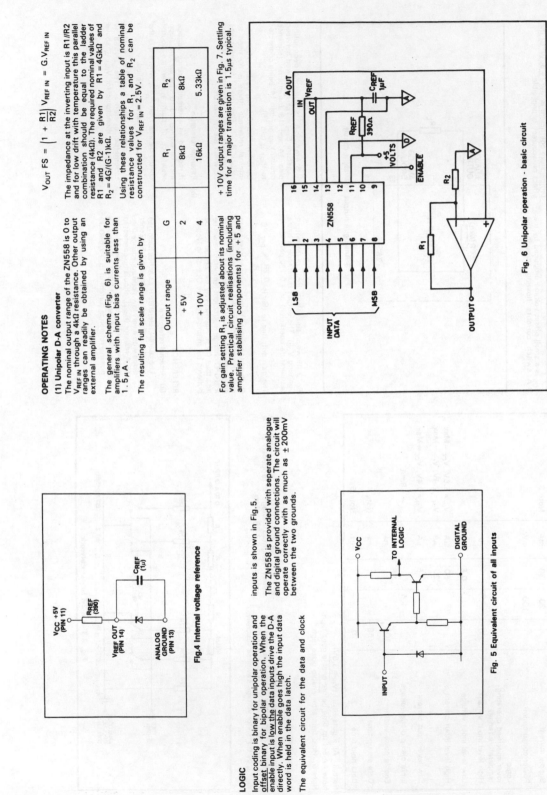

Fig. 6 Unipolar operation - basic circuit

## LOGIC

Input coding is binary for unipolar operation and offset binary for bipolar operation. When the enable input is low the data inputs drive the D-A directly. When enable goes high the input data word is held in the data latch.

The equivalent circuit for the data and clock inputs is shown in Fig.5.

The ZN558 is provided with seperate analogue and digital ground connections. The circuit will operate correctly with as much as ±200mV between the two grounds.

Fig.4 Internal voltage reference

Fig. 5 Equivalent circuit of all inputs

## (2) Bipolar D-A converter

For bipolar operation the output from the ZN558 is offset by half full scale by connecting a resistor $R_3$ between $V_{REF\ IN}$ and the inverting input of the buffer amplifier (Fig. 8).

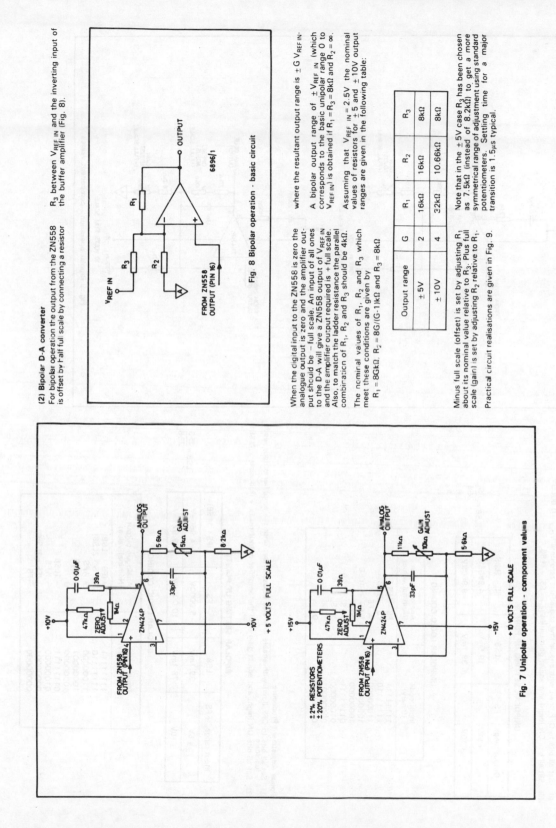

Fig. 8 Bipolar operation - basic circuit

When the digital input to the ZN558 is zero the analogue output is zero and the amplifier output should be − full scale. An input of all ones to the D-A will give a ZN558 output of $V_{REF\ IN}$ and the amplifier output required is + full scale. Also, to match the ladder resistance the parallel combination of $R_1$, $R_2$ and $R_3$ should be 4kΩ.

The nominal values of $R_1$, $R_2$ and $R_3$ which meet these conditions are given by

$$R_1 = 8G\ k\Omega, \quad R_2 = 8G/(G-1)\ k\Omega \quad \text{and} \quad R_3 = 8k\Omega$$

where the resultant output range is $\pm G\ V_{REF\ IN}$.

A bipolar output range of $\pm V_{REF\ IN}$ (which corresponds to the basic unipolar range 0 to $V_{REF\ IN}$) is obtained if $R_1 = R_3 = 8k\Omega$ and $R_2 = \infty$.

Assuming that $V_{REF\ IN} = 2.5V$ the nominal values of resistors for ±5 and ±10V output ranges are given in the following table:

| Output range | G | $R_1$ | $R_2$ | $R_3$ |
|---|---|---|---|---|
| ±5V | 2 | 16kΩ | 16kΩ | 8kΩ |
| ±10V | 4 | 32kΩ | 10.66kΩ | 8kΩ |

Note that in the ±5V case $R_2$ has been chosen as 7.5kΩ (instead of 8.2kΩ) to get a more symmetrical range of adjustment using standard potentiometers. Settling time for a major transition is 1.5µs typical.

Minus full scale (offset) is set by adjusting $R_1$ about its nominal value relative to $R_3$. Plus full scale (gain) is set by adjusting $R_2$ relative to $R_1$.

Practical circuit realisations are given in Fig. 9.

Fig. 7 Unipolar operation - component values

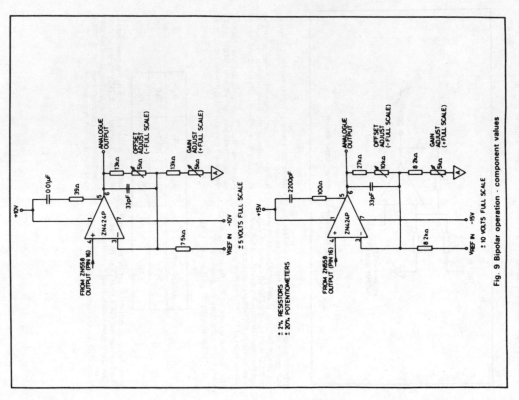

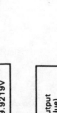

Fig. 9 Bipolar operation - component values

± 2% RESISTORS
± 20% POTENTIOMETERS

**UNIPOLAR ADJUSTMENT PROCEDURE**

(i) Set all bits to OFF (low) with enable low and adjust zero until $V_{OUT} = 0.0000V$.

(ii) Set all bits ON (high) and adjust gain until $V_{OUT} = FS - 1LSB$.

**UNIPOLAR SETTING UP POINTS**

| Output range, +FS | LSB | FS - 1LSB |
|---|---|---|
| +5V | 19.5mV | 4.9805V |
| +10V | 39.1mV | 9.9609V |

$$1LSB = \frac{FS}{256}$$

**UNIPOLAR LOGIC CODING**

| Input code (Binary) | Analog output (Nominal value) |
|---|---|
| 11111111 | FS - 1LSB |
| 11111110 | FS - 2LSB |
| 11000000 | ¾ FS |
| 10000001 | ½FS + 1LSB |
| 10000000 | ½FS |
| 01111111 | ½FS - 1LSB |
| 01000000 | ¼FS |
| 00000001 | 1LSB |
| 00000000 | 0 |

**Bipolar Adjustment Procedure**

(1) Set all bits to OFF (low) with enable low and adjust offset until the amplifier output reads − full scale.

(2) Set all bits ON (high) and adjust gain until the amplifier output reads + (full scale - 1LSB).

**BIPOLAR SETTING UP POINTS**

| Input range, ±FS | LSB | −FS | + (FS - 1LSB) |
|---|---|---|---|
| ±5V | 39.1mV | − 5.0000V | + 4.9609V |
| ±10V | 78.1mV | − 10.0000V | + 9.9219V |

$$1LSB = \frac{2FS}{256}$$

**BIPOLAR LOGIC CODING**

| Input code (Offset binary) | Analogue output (Nominal value) |
|---|---|
| 11111111 | + (FS - 1LSB) |
| 11111110 | + (FS - 2LSB) |
| 11000000 | + ½FS |
| 10000001 | + 1LSB |
| 10000000 | 0 |
| 01111111 | − 1LSB |
| 01000000 | − ½FS |
| 00000001 | − (FS - 1LSB) |
| 00000000 | − FS |

**Digital Integrated Circuits**

Most data sheets for digital ICs follow a similar pattern to the data sheets for analogue circuits. The first section of a sheet gives a general description of the device. The information given may include: (*a*) the number and title of the IC; (*b*) an outline of its main features; (*c*) package options and pinouts; (*d*) a description of the circuit and an outline description of its operation; (*e*) a function table; (*f*) the logic symbol for the device. The next part usually presents the absolute maximum ratings of the IC; these include the current, temperature and voltage figures that must not be exceeded during the operation of the circuit if the IC is not to suffer damage. The recommended operating conditions are then given which specify recommended figures for input, output, and supply currents and voltages, and for the ambient temperature. The maximum allowable rate of change of the input voltage waveform, both low to high and high to low, is often also given. The third part of the data sheet gives the guaranteed electrical characteristic limits of the device when it is operated under the recommended conditions. These electrical characteristics include the minimum high-level and the maximum low-level, input/output voltages and the input/output currents and capacitances.

The data sheets of sequential logic devices will then have sections that give the timing requirements and the switching characteristics. The timing requirements specify: (*a*) the maximum clock frequency; (*b*) the minimum pulse duration; (*c*) the minimum hold and set-up times. The switching characteristics give the maximum propagation delays for various conditions. A relatively simple example is shown by Fig. 1.15 and it is the data sheet for the Motorola MC 74HC573 octal three-state D-type latch. More data sheets for digital circuits are given in Chapter 6.

**Components**

Once a circuit has been designed and the necessary ICs and transistors have been selected some thought must then be given to the passive components employed in the circuit. Mainly, these are resistors and capacitors since the use of inductors is avoided wherever possible.

**Resistors**

The basic considerations that govern the choice of a resistor for an electronic circuit are: (*a*) its resistance value; (*b*) its power rating; (*c*) its physical dimensions; (*d*) the required characteristics.

*Resistance Values and Tolerances*

Resistors are offered by their manufacturers in a number of *preferred values*. There are three main series in use that have, respectively, tolerances of $\pm$ 10, $\pm$ 5 and $\pm$ 2%; smaller tolerances such as $\pm$ 1% are also available but these are considerably more expensive. Table

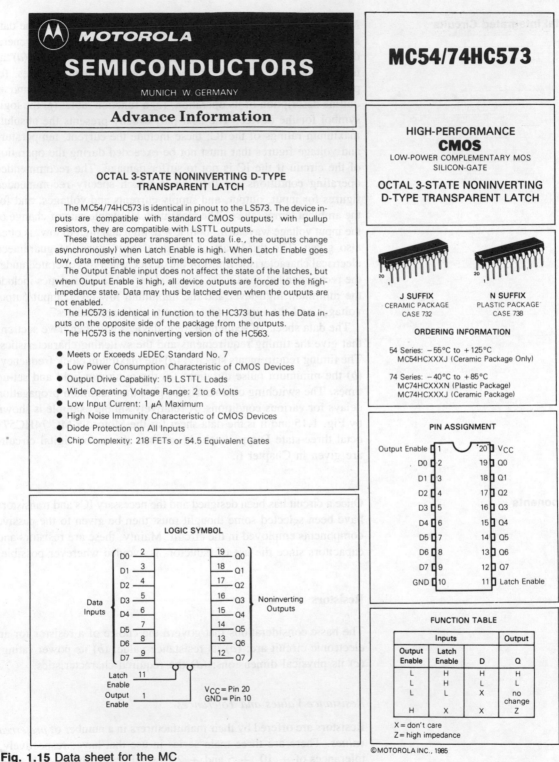

**MOTOROLA SEMICONDUCTORS**
MUNICH W. GERMANY

## Advance Information

**MC54/74HC573**

### OCTAL 3-STATE NONINVERTING D-TYPE TRANSPARENT LATCH

The MC54/74HC573 is identical in pinout to the LS573. The device inputs are compatible with standard CMOS outputs; with pullup resistors, they are compatible with LSTTL outputs.

These latches appear transparent to data (i.e., the outputs change asynchronously) when Latch Enable is high. When Latch Enable goes low, data meeting the setup time becomes latched.

The Output Enable input does not affect the state of the latches, but when Output Enable is high, all device outputs are forced to the high-impedance state. Data may thus be latched even when the outputs are not enabled.

The HC573 is identical in function to the HC373 but has the Data inputs on the opposite side of the package from the outputs.

The HC573 is the noninverting version of the HC563.

- Meets or Exceeds JEDEC Standard No. 7
- Low Power Consumption Characteristic of CMOS Devices
- Output Drive Capability: 15 LSTTL Loads
- Wide Operating Voltage Range: 2 to 6 Volts
- Low Input Current: 1 μA Maximum
- High Noise Immunity Characteristic of CMOS Devices
- Diode Protection on All Inputs
- Chip Complexity: 218 FETs or 54.5 Equivalent Gates

### HIGH-PERFORMANCE CMOS
LOW-POWER COMPLEMENTARY MOS
SILICON-GATE

### OCTAL 3-STATE NONINVERTING D-TYPE TRANSPARENT LATCH

**J SUFFIX**
CERAMIC PACKAGE
CASE 732

**N SUFFIX**
PLASTIC PACKAGE
CASE 738

**ORDERING INFORMATION**

54 Series: −55°C to +125°C
MC54HCXXXJ (Ceramic Package Only)

74 Series: −40°C to +85°C
MC74HCXXXN (Plastic Package)
MC74HCXXXJ (Ceramic Package)

### PIN ASSIGNMENT

| | | | |
|---|---|---|---|
| Output Enable | 1 | 20 | $V_{CC}$ |
| D0 | 2 | 19 | Q0 |
| D1 | 3 | 18 | Q1 |
| D2 | 4 | 17 | Q2 |
| D3 | 5 | 16 | Q3 |
| D4 | 6 | 15 | Q4 |
| D5 | 7 | 14 | Q5 |
| D6 | 8 | 13 | Q6 |
| D7 | 9 | 12 | Q7 |
| GND | 10 | 11 | Latch Enable |

### LOGIC SYMBOL

Data Inputs: D0 (2), D1 (3), D2 (4), D3 (5), D4 (6), D5 (7), D6 (8), D7 (9)

Noninverting Outputs: Q0 (19), Q1 (18), Q2 (17), Q3 (16), Q4 (15), Q5 (14), Q6 (13), Q7 (12)

Latch Enable (11)
Output Enable (1)

$V_{CC}$ = Pin 20
GND = Pin 10

### FUNCTION TABLE

| Inputs | | | Output |
|---|---|---|---|
| Output Enable | Latch Enable | D | Q |
| L | H | H | H |
| L | H | L | L |
| L | L | X | no change |
| H | X | X | Z |

X = don't care
Z = high impedance

© MOTOROLA INC., 1985

**Fig. 1.15** Data sheet for the MC 74HC573 octal three-state D latch (*courtesy of Motorola*)

## MAXIMUM RATINGS*

| Symbol | Parameter | Value | Unit |
|--------|-----------|-------|------|
| $V_{CC}$ | DC Supply Voltage (Referenced to GND) | $-0.5$ to $+7.0$ | V |
| $V_{in}$ | DC Input Voltage (Referenced to GND) | $-1.5$ to $V_{CC}+1.5$ | V |
| $V_{out}$ | DC Output Voltage (Referenced to GND) | $-0.5$ to $V_{CC}+0.5$ | V |
| $I_{in}$ | DC Input Current, per Pin | $\pm 20$ | mA |
| $I_{out}$ | DC Output Current, per Pin | $\pm 35$ | mA |
| $I_{CC}$ | DC Supply Current, $V_{CC}$ and GND Pins | $\pm 75$ | mA |
| $P_D$ | Power Dissipation, per Package† | 500 | mW |
| $T_{stg}$ | Storage Temperature | $-65$ to $+150$ | °C |
| $T_L$ | Lead Temperature (8-Second Soldering) | 260 | °C |

This device contains protection circuitry to guard against damage due to high static voltages or electric fields. However, precautions must be taken to avoid applications of any voltage higher than maximum rated voltages to this high-impedance circuit. For proper operation, $V_{in}$ and $V_{out}$ should be constrained to the range $GND \leq (V_{in}$ or $V_{out}) \leq V_{CC}$.

Unused inputs must always be tied to an appropriate logic voltage level (e.g., either GND or $V_{CC}$). Unused outputs must be left open.

*Maximum Ratings are those values beyond which damage to the device may occur.
Functional operation should be restricted to the Recommended Operating Conditions.
†Power Dissipation Temperature Derating:
    Plastic "N" Package: $-12$mW/°C from 65°C to 85°C
    Ceramic "J" Package: $-12$mW/°C from 100°C to 125°C

## RECOMMENDED OPERATING CONDITIONS

| Symbol | Parameter | | Min | Max | Unit |
|--------|-----------|--|-----|-----|------|
| $V_{CC}$ | DC Supply Voltage (Referenced to GND) | | 2.0 | 6.0 | V |
| $V_{in}, V_{out}$ | DC Input Voltage, Output Voltage (Referenced to GND) | | 0 | $V_{CC}$ | V |
| $T_A$ | Operating Temperature | 74HC Series | $-40$ | $+85$ | °C |
|  |  | 54HC Series | $-55$ | $+125$ | |
| $t_r, t_f$ | Input Rise and Fall Time (Figure 1) | $V_{CC}=2.0$ V | 0 | 1000 | ns |
|  |  | $V_{CC}=4.5$ V | 0 | 500 | |
|  |  | $V_{CC}=6.0$ V | 0 | 400 | |

## ELECTRICAL CHARACTERISTICS (Voltages Referenced to GND)

| Symbol | Parameter | Test Conditions | $V_{CC}$ | 25°C ★ 54HC and 74HC Typical* | 25°C ★ 54HC and 74HC Guaranteed Limit | 85°C 74HC | 125°C 54HC | Unit |
|--------|-----------|-----------------|----------|------|------|------|------|------|
| $V_{IH}$ | Minimum High Level Input Voltage | $V_{out}=0.1$ V or $V_{CC}-0.1$ V $\|I_{out}\|=20\ \mu A$ | 2.0 | 1.2 | 1.5 | 1.5 | 1.5 | V |
|  |  |  | 4.5 | 2.4 | 3.15 | 3.15 | 3.15 | |
|  |  |  | 6.0 | 3.2 | 4.2 | 4.2 | 4.2 | |
| $V_{IL}$ | Maximum Low-Level Input Voltage | $V_{out}=0.1$ V or $V_{CC}-0.1$ V $\|I_{out}\|=20\ \mu A$ | 2.0 | 0.6 | 0.3 | 0.3 | 0.3 | V |
|  |  |  | 4.5 | 1.8 | 0.9 | 0.9 | 0.9 | |
|  |  |  | 6.0 | 2.4 | 1.2 | 1.2 | 1.2 | |
| $V_{OH}$ | Minimum High-Level Output Voltage | $V_{in}=V_{IH}$ or $V_{IL}$ $I_{out}=-20\ \mu A$ | 2.0 | 1.999 | 1.9 | 1.9 | 1.9 | V |
|  |  |  | 4.5 | 4.499 | 4.4 | 4.4 | 4.4 | |
|  |  |  | 6.0 | 5.999 | 5.9 | 5.9 | 5.9 | |
|  |  | $V_{in}=V_{IH}$ or $V_{IL}$ $I_{out}=-6.0$ mA | 4.5 | 4.20 | 3.98 | 3.84 | 3.70 | |
|  |  | $I_{out}=-7.8$ mA | 6.0 | 5.80 | 5.48 | 5.34 | 5.20 | |
| $V_{OL}$ | Maximum Low-Level Output Voltage | $V_{in}=V_{IH}$ or $V_{IL}$ $I_{out}=20\ \mu A$ | 2.0 | 0.001 | 0.1 | 0.1 | 0.1 | V |
|  |  |  | 4.5 | 0.001 | 0.1 | 0.1 | 0.1 | |
|  |  |  | 6.0 | 0.001 | 0.1 | 0.1 | 0.1 | |
|  |  | $V_{in}=V_{IH}$ or $V_{IL}$ $I_{out}=6.0$ mA | 4.5 | 0.20 | 0.26 | 0.33 | 0.40 | |
|  |  | $I_{out}=7.8$ mA | 6.0 | 0.20 | 0.26 | 0.33 | 0.40 | |
| $I_{in}$ | Maximum Input Leakage Current | $V_{in}=V_{CC}$ or GND | 6.0 | $\pm 0.00001$ | $\pm 0.1$ | $\pm 1.0$ | $\pm 1.0$ | $\mu A$ |
| $I_{OZ}$ | Maximum Three-State Leakage Current | Output Enable = $V_{IH}$ $V_{out}=V_{CC}$ or GND | 6.0 | $\pm 0.005$ | $\pm 0.5$ | $\pm 5.0$ | $\pm 10.0$ | $\mu A$ |
| $I_{CC}$ | Maximum Quiescent Supply Current (Per Package) | $V_{in}=V_{CC}$ or GND $I_{out}=0\ \mu A$ | 6.0 | 0.04 | 8 | 80 | 160 | $\mu A$ |

★ The 25°C Guaranteed Limits are also valid for the 74HC series at $-40$°C and the 54HC series at $-55$°C.
* Data labeled "Typical" is not to be used for design purposes, but is intended as an indication of the IC's potential performance.

### SWITCHING CHARACTERISTICS (Input $t_r = t_f = 6$ ns)

| Symbol | Parameter | | $V_{CC}$ | 25°C ★ 54HC and 74HC Typical* | Guaranteed Limit 25°C | 85°C 74HC | 125°C 54HC | Unit |
|---|---|---|---|---|---|---|---|---|
| $t_{PLH}$, $t_{PHL}$ | Maximum Propagation Delay, Input D to Q (Figures 1 and 5) | $C_L = 50$ pF | 2.0 | 70 | 120 | 150 | 180 | ns |
| | | $C_L = 150$ pF | | 100 | 170 | 215 | 255 | |
| | | $C_L = 50$ pF | 4.5 | 17 | 24 | 30 | 36 | |
| | | $C_L = 150$ pF | | 24 | 34 | 43 | 51 | |
| | | $C_L = 50$ pF | 6.0 | 13 | 20 | 26 | 30 | |
| | | $C_L = 150$ pF | | 18 | 29 | 36 | 43 | |
| $t_{PLH}$, $t_{PHL}$ | Maximum Propagation Delay, Latch Enable to Q (Figures 2 and 5) | $C_L = 50$ pF | 2.0 | 70 | 115 | 145 | 170 | ns |
| | | $C_L = 150$ pF | | 100 | 165 | 210 | 245 | |
| | | $C_L = 50$ pF | 4.5 | 16 | 23 | 29 | 34 | |
| | | $C_L = 150$ pF | | 23 | 33 | 42 | 49 | |
| | | $C_L = 50$ pF | 6.0 | 12 | 20 | 25 | 29 | |
| | | $C_L = 150$ pF | | 17 | 28 | 35 | 42 | |
| $t_{PLZ}$, $t_{PHZ}$ | Maximum Propagation Delay, Output Enable to Q (Figures 3 and 6) | $C_L = 50$ pF | 2.0 | 40 | 125 | 160 | 185 | ns |
| | | | 4.5 | 15 | 25 | 32 | 37 | |
| | | | 6.0 | 12 | 21 | 27 | 32 | |
| $t_{PZL}$, $t_{PZH}$ | Maximum Propagation Delay, Output Enable to Q (Figures 3 and 6) | $C_L = 50$ pF | 2.0 | 70 | 140 | 175 | 210 | ns |
| | | $C_L = 150$ pF | | 100 | 190 | 240 | 285 | |
| | | $C_L = 50$ pF | 4.5 | 18 | 28 | 35 | 42 | |
| | | $C_L = 150$ pF | | 25 | 38 | 48 | 57 | |
| | | $C_L = 50$ pF | 6.0 | 15 | 24 | 30 | 35 | |
| | | $C_L = 150$ pF | | 20 | 32 | 41 | 48 | |
| $t_{TLH}$, $t_{THL}$ | Maximum Output Transition Time, Any Output (Figures 1 and 5) | $C_L = 50$ pF | 2.0 | 30 | 60 | 75 | 90 | ns |
| | | | 4.5 | 6 | 12 | 15 | 18 | |
| | | | 6.0 | 5 | 10 | 13 | 15 | |
| $C_{out}$ | Three-State Output Capacitance (Output Enable = $V_{CC}$) | | — | 7.5 | 15 | 15 | 15 | pF |
| $C_{in}$ | Input Capacitance | | — | 5 | 10 | 10 | 10 | pF |
| $C_{PD}$ | Power Dissipation Capacitance Used to determine the no-load dynamic power consumption: $P_D = C_{PD} V_{CC}^2 f + I_{CC} V_{CC}$ | | — | 200 | — | — | — | pF |

★ The 25°C Guaranteed Limits are also valid for the 74HC series at −40°C and the 54HC series at −55°C.
* Data labeled "Typical" is not to be used for design purposes, but is intended as an indication of the IC's potential performance.

### TIMING REQUIREMENTS (Input $t_r = t_f = 6$ ns)

| Symbol | Parameter | $V_{CC}$ | 25°C ★ 54HC and 74HC Typical* | Guaranteed Limit 25°C | 85°C 74HC | 125°C 54HC | Unit |
|---|---|---|---|---|---|---|---|
| $t_{su}$ | Minimum Setup Time, Input D to Latch Enable (Figure 4) | 2.0 | 14 | 75 | 95 | 110 | ns |
| | | 4.5 | 4 | 15 | 19 | 22 | |
| | | 6.0 | 3 | 13 | 16 | 19 | |
| $t_h$ | Minimum Hold Time, Latch Enable to Input D (Figure 4) | 2.0 | −15 | 5 | 5 | 5 | ns |
| | | 4.5 | −1 | 5 | 5 | 5 | |
| | | 6.0 | −1 | 5 | 5 | 5 | |
| $t_w$ | Minimum Pulse Width, Latch Enable (Figure 2) | 2.0 | 20 | 80 | 100 | 120 | ns |
| | | 4.5 | 8 | 16 | 20 | 24 | |
| | | 6.0 | 7 | 14 | 17 | 20 | |
| $t_r$, $t_f$ | Maximum Input Rise and Fall Times (Figure 1) | 2.0 | 2000 | 1000 | 1000 | 1000 | ns |
| | | 4.5 | 1000 | 500 | 500 | 500 | |
| | | 6.0 | 800 | 400 | 400 | 400 | |

★ The 25°C Guaranteed Limits are also valid for the 74HC series at −40°C and the 54HC series at −55°C.
* Data labeled "Typical" is not to be used for design purposes, but is intended as an indication of the IC's potential performance.

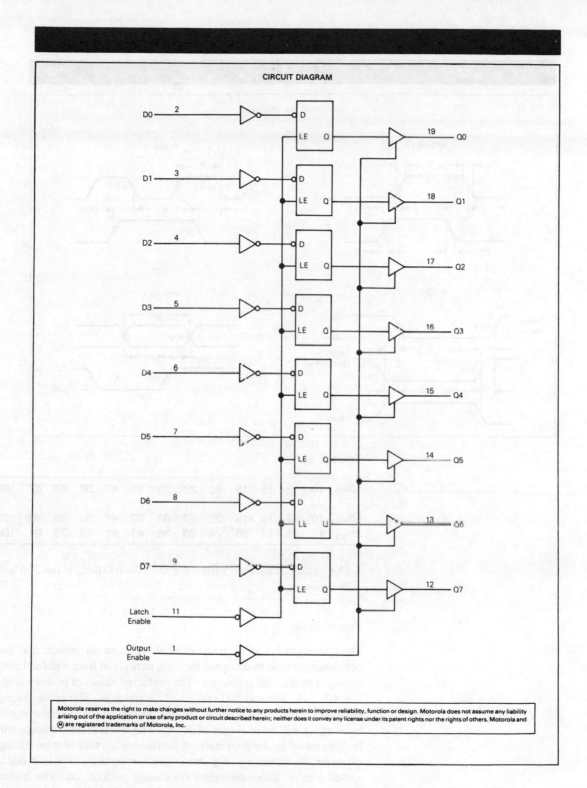

CIRCUIT DIAGRAM

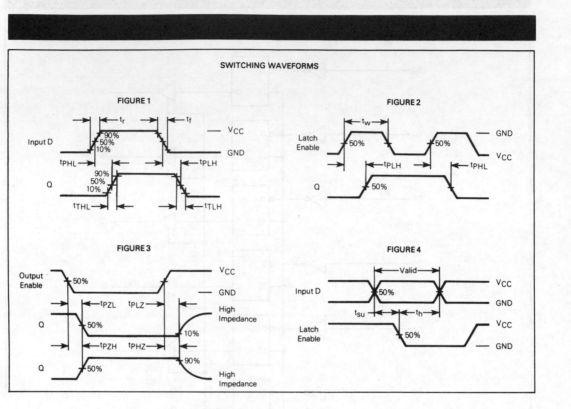

SWITCHING WAVEFORMS

**Table 1.14**

| 10% | 10 | 12 | 15 | 18 | 22 | 27 | 33 | 39 | 47 | 56 | 68 | 82 | 100 |
|-----|----|----|----|----|----|----|----|----|----|----|----|----|-----|
| 5% ⎫ | 10 | 12 | 15 | 18 | 22 | 27 | 33 | 39 | 47 | 56 | 68 | 82 | 100 |
| 2% ⎭ | 11 | 13 | 16 | 20 | 24 | 30 | 36 | 43 | 51 | 62 | 75 | 91 | 100 |

1.14 gives the resistance values for the 10−100 decade for 2, 5 and 10% resistors.

### Power Rating

The power rating of a resistor is the maximum power that the component is able to dissipate for long periods of time without being damaged by the heat developed. The preferred values of power ratings are: 63, 125, 250 and 500 mW, 1, 2, 3 and 4 W. The power rating must be derated if the ambient temperature in which the resistor will be working will be in excess of 70 °C. The ambient temperature will be determined by the proximity of the resistor to other heat-producing elements, its mounting, e.g. horizontal or vertical, if it is inside a closed area, if the components are closely packed, etc. The higher the power rating the larger will be the physical dimensions of the resistor and the more space the resistor will both occupy, and need to be left clear around it.

## Temperature Coefficient

The resistance of a resistor may change with any change in the ambient temperature. Its *temperature coefficient* is the relative variation of the resistance value between two given temperatures divided by the difference between the two temperatures, and it is usually expressed in parts per million per degree Celsius (ppm/°C).

## Stability

The stability of a resistor is its ability to keep the same resistance value over a period of time. The higher the power rating the lower will be the stability.

## Noise

The noise generated within a resistor is only of importance for those resistors that are to be used in the input stages of a low-noise circuit, such as a preamplifier or a sensitive measurement instrument.

## Applications

The four main kinds of resistor used in modern electronic circuits are the carbon film, the metal film, the metal glaze and the metal oxide. Wirewound resistors are only employed in applications where a high power rating is required and carbon composition resistors, once the most common type, are now rarely used. Table 1.15 gives a comparison between the four types of resistor from which it should be possible to select the best type for a given application.

For general-purpose use, where the parameters are not critical, the first choice should be the carbon film type of resistor since it is the cheapest. When long-term stability is important either a metal film or metal oxide type should be chosen. If a low temperature coefficient is wanted then a precision metal film resistor must be employed, and if voltage surges are to be withstood a metal glaze type is

**Table 1.15**

| Type | Tolerance (± %) | Power dissipation (W) | Stability | Temperature coefficient (ppm/°C) | Resistance range (Ω) | High voltage | High frequency | Noise | Cost |
|------|------|------|------|------|------|------|------|------|------|
| Carbon film | 2.5 | 2 | Average | −1500 | 1–1M | Good | Average | Poor | Low |
| Metal film | 1.2 | 0.5 | Good | ±100 | 0.1–1M | Average | Average | Good | Average |
| Metal glaze | 2.5 | 0.5 | Very good | ±20 | 1–100M | Good | Average | Average | High |
| Metal oxide | 2.5 | 0.6 | Average | ±300 | 0.1–240K | Good | Average | Average | Average |

recommended since voltage surges may leap from one spiral to another in a film resistor. Low-noise applications are best served by choosing a metal film resistor.

Resistor networks using either thick or thin film techniques are also employed, particularly for surface-mounted components. Thick films are composed of a mixture of glass and metal, such as ruthenium oxide, that are connected to the outside circuitry by silver, or palladium silver, contacts. The resistors are laser trimmed to give a resistance accuracy of $\pm 2\%$ or less with a temperature coefficient of typically $\pm 100\ \text{ppm}/^\circ\text{C}$. Thin film resistors use nichrome or tantalum nitride as the resistive material. Laser trimming gives accuracies as good as $\pm 0.1\%$.

## Capacitors

When a capacitor is chosen for a particular circuit application a number of factors ought to be considered. The main factors are, of course, the capacitance value and the tolerance. If the capacitance value needed is greater than about $2.2\ \mu\text{F}$ then an electrolytic type of capacitor will probably have to be used. Aluminium electrolytic capacitors can be obtained in values ranging from about 1 up to $68\,000\ \mu\text{F}$ with tolerances that may, typically, be $-10$, $+50$ or $\pm 20\%$. Tantalum capacitors are available in values $4.7-100\ \mu\text{F}$. Several different kinds of plastic capacitor are in existence, namely polypropylene, polycarbonate, polyester and polystyrene, and these can be obtained in a wide range of values, e.g. polypropylene $100\ \text{pF}$ to $1000\ \mu\text{F}$, polycarbonate $100\ \text{pF}$ to $10\ \mu\text{F}$, polyester $4.7\ \text{nF}$ to $4.7\ \mu\text{F}$, and polystyrene $10\ \text{pF}$ to $39\ \text{nF}$ with tolerances that vary from $\pm 2$ to $\pm 20\%$ according to to the type considered. Silvered mica capacitors vary from $2.2\ \text{pF}$ to $10\ \text{nF} \pm 1\%$, and ceramic capacitors from $10\ \text{pF}$ to $1\ \mu\text{F}$. The tolerance of a ceramic capacitor varies considerably with the type, i.e monolithic, disc, plate or dil.

Other capacitor parameters of importance are the voltage rating, the power factor, the series loss resistance, the insulation resistance, the variation of capacitance with temperature, voltage, frequency and time (i.e. the stability), and current ratings. Table 1.16 gives a guide to the type(s) of capacitors that should be used for various common electronic circuit functions.

**Table 1.16**

| Application | Ceramic | Mica | Aluminium | Tantalum | Polypropylene | Polycarbonate | Polyester | Polystyrene |
|---|---|---|---|---|---|---|---|---|
| d.c. blocking | | | X | X | | | X | X |
| Decoupling | X | | X | X | | | X | X |
| Coupling | X | | X | X | | | X | |
| Smoothing | X | | X | X | | | | |
| Timing | | X | | | X | X | | X |
| Tuning | | X | | | X | X | | X |

*Notes*

(*a*) To improve reliability capacitors should be operated at a voltage lower than their voltage rating.

(*b*) Non-polarized electrolytic capacitors can be obtained.

(*c*) An electrolytic component employed as the reservoir capacitor in a power supply must have a sufficiently high ripple current rating.

## Transducers

The term *transducer* is applied to any device that is able to convert energy from one form to another. An electrical transducer senses a change in some non-electrical quantity and converts it into a corresponding change in an electrical signal; examples of this are the telephone microphone and a pressure sensor. Other transducers convert an electrical signal into the corresponding non-electrical signal and examples of these include an electric light and a radio loudspeaker. A wide variety of different kinds of transducer are available to perform many varied tasks and only a few of them will be mentioned in this chapter. Whatever the type of transducer that is required for a particular application there are a number of common factors that are usually of importance and are often specified in the manufacturer's data sheet for the device. These factors include: (*a*) accuracy; (*b*) frequency range; (*c*) ease of use; (*d*) linearity; (*e*) physical dimensions and weight; (*f*) range of values, both electrical and non-electrical; (*g*) reliability; (*h*) resolution; (*i*) sensitivity. In addition, the cost and availability of a transducer should be considered before it is selected for a particular purpose.

## Pressure Sensors

A pressure sensor converts pressure into a proportional d.c. voltage. Traditionally, pressure has been sensed by a strain gauge but more recent sensors use piezo-electric resistance. Piezo-resistance is the change in the resistance of a piezo-electric substance when it is subjected to an applied strain. A piezo-electric pressure sensor consists of four identical piezo-electric resistors which are attached to the surface of a diaphragm. When pressure is exerted upon the diaphragm, the diaphragm is caused to flex and this induces a strain into the resistors which, in turn, alters their resistances. In this way a change in pressure is converted into a change in electrical resistance. The four resistances are bridge-connected, as shown by Fig. 1.16; if $R$ is the piezo-resistance for zero pressure on the diaphragm, then $R + \Delta R$ and $R - \Delta R$ are the actual resistances of the resistors when a particular value of pressure is exerted upon the diaphragm. All four resistors change their resistance by approximately the same amount, two resistors increasing in value and the other two decreasing in value.

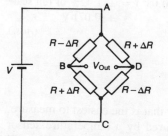

**Fig. 1.16** Pressure sensor bridge circuit

When a positive pressure is applied to port B the voltage at that point, with respect to earth, increases and the voltage at port D decreases. This results in an increase in the differential voltage $V_B - V_D$. The d.c. output voltage of the sensor is directly proportional to both the supply voltage $V$ and the applied pressure.

Among the terms employed in pressure sensor data sheets are:

(a) *Null effect*: this is the output voltage obtained when equal pressures are applied to the two measurement ports and there ought to be zero output.

(b) *Linearity error*: this is the deviation of the sensor's output voltage from a straight-line graph as the applied pressure is varied over the designed-for range.

(c) *Repeatability error*: this is the deviation between the output voltages obtained for successive applications of the same pressure.

(d) *Mechanical hysteresis*: this is the difference between the output voltage obtained when a certain pressure is approached first with increasing pressure and then with decreasing pressure.

(e) *Temperature hysteresis*: this is the difference between the output voltages obtained for the same pressure at two different temperatures.

(f) *Span*: this is the arithmetic difference between the output voltage obtained for the specified maximum and minimum operating pressures.

(g) *Sensitivity*: this is the ratio (span)/(specified pressure range).

The total error in the output voltage is given by the root mean square (r.m.s.) value of the individual errors.

### Example 1.2

A pressure sensor has a full-scale output voltage of 5 V. Its inherent errors are: (a) null offset $\pm$ 0.05 V; (b) linearity $\pm$ 1% of full scale; (c) total hysteresis error $\pm$ 0.08% of full scale; (d) repeatability error $\pm$ 0.08% of full scale. Calculate the total error.

### Solution

$\pm$ 0.05 V = $\pm$ 1% of full scale. Hence

$$\text{Total error} = \sqrt{(1^2 + 1^2 + 0.08^2 + 0.08^2)} = \pm 1.42\% \text{ of full scale}$$
$$= \pm 0.071 \text{ V}$$

(*Ans.*)

### Temperature Measurement

Temperature is the physical parameter that is the easiest to measure. Temperature can be accurately measured by a temperature sensor based upon either the thermoelectric effect or upon the variation of resistance with temperature. The main characteristics of the different kinds of temperature sensor are listed in Table 1.17.

**Table 1.17**

| Type | Temperature range (°C) | Sensitivity at 25 °C | Accuracy | Linearity | Cost |
|------|------------------------|----------------------|----------|-----------|------|
| Thermocouple | −270 to 1800 | <50 V/°C | ±0.5 °C | Poor | Dearer |
| Thermistor | −100 to 450 | 5%/°C | ±0.1 °C | ± 0.2 °C | Cheaper |
| PRW | −250 to 900 | 0.5%/°C | ±0.1 °C | Very good | Dearest |
| Diodes and transistors | −270 to 175 | −2.2 mV/°C | ±2 °C | Average | Cheapest |

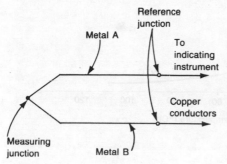

**Fig. 1.17** Basic thermocouple

Although a number of different metals could be employed, most resistance thermometers use platinum resistance wire (PRW) since it is stable and has a linear temperature coefficient over a wide range of temperatures. The standard platinum resistance thermometer has a resistance of 100 Ω at 0 °C and produces a resistance change of 38.5 Ω over a temperature range of 0−100 °C. This is known as the *fundamental interval*.

The action of a thermocouple is based upon the generation of an electromotive force (e.m.f.) at the point where two different metals are joined together. This thermal e.m.f. is directly proportional to the temperature of the junction. If, as shown by Fig. 1.17, the two metals are joined together at one point and are each connected either together or (as shown) to copper conductors and the junctions are at differing temperatures, a potential difference will exist. The magnitude of the generated e.m.f. is a function of the difference between the temperatures of the two junctions. In practice, the reference junction is held at a constant, known, temperature and the potential difference that is generated is proportional to the temperature of the measuring junction. A number of different types of thermocouple can be obtained; (*a*) type J: uses iron and copper−nickel and has a temperature range of −200 to +850 °C; (*b*) type K: uses nickel alloys, (−200 to +1000 °C); (*c*) type N: uses nickel alloys as well but the exact composition of the conductors is specified (−230 to +1230 °C); (*d*) type R: uses platinum and rhodium (up to 1350 °C); (*e*) type T: uses copper and copper−nickel (−200 to +400 °C).

The thermocouple offers the advantages of cheapness, versatility, ruggedness and reasonable sensitivity. The platinum resistance thermometer, although more expensive, offers greater accuracy and stability and better resolution.

*Thermistors*

The *thermistor* is a device whose resistance changes with change in temperature and Fig. 1.18 shows a typical characteristic. Other types of thermistor may have a temperature range as large as −80 to +150 °C with a resistance as high as, say, 100 kΩ. Thermistors can be obtained with either a positive, or a negative, temperature coefficient of resistance, known respectively as PTC and NTC devices. The short-term data for thermistors give typical resistance values at

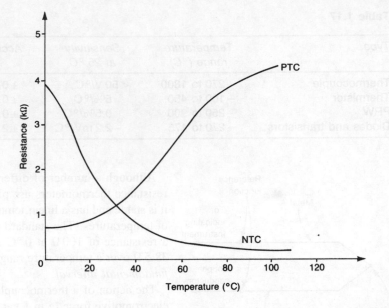

**Fig. 1.18** Resistance–temperature characteristic of a thermistor

a quoted temperature — usually 20 or 25 °C — and the temperature range.

Light-dependent resistors are also employed in electronic circuits. These devices have a *dark* resistance that falls linearly as the illumination, quoted in lux, incident upon them increases. Typically, the change in resistance may be from 100 kΩ to 10 Ω for a change in lux of from 1 to $10^4$.

## Quartz Crystals

A quartz crystal is a piezo-electric device that if subjected to a mechanical stress has an electric potential developed across it, if the direction of the stress is reversed the polarity of the voltage is also reversed. Conversely, if a voltage is applied across a quartz crystal the shape of the crystal will change, causing a stress to be applied to it in a direction determined by the polarity of the voltage. If the applied voltage is alternating the crystal will be subjected to an alternating stress and it will tend to vibrate at its natural frequency. Crystal units are cut from the parent crystal at a precise angle relative to the crystal's axis and then the unit is mounted in a suitable holder and silver plated on each side to provide electrical connections. The natural frequency of each crystal unit is determined by its physical dimensions, its mode of vibration, and its *cut*. A large number of different cuts are possible and Table 1.18 lists the more commonly employed.

The modes of vibration are extensional, face shear, flexural and thickness and these modes are illustrated by Fig. 1.19. Other than

**Table 1.18**

| Cut | Mode of vibration | Frequency range |
|-----|-------------------|-----------------|
| AT | Thickness (fundamental) | 800 kHz to 30 MHz |
| | (3rd overtone) | 4–90 MHz |
| | (5th overtone) | 4–150 MHz |
| | (7th overtone) | 100–200 MHz |
| BT | Thickness | 1–3 MHz |
| CT | Face shear | 300–850 kHz |
| DT | Face shear | 100–800 kHz |
| GT | Thickness | 1–3 MHz |
| J | Flexural | Below 10 kHz |
| NT | Flexural | 10–100 kHz |
| X | Extensional | 40–200 kHz |
| XY | Extensional | 3–85 kHz |

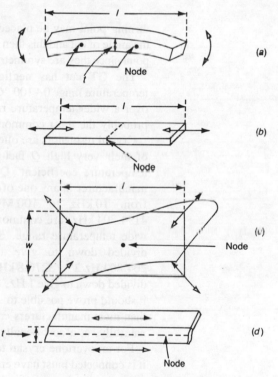

**Fig. 1.19** Quartz crystal vibration modes: (a) thickness flexural, $f = k_a t/l^2$; (b) length extensional, $f = k_b/l$; (c) face shear, $f = k_c/(1/l^2 + 1/w^2)$; (d) thickness shear, $f = k_d/t$.

the obvious choice of a crystal to give the wanted frequency, usually the main requirement is for temperature stability. This is the relative frequency variation $\Delta f/f$ as a function of temperature of the natural frequency of a crystal, where $f$ is the frequency at the reference temperature. Typical frequency/temperature curves for some commonly employed crystals are shown in Fig. 1.20. The points of zero temperature coefficient are known as the *turning-points*. One

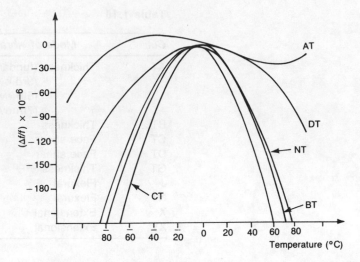

**Fig. 1.20** Frequency–temperature characteristics of various crystal cuts

turning-point may be placed wherever desired by suitable choice of the angle of cut and this then also fixes the position of the other turning point since they are symmetrical about a point in the 20–30 °C range.

The GT cut has negligible temperature coefficient over the temperature range 0–100 °C but the AT cut has the best performance over a wider temperature range such as −55 to +100 °C and it is probably the most commonly employed.

Crystal resonators are often employed in electronic circuits because of their very high $Q$ factor, relatively small size and very good temperature coefficient. Quartz crystals can be obtained from a manufacturer in any one of nearly 100 standard frequencies ranging from 10 kHz to 100 MHz. Two frequencies, 32.768 and 4194.304 kHz, are commonly used since they perform well over a wide temperature range. Some of the crystal frequencies can be divided down to give the mains frequency of 50 Hz, e.g. 204.80 kHz/$2^{12}$, 3276.8 kHz/$2^{16}$; other crystal frequencies can be divided down to give 1 Hz, 1 kHz, 10 kHz, etc. For most applications it should prove possible to select one of the standard crystals; if not, then most manufacturers can supply specified crystals to order but, naturally, at a cost penalty.

For an overtone crystal to work satisfactorily the circuit in which it is connected must have enough selectivity to ensure that the crystal cannot resonate at its fundamental frequency.

## Optical Transducers

The operation of many electronic systems relies on the conversion of light energy into an electrical signal. There are three basic photo-electric effects:

(*a*) *Photoconductivity* in which the conductivity of a semiconductor material is altered by incident light energy;

(*b*) *Photovoltaic* in which an e.m.f. is generated that is directly proportional to the intensity of the incident light;

(*c*) *Photoemissivity* in which a current flowing in a semiconductor material causes it to radiate light energy in the visible part of the frequency spectrum.

Light energy is measured in terms of units known as the *lumen* (lm), the *candela* (cd), the *lux* (lx), and *candela/metre* (cd/m). The total light energy emitted, or received, by a surface is known as its *luminous flux*, measured in lumens. One lumen is the luminous flux emitted by a source of 1 candela within a solid angle of 1 steradian. *Luminous intensity* is the strength of a light source in a given direction and it is measured in candela, i.e. candela = lumens/steradian. The luminous intensity of a light source may vary with the angle from which the source is viewed. When the angle is large the term *luminance*, in candela/metre, is employed. Lastly, the amount of luminous flux incident upon a surface is measured in lux. One lux is the illumination produced when 1 lumen of light flux is incident upon a surface of area $1 \text{ m}^2$. An illumination of 1 lx is produced in an area of $1 \text{ m}^2$ at a distance of 1 m by a source of 1 cd. The *dark current* is the current that flows in a photo device when it is not illuminated and the *light current* is the current that flows when it is illuminated.

A *photoconductive cell*, or *light-dependent resistor*, is a device that has a resistance that decreases with increase in the illumination incident upon the cell. Its most important parameters are its illumination sensitivity, i.e. the ratio [light current (A)]/[incident illumination (lx)], and its spectral response, i.e. its sensitivity plotted against the wavelength of the incident light.

A *photodiode* acts like a photoconductive device when it is reverse-biased, since then the leakage current increases in direct proportion to the incident illumination. When the diode is forward-biased it acts like a photovoltaic device. The data sheet for a photodiode gives figures for: (*a*) its sensitivity in nanoamps/lux when reverse-biased; (*b*) its maximum reverse bias voltage; (*c*) its maximum forward- and reverse-bias currents; (*d*) its maximum power dissipation; (*e*) its spectral response.

A *photovoltaic cell* (solar cell) is essentially a photodiode that has been manufactured to develop the maximum output power.

A *phototransistor* is a bipolar transistor with normal collector and emitter terminals but no base terminal. When the base region of the phototransistor is illuminated a collector current flows that is proportional to the luminous flux in lux. The output characteristics of a phototransistor are very similar to that of a bipolar transistor except that the curves are labelled with lux values instead of values of base current. The data sheet will give the usual transistor data plus figures for the maximum dark and light currents.

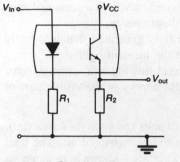

**Fig. 1.21** Opto-coupler

A *light-emitting diode* (LED) radiates light energy in the visible part of the frequency spectrum whenever it is turned ON by a forward-bias voltage.

An *opto-isolator*, or *photocoupler*, consists of an LED and a phototransistor in the same package. When a current is passed through the LED it emits light energy which illuminates the phototransistor, the phototransistor then conducts and an output voltage is developed across the load. The basic principle is illustrated by Fig. 1.21. The data sheet of an opto-isolator includes such data as the isolation voltage in kilovolts, the current transfer ratio as a percentage, the breakdown voltage in volts, and the switching time in microseconds.

# 2 CR Circuits and Attenuators

Capacitor–resistor (CR) circuits are widely employed in analogue electronic circuits, and less often in digital electronic circuits, for a wide variety of purposes. Three of the many possible examples are first-order filters, differentiating and integrating circuits, and inter-stage coupling circuits. Attenuators are often used in electronic circuitry to reduce the level of a signal applied to an amplifier in order to avoid overloading and distortion. For a specified loss to be obtained, a passive attenuator ought to be matched at its input terminals to the source resistance and at its output terminals to the load resistance. When the source and the load resistances are equal to one another a symmetrical attenuator can be employed, but when the resistances differ a non-symmetrical attenuator is required. This will usually provide matching as well as attenuation. Attenuators can always be connected in cascade to obtain a larger attenuation than is provided by each attenuator alone. Active attenuators are also available and some of these have an attenuation that can be programmed by an applied control voltage. Loss-less matching networks are employed at radio frequencies to match a source to a load without the introduction of power losses and these networks also provide some degree of filtering.

## The Low-pass Filter

Figure 2.1 shows that the circuit of the basic CR low-pass filter consists merely of a series resistor R and a parallel capacitor C. The voltage source is supposed to be of very low resistance and the load connected across the output terminals is supposed to be of high resistance, so that the resistances do not affect the operation of the circuit.

The output voltage $V_{out}$ of the filter is

$$V_{out} = V_{in}jX_C/(R + jX_C)$$
$$= (V_{in}/j\omega C)/(R + 1/j\omega C)$$
$$= V_{in}/(1 + j\omega CR)$$

The magnitude of the output voltage is

$$|V_{out}| = V_{in}/\sqrt{(1 + \omega^2 C^2 R^2)} \qquad (2.1)$$

and its phase angle $\phi$, relative to the input voltage, is

$$\phi = -\tan^{-1}\omega CR \qquad (2.2)$$

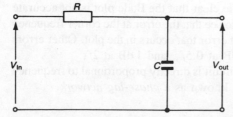

**Fig. 2.1** CR low-pass filter

Clearly, both the magnitude and the phase of the output voltage vary with the frequency of the input signal.

At the *corner frequency* $f_3$ the attenuation of the filter is 3 dB or $1/\sqrt{2}$. Hence

$$|V_{out}/V_{in}| = 1/\sqrt{2} = 1/\sqrt{(1 + \omega_3^2 C^2 R^2)}$$
$$1 = \omega_3^2 C^2 R^2 \quad \text{or} \quad \omega_3 = 1/CR \qquad (2.3)$$

and

$$f_3 = 1/2\pi CR \text{ Hz} \qquad (2.4)$$

This relationship enables equation (2.1) to be written in terms of the 3 dB frequency as shown by equation (2.5), i.e.

$$|V_{out}/V_{in}| = 1/\sqrt{(1 + \omega^2/\omega_3^2)} \qquad (2.5)$$

Also, the phase shift $\phi$ introduced by the filter can be written in the form

$$\phi = -\tan^{-1}(\omega/\omega_3) \qquad (2.6)$$

At a frequency equal to twice the corner frequency the amplitude ratio of the circuit is

$$|V_{out}/V_{in}| = 1/\sqrt{[1 + (2\omega_3)^2 C^2 R^2]} = 1/\sqrt{(1^2 + 2^2)}$$
$$= 1/\sqrt{5} = -7 \text{ dB}$$

Similarly, at frequency $4f_3$

$$|V_{out}/V_{in}| = 1/\sqrt{(1 + 4^2)} = 1/\sqrt{17} = -12.3 \text{ dB}$$

at frequency $8f_3$

$$|V_{out}/V_{in}| = 1/\sqrt{(1 + 8^2)} = 1/\sqrt{65} = -18.1 \text{ dB}$$

at frequency $16f_3$

$$|V_{out}/V_{in}| = 1/\sqrt{(1 + 16^2)} = 1/\sqrt{257} = -24.1 \text{ dB}$$

and so on. The loss−frequency curve of the circuit is shown plotted in Fig. 2.2(a). Note that at frequencies higher than $2f_3$ the attenuation of the circuit increases at the constant rate of 6 dB/octave, or 20 dB/decade.

A *Bode plot* of the amplitude response of the filter assumes that the attenuation is constant at 0 dB from 0 Hz up to the corner frequency $f_3$, whereafter it increases at 6 dB/octave. The Bode plot of the *CR* low-pass filter is shown by Fig. 2.2(b) and can be seen to consist of two straight lines that intersect at the corner frequency. Comparing Figs 2.2(a) and (b) it is quite clear that the Bode plot is not accurate at all frequencies. It is easy to see that the error at the corner frequency is 3 dB and this is the largest error that occurs in the plot. Other errors are: 0.53 dB at $0.25f_3$, 1 dB at $0.5f_3$, and 1 dB at $2f_3$.

The phase shift $\phi$ of the circuit is directly proportional to frequency and so this circuit is often known as a *phase-lag network*.

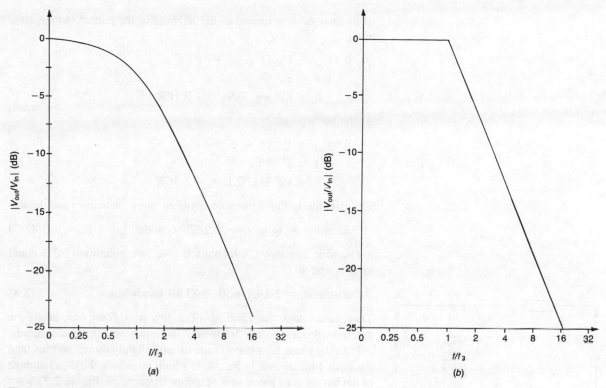

**Fig. 2.2** (*a*) Loss—frequency
characteristic of a *CR* low-pass filter;
(*b*) Bode plot of (*a*)

### Example 2.1

A low-pass filter is to have a corner frequency of 4000 Hz. Determine the
frequency at which its loss is (*a*) 8 dB and (*b*) 20 dB.

*Solution*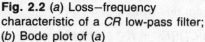

(*a*) 8 dB is a voltage ratio of 2.51 : 1 and hence
$$1/2.51 = 1/\sqrt{(1 + \omega^2/\omega_3^2)} \quad \text{and} \quad \omega/\omega_3 = \sqrt{5.3}$$
Therefore
$$f = 4000\ \sqrt{5.3} = 9209\ \text{Hz} \quad (Ans.)$$
(*b*) 20 dB is a voltage ratio of 10 : 1 and hence
$$\omega/\omega_3 = 1/\sqrt{(1 + \omega^2/\omega_3^2)} = \sqrt{99}$$
Therefore
$$f = \sqrt{99} \times 4000 = 39.8\ \text{kHz} \quad (Ans.)$$

### Risetime

When a rectangular pulse is applied to the input terminals of a low-
pass filter the output voltage does not change its value instantaneously.
The time taken for the output voltage to increase from 10 to 90%

of its final value is known as the *risetime* of the circuit. At the 10% point,

$$0.1V_{out} = V_{out}(1 - e^{-t_1/CR})$$
$$0.9 = e^{-t_1/CR}$$
$$t_1 = CR \log_e 0.9 = -0.1CR$$

At the 90% point,

$$0.9V_{out} = V_{out}(1 - e^{-t_2/CR})$$
$$0.1 = e^{-t_2/CR}$$
$$t_2 = CR \log_e 0.1 = -2.3CR$$

The risetime is the difference between these two times and hence

$$\text{Risetime} = t_2 - t_1 = 2.2CR \text{ seconds} \qquad (2.7)$$

The time constant $CR$ seconds is also (see equation (2.3)) equal to $1/\omega_3$ and so

$$\text{Risetime} = 2.2/\omega_3 = 0.35/(3 \text{ dB bandwidth}) \qquad (2.8)$$

This means that the risetime of a low-pass filter can easily be determined if its corner frequency is known or can be measured.

Practical phase lag networks are usually slightly more complex than the basic filter shown in Fig. 2.1, which possesses the disadvantage of not having zero phase shift at higher frequencies. Figure 2.3 shows a typical phase-lag circuit. From this figure,

$$V_{out} = V_{in}[R_2 + 1/j\omega C_1]/[R_1 + R_2 + 1/j\omega C_1]$$
$$V_{out}/V_{in} = [1 + j\omega C_1 R_2]/[1 + j\omega C_1(R_1 + R_2)]$$
$$= [1 + j\omega(R_2/(R_1 + R_2)C_1(R_1 + R_2))]/$$
$$[1 + j\omega C_1(R_1 + R_2)]$$

Let $\alpha = R_2/(R_1 + R_2)$ and $\tau = C_1(R_1 + R_2)$, then

$$V_{out}/V_{in} = (1 + j\alpha\omega\tau)/(1 + j\omega\tau) \qquad (2.9)$$

The magnitude $|V_{out}/V_{in}|$ of the network's response is

$$20 \log_{10} [\sqrt{(1 + \alpha^2\omega^2\tau^2)}] - 20 \log_{10} [\sqrt{(1 + \omega^2\tau^2)}].$$

At low frequencies where $\omega\tau \ll 1$ the magnitude of the response is 0 dB, and at high frequencies where $\omega\tau \gg 1$ the voltage ratio is

$$20 \log_{10} [(\alpha\omega\tau)/(\omega\tau)] = 20 \log_{10} \alpha \text{ decibels}$$

and this is always a negative value. Corner frequencies occur at $\omega_1 = 1/\tau$ and at $\omega_2 = 1/\alpha\tau$; between these two corner frequencies the slope of the network's loss—frequency characteristic is 6 dB/octave or 20 dB/decade. The Bode plot of the amplitude response of the phase-lag network is given by Fig. 2.4.

The phase lag $\phi$ introduced by the network is

$$\phi = \tan^{-1}(\alpha\omega\tau) - \tan^{-1}(\omega\tau) = \phi_1 - \phi_2 \qquad (2.10)$$

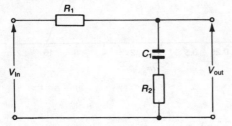

**Fig. 2.3** Phase-lag circuit

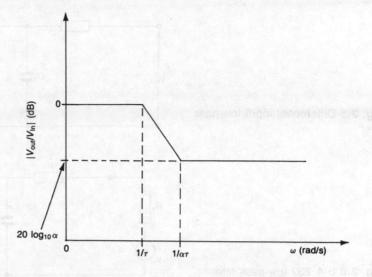

**Fig. 2.4** Bode plot of phase-lag circuit

$$\tan \phi = (\tan \phi_1 - \tan \phi_2)/(1 + \tan \phi_1 \tan \phi_2)$$

$$-\frac{(\omega \alpha \tau - \omega \tau)}{(1 + \omega \alpha \tau \, \omega \tau)} = \frac{\omega \tau (\alpha - 1)}{(1 + \alpha \omega^2 \tau^2)} \quad (2.11)$$

The maximum value of the phase shift $\phi$ occurs when $\tan \phi$ is maximum since $\phi < 90°$. Differentiating

$$\frac{d(\tan \phi)}{d\omega} = \frac{[\tau(\alpha - 1)(1 + \alpha \omega^2 \tau^2) - \omega \tau(\alpha - 1) 2 \alpha \omega \tau^2]}{(1 + \alpha \omega^2 \tau^2)^2}$$

For a maximum $d(\tan \phi)/d\omega - 0$. Hence

$$1 = \alpha \omega^2 \tau^2 \quad \text{or} \quad \omega = \frac{1}{\tau \sqrt{\alpha}}$$

Therefore

$$\phi_{max} = \tan^{-1}[(\alpha - 1)/\sqrt{\alpha}]/[1 + \alpha/\alpha] = \tan^{-1}[(\alpha - 1)/2\sqrt{\alpha}]$$
$$= \sin^{-1}[(\alpha - 1)/(\alpha + 1)] \quad (2.12)$$

*Example 2.2*

In the phase lag circuit of Fig. 2.3 $R_1 = 10$ k$\Omega$, $R_2 = 22$ k$\Omega$, and $C_1 = 0.01$ $\mu$F. Calculate (a) the corner frequencies of the network, and (b) the phase shift at each of these frequencies.

*Solution*
$\alpha = 22/32 = 0.69$. $\tau = 1 \times 10^{-8} \times 32 \times 10^3 = 320$ $\mu$s.
(a) $f_1 = (1 \times 10^6)/(2\pi \times 320) = 497$ Hz     (*Ans.*)
$f_2 = (1 \times 10^6)/(2\pi \times 320 \times 0.69) = 721$ Hz     (*Ans.*)

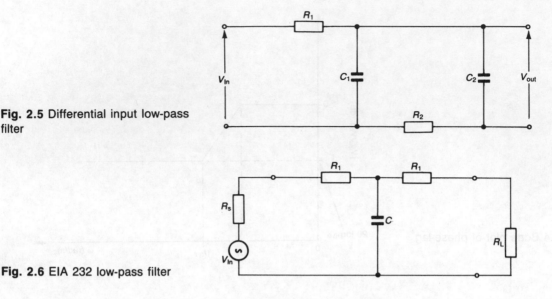

**Fig. 2.5** Differential input low-pass filter

**Fig. 2.6** EIA 232 low-pass filter

(*b*) At 497 Hz $\phi_1 = 34.6°$ and $\phi_2 = 44.5°$
and phase lag = 9.9° (*Ans.*)
At 721 Hz $\phi_1 = 45°$ and $\phi_2 = 55.4°$
and phase lag = 10.4° (*Ans.*)

The circuit of a differential input low-pass filter is shown by Fig. 2.5. This type of filter is used in data acquisition circuits.

Figure 2.6 shows the kind of low-pass filter that is often fitted on circuit boards to filter EIA 232 lines. The resistors $R_1$ and $R_2$ are of equal value $R$ so that the filter has the same attenuation in both directions of transmission. The transfer function of the circuit is

$$V_{out}/V_{in} = R_L/[R_s + R_L + 2R + j\omega C(R_s + R)(R_L + R)] \tag{2.13}$$

and the 3 dB frequency $f_{3dB}$ is

$$f_{3dB} = [2R + R_s + R_L]/[2\pi C(R_s + R)(R_L + R)] \tag{2.14}$$

**The High-pass Filter**

The circuit of the basic *CR* high-pass filter is shown in Fig. 2.7. The output voltage $V_{out}$ of the circuit is taken from the resistor $R$ and is

$$V_{out} = V_{in}/(1 + 1/j\omega CR) = V_{in}/(1 - j/\omega CR).$$

The magnitude of the output voltage is

$$|V_{out}| = V_{in}/\sqrt{[1 + (1/\omega^2 C^2 R^2)]}. \tag{2.15}$$

At the corner frequency $f_3$ the ratio $|V_{out}/V_{in}|$ is 3 dB down on the value at 0 Hz. Hence

$$|V_{out}/V_{in}| = 1/\sqrt{2} = 1/\sqrt{[1 + (1/\omega_3^2 C^2 R^2)]}$$
$$\omega_3 = 1/CR \quad \text{and} \quad f_3 = 1/(2\pi CR) \text{ hertz} \tag{2.16}$$

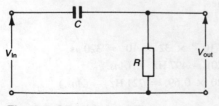

**Fig. 2.7** *CR* high-pass filter

The expression for the magnitude of the output voltage can be written in the form

$$|V_{out}/V_{in}| = 1/\sqrt{(1 + \omega_3^2/\omega^2)} \tag{2.17}$$

At frequencies higher than the corner frequency the ratio $\omega_3^2/\omega^2$ is small and then the ratio $|V_{out}/V_{in}|$ approaches unity. At frequencies below the corner frequency the ratio $\omega_3^2/\omega^2$ becomes increasingly large and this means that the output falls with decreasing frequency. Hence the circuit acts like a high-pass filter. At frequency $f = f_3/2$

$$|V_{out}/V_{in}| = 1/\sqrt{(1 + 2^2)} = 1/\sqrt{5} = -7\,\text{dB}$$

At frequency $f = f_3/4$

$$|V_{out}/V_{in}| = 1/\sqrt{(1 + 4^2)} = 1/\sqrt{17} = -12.3\,\text{dB etc.}$$

The figures should be compared with those obtained for the low-pass filter, they will be seen to be exactly the same as the figures for $2f_3$, $4f_3$, etc. Hence, the loss–frequency characteristic of a high-pass filter can be drawn by plotting $|V_{out}/V_{in}|$ in decibels against frequency and also by the use of a Bode plot and these are shown in Fig. 2.8. The errors inherent in the Bode plot are the same as those obtained in the Bode plot of a low-pass filter.

The high-pass filter introduces a phase shift $\phi$ equal to

$$-\tan^{-1}(-1/\omega CR) = -\tan^{-1}(-\omega_3/\omega).$$

At the corner frequency $f_3$

$$\phi = -\tan^{-1}(-1) = +45°$$

at frequency $f_3/2$

$$\phi = -\tan^{-1}(-2) = +63.4°$$

and at frequency $2f_3$

$$\phi = -\tan^{-1}(-1/2) = +26.6°.$$

This means that the high-pass filter acts as a *phase-advance* circuit.

### Example 2.3

A phase-advance circuit uses a resistor of 12 kΩ and a capacitor of 0.1 μF. At what frequency is the phase shift equal to 30°?

*Solution*
$CR = 0.1 \times 10^{-6} \times 12 \times 10^3 = 1.2\,\text{ms}$
$f_3 = 1/(2\pi \times 1.2 \times 10^{-3}) = 133\,\text{Hz}$
$30 = -\tan^{-1}(-133/f) \quad \tan 30° = 0.577 = 133/f$
$f = 133/0.577 = 230\,\text{Hz} \quad (Ans.)$

A more complex phase advance circuit is shown in Fig. 2.9.

$$V_{out} = V_{in}R_2/[R_2 + R_1/(1 + j\omega C_1 R_1)]$$

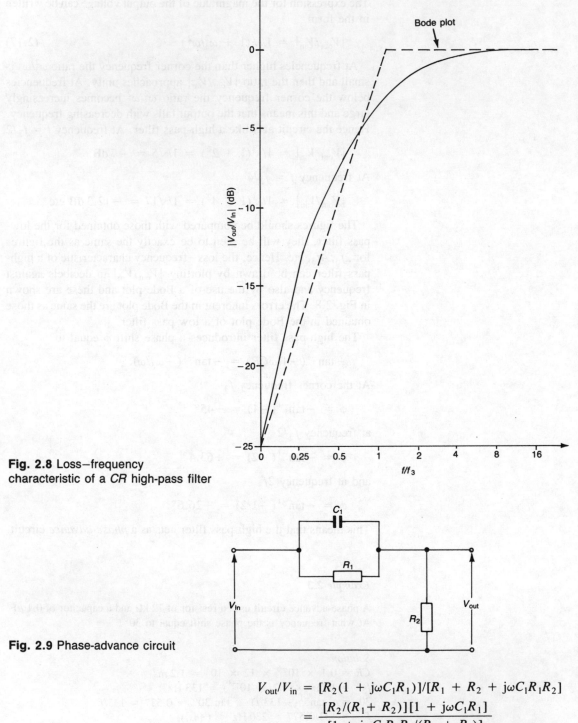

**Fig. 2.8** Loss−frequency characteristic of a *CR* high-pass filter

**Fig. 2.9** Phase-advance circuit

$$V_{out}/V_{in} = [R_2(1 + j\omega C_1 R_1)]/[R_1 + R_2 + j\omega C_1 R_1 R_2]$$

$$= \frac{[R_2/(R_1 + R_2)][1 + j\omega C_1 R_1]}{[1 + j\omega C_1 R_1 R_2/(R_1 + R_2)]}$$

$$= \alpha(1 + j\omega\tau)/(1 + j\alpha\omega\tau) \qquad (2.18)$$

where $\alpha = R_2/(R_1 + R_2)$ and $\tau = C_1 R_1$.

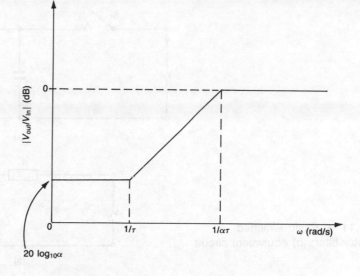

**Fig. 2.10** Amplitude–frequency characteristic of phase-advance circuit

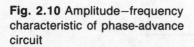

$20 \log_{10} \alpha$

The amplitude response of the circuit in decibels is

$$20 \log_{10}|V_{out}/V_{in}| = 20 \log_{10} \alpha + 20 \log_{10} \sqrt{(1 + \omega^2\tau^2)}$$
$$- 20 \log_{10} \sqrt{(1 + \omega^2\alpha^2\tau^2)} \text{ decibels}$$

At low frequencies where $\omega\tau \ll 1$ the second and third terms have a negligible contribution to the output voltage and so

$$|V_{out}/V_{in}| = 20 \log_{10} \alpha \text{ decibels}$$

which is a negative value. At high frequencies where $\omega\tau \gg 1$,

$$|V_{out}/V_{in}| = 20 \log_{10} \omega\tau - 20 \log_{10} \alpha\omega\tau = 20 \log_{10} (1/\alpha) \text{ decibels}.$$

Corner frequencies exist at two points: (*a*) at $\omega_1 = 1/\tau$ and (*b*) at $\omega_2 = 1/\alpha\tau$. In between these two points the amplitude characteristic has a slope of 6 dB/octave, or 20 dB/decade. The amplitude–frequency characteristic is shown plotted in Fig. 2.10.

The angle $\phi$ of phase advance provided by the circuit is given by equation (2.19), i.e.

$$\phi = \tan^{-1}\omega\tau - \tan^{-1}\alpha\omega\tau \tag{2.19}$$

At low frequencies both $\omega\tau$ and $\alpha\omega\tau$ are small and the phase advance angle $\phi$ is near 0°. At high frequencies $\omega\tau \simeq \alpha\omega\tau$ and again the phase angle is near 0°.

**Switched-capacitor Filters**

The standard active filters, see *Electronics IV*, consist of a combination of capacitors, resistors and op-amps. The resistance values needed are relatively large so that when the filter is implemented on-chip an uneconomically large chip area is required. The *switched-capacitor filter* overcomes this problem by simulating the required resistance(s). The elimination of physical resistance also reduces the power dissipation of a filter circuit. Figure 2.11(*a*) shows a simple circuit

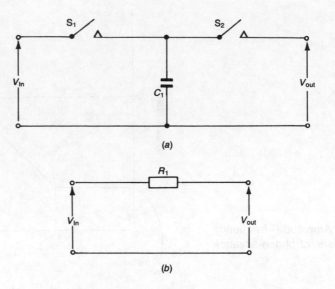

**Fig. 2.11** (a) Basic switched capacitor filter; (b) equivalent circuit of (a)

consisting of two switches $S_1$ and $S_2$ and a capacitor $C_1$. When switch $S_1$ is closed for $t_1$ seconds switch $S_2$ is open for the same period of time; conversely, when $S_1$ is open for $t_2$ seconds $S_2$ is closed. The sum of the two time periods is equal to the periodic time of the switching cycle.

When switch $S_1$ is closed and $S_2$ is open capacitor $C_1$ is charged to $V_{in}$ volts. At time $t_1$ $S_1$ is opened and $S_2$ is closed; this action causes capacitor $C_1$ to discharge to a lower voltage $V_{out}$. After time $t_1 + t_2$ seconds, switch $S_1$ closes and $S_2$ opens again and then the capacitor is again charged up to $V_{in}$ volts. During one switching cycle the charge moved from the input terminal of the circuit to the output terminal is $Q = C(V_{in} - V_{out})$. This movement of charge takes place in $T = t_1 + t_2$ seconds and so the equivalent current flow is

$$I_e = Q/T = C(V_{in} - V_{out})/T \qquad (2.20)$$

The equivalent resistance $R_1$ of the circuit (see Fig. 2.11(b)), is

$$R_1 = (V_{in} - V_{out})/I_e = T[V_{in} - V_{out}]/[C(V_{in} - V_{out})]$$
$$= T/C \qquad (2.21)$$

The switching frequency $f_s$ is the reciprocal of the switching period so that

$$R_1 = 1/Cf_s \qquad (2.22)$$

The switching frequency must be several times higher than the maximum frequency of the input signal. The two switches are usually provided by NMOS or CMOS devices and Fig. 2.12 shows a simple circuit. Switched capacitor filters are implemented as an integral part of many complex LSI/VSLI circuits and they are also available as discrete IC packages, e.g. the National Semiconductor MF 10.

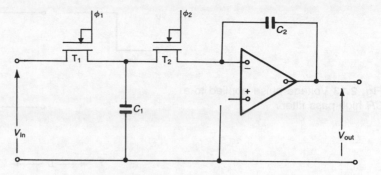

**Fig. 2.12** Switched capacitor filter

**Response of a CR Circuit to Pulse Waveforms**

The output voltage of the high-pass filter shown in Fig. 2.7 can be written as

$$V_{out}(S) = V_{in}(S)SCR/(1 + SCR)$$

where $S$ is the complex frequency.

If the input voltage to the filter is a step of amplitude $V$ volts then

$$V_{out}(S) = (V/S)[(SCR)/(1 + SCR)] = V\,CR/(1 + SCR)$$
$$= V/(S + 1/CR).$$

From Laplace transform tables the output voltage $V_{out}$ is

$$V_{out} = V\,e^{-t/CR} \tag{2.23}$$

Since, in general, the input voltage may not always start from 0 V the output voltage for any constant value of input voltage is of the form $V_{out} = A + Be^{-t/CR}$, where $A$ and $B$ are constants, one of which represents the initial value and the other represents the final value, of the output voltage. The initial value $V_I$ is the output voltage of the circuit at time $t = 0$ and the final value is the output voltage when the time $t$ is theoretically infinite but, in practice, is at least five times the time constant.

At time $t = 0$

$$V_I = A + B^0 = A + B$$

At time $t = \infty$

$$V_F = A$$

Hence $B = V_I - A = V_I - V_F$, and so the general expression for the output voltage of a $CR$ circuit is

$$V_{out} = V_F + (V_I - V_F)\,e^{-t/CR} \tag{2.24}$$

**High-pass Filter**

(a) When the input voltage applied to the circuit consists of a single step from zero to $V$ volts, as shown by Fig. 2.13, the initial output

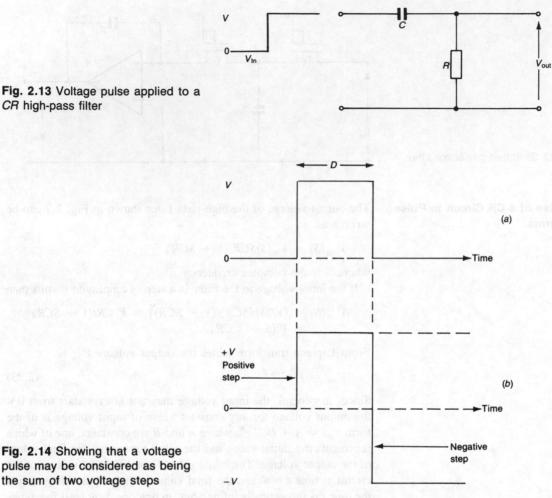

**Fig. 2.13** Voltage pulse applied to a *CR* high-pass filter

**Fig. 2.14** Showing that a voltage pulse may be considered as being the sum of two voltage steps

voltage is $V$ and the final output voltage is zero. Substituting into equation (2.24) gives the expression for the output voltage as

$$V_{out} = V \, e^{-t/CR} \qquad (2.23) \text{ (again)}$$

(b) When a single pulse of amplitude $V$ volts is applied to the high-pass filter it is best considered to consist of a positive step of $V$ volts followed after a time $D$, where $D$ is the pulse duration, by a negative step of $V$ volts. The concept is illustrated by Fig. 2.14(a) which shows the applied pulse and Fig. 2.14(b) which shows the pulse as the sum of two steps. In the time period from $t = 0$ to $t = D$ the output voltage of the circuit is given by equation (2.23). At the instant when the negative step is applied to the circuit the output voltage is

$$V_{out} = V \, e^{-D/CR} \text{ volts;}$$

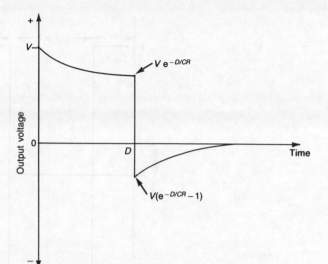

**Fig. 2.15** Output voltage of **Fig. 2.13**

this instantaneously changes in the negative direction by $V$ volts so that the initial voltage for the negative step is

$$V e^{-D/CR} - V = V(e^{-D/CR} - 1) \text{ volts}$$

The final value of the output voltage is zero and hence, from equation (2.24),

$$V_{\text{out}} = [V(e^{-D/CR} - 1)] e^{-(t-D)/CR} \tag{2.25}$$

The waveform of the output voltage is shown in Fig. 2.15.

(c) When a rectangular pulse waveform is applied to the input of a $CR$ high-pass filter some time elapses before the output voltage settles down to its *steady-state condition* when any change in the output voltage is equal to the change in the input voltage. Figure 2.16 shows a rectangular pulse waveform of amplitude $\pm V$ and periodic time $T$ that has unequal positive and negative time periods. The waveform of the output voltage is determined by the relationship between the time constant $CR$ of the circuit and the periodic time $T$ of the input voltage. When $T \gg CR$ the output voltage will have fallen to zero before the end of each pulse occurs and the output waveform will be of the general shape shown by Fig. 2.17(a). When $T \simeq CR$ the output voltage will have fallen some way towards zero when a pulse ends as shown by Fig. 2.17(b). Lastly, when $T \ll CR$ the output voltage will have hardly changed by the time each pulse comes to its end and the output voltage waveform will be more or less unchanged (see Fig. 2.17(c)).

Fig. 2.17(b) is shown, in a little more detail, in Fig. 2.18. Once the steady-state condition has been reached the output voltage must have an average value of zero since the series-connected capacitor

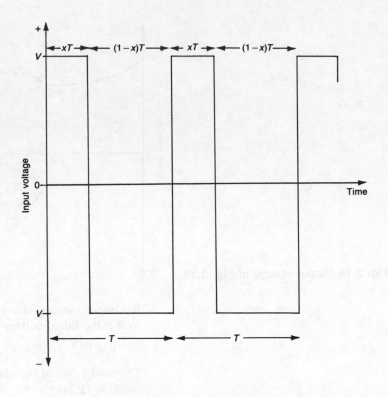

**Fig. 2.16** Rectangular pulse
waveform

cannot pass a d.c. component. This means that $V_1$ must be equal to $V_3$ and $V_2$ must be equal to $V_4$, and since the input voltage changes by $2V$ volts at the end of each positive and negative pulse,

$$2V = V_1 + V_2 = V_3 + V_4$$

During the time occupied by the positive pulse, $V_{out} = V\,e^{-t/CR}$ and for the time occupied by the negative pulse $V_{out} = -V\,e^{-(t-xT)/CR}$.
From Fig.2.18,

$$V_2 = V_1 + 2V$$
$$V_3 = V_2\,e^{-xT/CR}$$
$$V_4 = V_3 - 2V = V_2\,e^{-xT/CR} - 2V$$
$$V_1 = V_4\,e^{-(1-x)T/CR}$$

Consider $V_2$,

$$V_2 = V_4\,e^{-(1-x)T/CR} + 2V$$
$$= [V_2\,e^{-xT/CR} - 2V]\,e^{-(1-x)T/CR} + 2V$$
$$= V_2\,e^{-xT/CR}\,e^{-(1-x)T/CR} - 2V\,e^{-(1-x)T/CR}$$
$$\quad + 2V$$
$$= V_2\,e^{-T/CR} - 2V\,e^{-(1-x)T/CR} + 2V$$
$$V_2[1 - e^{-T/CR}] = 2V[1 - e^{-(1-x)T/CR}]$$
$$V_2 = 2V[1 - e^{-(1-x)T/CR}]/[1 - e^{-T/CR}]$$

$$(2.26)$$

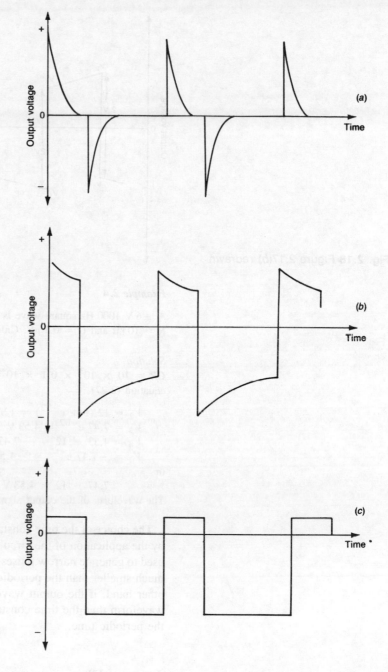

**Fig. 2.17** Output voltage of filter depends upon the time constant: (a) $T \gg CR$; (b) $T \simeq CR$; (c) $T \ll CR$

If the input waveform is square so that $x = \frac{1}{2}$ this expression becomes

$$V_2 = 2V[1 - e^{-T/2CR}]/[1 - e^{-T/CR}]$$
$$= 2V/[1 + e^{-T/2CR}] \qquad (2.27)$$

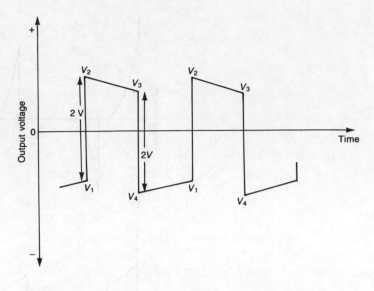

**Fig. 2.18** Figure 2.17(b) redrawn

*Example 2.4*

A ±6 V 1000 Hz square wave is applied to a high-pass *CR* filter that has $R = 10\,\text{k}\Omega$ and $C = 0.1\,\mu\text{F}$. Calculate and sketch the output waveform.

*Solution*

$CR = 10 \times 10^3 \times 0.1 \times 10^{-6} = 1$ ms, $T = 1/1000 = 1$ ms. From equation (2.27),

$$V_2 = 12/(1 + e^{-1/2}) = 12/1.607 = 7.47\,\text{V} \qquad (Ans.)$$
$$V_3 = 7.47\,e^{-1/2} = 4.53\,\text{V} \qquad (Ans.)$$
$$V_4 = 4.53 - 12 = -7.47\,\text{V} \qquad (Ans.)$$
$$V_1 = -7.47\,e^{-1/2} = -4.53\,\text{V} \qquad (Ans.).$$

or

$$7.47 - 12 = 4.53\,\text{V} \qquad (Ans.)$$

The waveform of the output signal is shown in Fig. 2.19.

The choice of the time constant for a high-pass filter is determined by the application of the circuit. If, for example, the circuit is to be used to generate narrow pulses then the time constant should be very much smaller than the periodic time of the input waveform. On the other hand, if the output waveform is to be the same as the input waveform then the time constant should be very much longer than the periodic time.

**Low-pass Filter**

(*a*) If a positive step of *V* volts is applied to a low-pass filter, as in Fig. 2.20, the initial value $V_I$ of the output voltage will be zero and the final value $V_F$ will be *V* volts. Then, from equation (2.24),

$$V_{\text{out}} = V - V\,e^{-t/CR} = V(1 - e^{-t/CR}) \qquad (2.28)$$

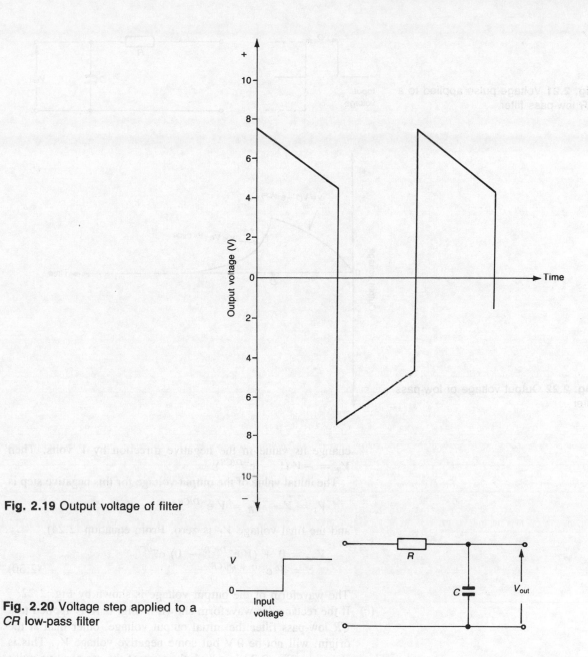

**Fig. 2.19** Output voltage of filter

**Fig. 2.20** Voltage step applied to a *CR* low-pass filter

(b) A single positive pulse of amplitude $V$ volts applied to the low-pass filter can be regarded as being the sum of a positive step followed, after a time $D$, by a negative step of the same voltage (see Fig. 2.21). At time $t = D$,

$$V_{out} = V(1 - e^{-D/CR}) \tag{2.29}$$

At this instant the voltage $V_R$ across the resistor $R$ is equal to $V - V_{out} = V e^{-D/CR}$. The application of the negative step after time $D$ causes the voltage across the resistor to instantaneously

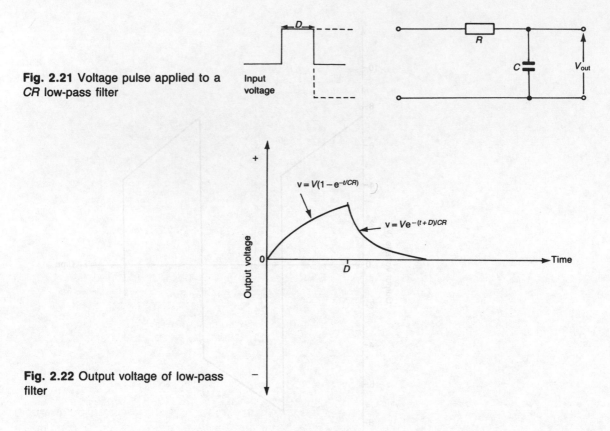

**Fig. 2.21** Voltage pulse applied to a *CR* low-pass filter

**Fig. 2.22** Output voltage of low-pass filter

change its value in the negative direction by $V$ volts. Then $V_R = -V (1 - e^{-D/CR})$.

The initial value of the output voltage for this negative step is

$$V_I = V - V_R = V e^{-D/CR}$$

and the final voltage $V_F$ is zero. From equation (2.24),

$$V_{out} = 0 + (V e^{-D/CR} - 0) e^{-t/CR}$$
$$= V e^{-(t + D)/CR} \qquad (2.30)$$

The waveform of the output voltage is shown by Fig. 2.22.

(*c*) If the rectangular waveform shown in Fig. 2.16 is applied to a *CR* low-pass filter the initial output voltage, other than at the origin, will not be $0 \, V$ but some negative voltage $V_1$. This is shown by Fig. 2.23 in which the start of the $n$th positive pulse has been labelled, for convenience, as $t = 0$. At time $t = 0$ the initial output voltage $V_I = V_1$ and the final output voltage $V_F = V$. Substituting these values into equation (2.24) gives

$$V_{out} = V + (V_1 - V) e^{-t/CR}$$

After $xT$ seconds, when the positive pulse ends, the output voltage is equal to

$$V_{out} = V_2 = V + (V_1 - V) e^{-xT/CR} \qquad (2.31)$$

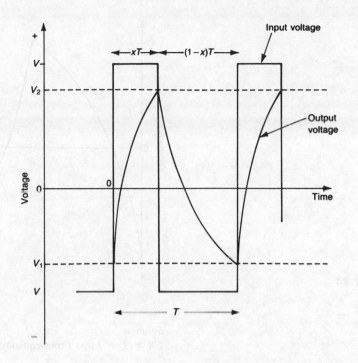

**Fig. 2.23** Output voltage of low-pass filter when the waveform of Fig. 2.16 is applied

For the time period from $xT$ to $T$, when the negative pulse is applied to the circuit, the initial output voltage $V_1$ is $V_2$ and the final output voltage $V_F$ is $-V$. From equation (2.24),

$$V_{out} = -V + (V_2 + V)\,e^{-(t-xT)/CR}$$

At the end of the negative pulse, when $t = T$, the output voltage is $V_1$, where

$$V_1 = -V + (V_2 + V)\,e^{-(1-x)T/CR} \tag{2.32}$$

If $V_1 = -V_2$ then

$$V_2 = V + (-V_2 - V)\,e^{-xT/CR}$$
$$= V - V_2\,e^{-xT/CR} - V\,e^{-xT/CR}$$
$$V_2(1 + e^{-xT/CR}) = V(1 - e^{-xT/CR})$$
$$V_2 = V(1 - e^{-xT/CR})/(1 + e^{-xT/CR}) \tag{2.33}$$

If the input waveform is square so that $x = \frac{1}{2}$,

$$V_2 = \frac{V\,(1 - e^{-T/2CR})}{(1 + e^{-T/2CR})} \tag{2.34}$$

### Example 2.5

The waveform in Example 2.4 is now applied to a *CR* low-pass filter with the same component values. Calculate and sketch the output waveform.

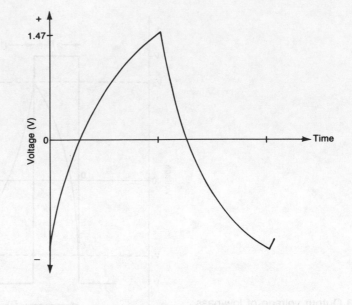

**Fig. 2.24**

*Solution*

$CR = T = 1$ ms. From equation (2.34),

$$V_2 = 6(1 - e^{-1/2})/(1 + e^{-1/2}) = 6 \times 0.393/1.607$$
$$= 1.47 \text{ V} \quad (Ans.).$$

Note that $1.47 = 7.47 - 6$. The waveform is shown in Fig. 2.24.

## Attenuators

An attenuator is a purely resistive network that has been designed to introduce a specified amount of attenuation into a circuit when it is matched to its source and load resistances. Symmetrical networks are used when the load and source resistances are of equal value, known as the characteristic resistance $R_0$, but when the resistances differ from one another a non-symmetrical network must be employed.

## Symmetrical Attenuators

Figure 2.25 shows a T-attenuator connected between a source and a load that are both of resistance $R_0$. From Fig. 2.25,

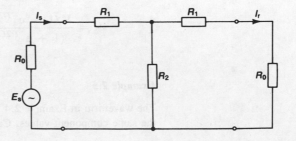

**Fig. 2.25** Symmetrical T-attenuator

$$I_r = I_s R_2 / (R_1 + R_2 + R_0)$$
$$I_s / I_r = M = (R_1 + R_2 + R_0)/R_2 \qquad (2.35)$$

Also,

$$R_0 = R_1 + R_2(R_1 + R_0)/(R_1 + R_2 + R_0)$$
$$= R_1 + (R_1 + R_0)/M$$
$$R_0 M = R_1 M + R_1 + R_0 \qquad R_0(M - 1) = R_1(M + 1)$$

or

$$R_1 = R_0(M - 1)/(M + 1) \qquad (2.36)$$

From equation (2.35),

$$R_2 M = R_1 + R_2 + R_0$$
$$R_2(M - 1) = R_1 + R_0$$
$$= R_0(M - 1)/(M + 1) + R_0$$
$$R_2(M - 1)(M + 1) = R_0(M - 1) + R_0(M + 1) = 2R_0 M$$

or

$$R_2 = 2R_0 M/(M^2 - 1) \qquad (2.37)$$

### Example 2.6

Design a T-attenuator to have $R_0 = 2000\ \Omega$ and an attenuation of 10 dB.

### Solution

10 dB is a voltage ratio of 3.162. Hence

$$R_1 = 2000 \times 2.162/4.162 = 1039\ \Omega \qquad (Ans.)$$
$$R_2 = 2 \times 2000 \times 3.162/9 = 1405\ \Omega \qquad (Ans.)$$

Figure 2.26 shows a symmetrical $\pi$-attenuator connected between load and source resistances that are matched to the characteristic resistance $R_0$ of the attenuator. From the circuit,

$$I_s = V_s/R_0 = V_s/R_2 + V_r/R_2 + V_r/R_0$$

Dividing throughout by $V_r$ gives

$$M/R_0 = M/R_2 + 1/R_2 + 1/R_0$$
$$(1/R_0)(M - 1) = (1/R_2)(M + 1)$$
$$R_2 = R_0(M + 1)/(M - 1) \qquad (2.38)$$

Also,

$$I_s = V_s/R_0 = V_s/R_2 + (V_s - V_r)/R_1$$

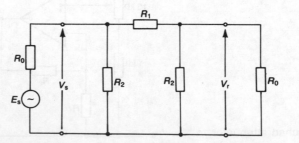

**Fig. 2.26** Symmetrical $\pi$-attenuator

Dividing throughout by $V_r$ gives

$$(1/R_1)(M - 1) = M/R_0 + M/R_1 - 1/R_1$$
$$(1/R_1)(M - 1) = M/R_0 - M/R_2 = M/R_0 -$$
$$M/[R_0(M + 1)/(M - 1)]$$
$$= (1/R_0)[M - M(M - 1)/(M + 1)]$$
$$= (1/R_0)[2M/(M + 1)]$$
$$1/R_1 = (1/R_0[2M/(M + 1)(M - 1)]$$
$$R_1 = R_0(M^2 - 1)/2M \tag{2.39}$$

**Example 2.7**

Design a $\pi$-attenuator to have $R_0 = 2000\,\Omega$ and 10 dB loss.

*Solution*
$R_1 = 2000 \times 9/(2 \times 3.162) = 2846\,\Omega$ (*Ans.*)
$R_2 = 2000 \times 4.162/2.162 = 3850\,\Omega$ (*Ans.*)

The circuit of a bridged T-attenuator is shown in Fig. 2.29. The values of the components can be calculated from:

$$R_1 = R_0(M - 1) \tag{2.40}$$
$$R_2 = R_0/(M - 1) \tag{2.41}$$

A ladder network is employed when it is necessary to be able to switch different values of attenuation into a circuit. The best accuracy can be obtained if the ladder network is supplied by a non-inverting op-amp as shown by Fig. 2.27. When the range switch is in the 0 dB position the load voltage is $0.5V_{out}$, where $V_{out}$ is the output voltage of the op-amp. The gain of the op-amp is set to 2 so that the attenuator will have 0 dB loss when the 0–20 dB potentiometer is set to its top position.

When the range switch is in position 2 the circuit is that shown by Fig. 2.28(*a*). From this,

$$V_L = (330V_{out})|(6004 + 330) = 0.05V_O$$

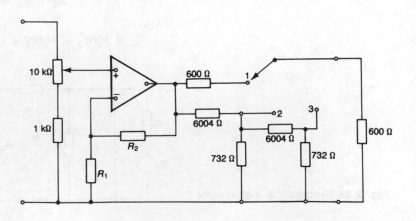

**Fig. 2.27** Switched attenuator

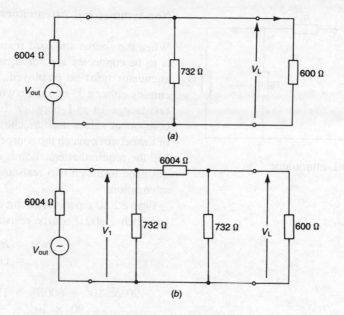

**Fig. 2.28**

and hence the insertion loss is

$$20 \log_{10}[(0.5V_{out})/(0.05V_{out})] = 20 \,\text{dB}$$

With the range switch in position 3 the circuit is that given by Fig. 2.28(*b*). From this circuit,

$$V_1 = V_{out} \frac{[(732 \times 6334)/(732 + 6334)]}{[6004 + (732 \times 6334)/(732 + 6334)]}$$
$$= 0.1V_{out}.$$

Hence

$$V_L = 330(0.1V_{out})/6334 = 5.2 \times 10^{-3} V_0.$$

The insertion loss is now

$$20 \log_{10}[0.5V_{out}/(5.2 \times 10^{-3}V_{out})] = 40 \,\text{dB}$$

Figure 2.29 shows the circuit of a bridged attenuator.

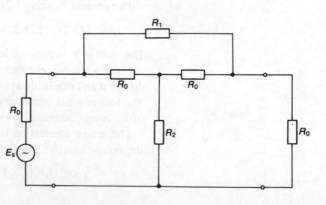

**Fig. 2.29** Bridged attenuator

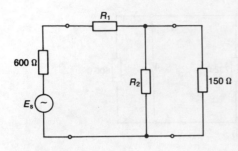

**Fig. 2.30** L-attenuator

### Non-symmetrical Attenuators

When the source and load resistances between which an attenuator is to be connected are *not* equal to one another a non-symmetrical attenuator must be employed. A non-symmetrical attenuator may employ either a T- or a $\pi$-network, with different series, or parallel, resistances, or an L-network. In both the T- and the $\pi$-networks the component values must be chosen so that the two image resistances of the network match the source and the load resistances respectively and the required attenuation is obtained. An L-network can only be designed to match two resistances and then it will give a particular attenuation.

Figure 2.30 shows the circuit of an L-network that is to be employed to match a $600\,\Omega$ source resistance to a $150\,\Omega$ load resistance.

$$600 = R_1 + 150R_2/(150 + R_2)$$
$$= (150R_1 + R_1R_2 + 150R_2)/$$
$$(150 + R_2)$$
$$90 \times 10^3 + 600R_2 = 150R_1 + R_1R_2 + 150R_2$$
$$90 \times 10^3 = -450R_2 + 150R_1 + R_1R_2 \qquad (2.42)$$

$$150 = \frac{R_2(R_1 + 600)}{(R_1 + R_2 + 600)}$$
$$150R_1 + 150R_2 + 90 \times 10^3 = R_1R_2 + 600R_2$$
$$90 \times 10^3 = 450R_2 - 150R_1 + R_1R_2 \qquad (2.43)$$

Adding equations (2.42) and (2.43) gives

$$180 \times 10^3 = 2R_1R_2 \qquad R_2 = (90 \times 10^3)/R_1$$

Subtracting equation (2.42) from (2.43) gives

$$0 = -900R_2 + 300R_1 \qquad R_2 = 300R_1/900 = R_1/3$$

Therefore

$$R_1/3 = (90 \times 10^3)/R_1 \quad \text{and} \quad R_1 = 519.6\,\Omega$$
$$R_2 = 519.6/3 = 173.2\,\Omega.$$

The current flowing in the load resistance is

$$I_r = 173.2I_s/(173.2 + 150) = 0.536I_s$$

The voltage $V_r$ across the load is $80.4V_s/(519.6 + 80.4) = 0.134V_s$.

Clearly the current and voltage ratios of the attenuator are quite different and this is always the case with a non-symmetrical network. The loss of a non-symmetrical attenuator must be expressed in terms of its *image attenuation coefficient* $\alpha$.

The image attenuation coefficient $\alpha$ is the real part, in nepers, of the *image transfer coefficient* $\gamma$, and $\gamma$ is defined by

$$\gamma = \tfrac{1}{2}\log_e(I_sV_s/I_rV_r) \qquad (2.44)$$

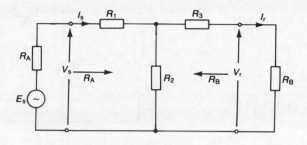

**Fig. 2.31** Non-symmetrical T-attenuator

$$= \log_e[(I_s/I_r)\sqrt{(Z_A/Z_B)}] \tag{2.45}$$

$$= \log_e[(V_s/V_r)\sqrt{(Z_B/Z_A)}] \tag{2.46}$$

where $Z_A$ and $Z_B$ are the two image impedances of the network.

When it is required that the matching network provide a specified current or voltage attenuation it becomes necessary to use either a T- or a $\pi$-network. Figure 2.31 shows a non-symmetrical T-attenuator connected between a source of resistance $R_s$ and a load of resistance $R_L$. For the source and the load to be matched to the network it is necessary that the image resistances of the network are $R_A = R_s$ and $R_B = R_L$. The design equations for the attenuator, assuming that $R_s > R_L$, are given by equations (2.47), (2.48), and (2.49).

$$R_2 = [2\sqrt{(R_A R_B)}]/[e^\alpha - e^{-\alpha}] \tag{2.47}$$

$$R_1 = R_A[(e^\alpha + e^{-\alpha})/(e^\alpha - e^{-\alpha})] - R_2 \tag{2.48}$$

$$R_3 = R_B[(e^\alpha + e^{-\alpha})/(e^\alpha - e^{-\alpha})] - R_2 \tag{2.49}$$

It will be found that there is a minimum loss that can be obtained for a given matching ratio, any lesser loss will be found to require a negative resistance value; this minimum will differ for current and for voltage ratios.

### Example 2.8

Design a T-attenuator to have a current loss of 10 dB when connected between a source of 600 Ω resistance and a 150 Ω load.

*Solution*

Here 10 dB is a current ratio of 3.162 : 1.

$\alpha = \log_e[3.162 \sqrt{(600/150)}] = \log_e 6.324 = 1.844$

$e^\alpha = 6.324$ and $e^{-\alpha} = 0.158$

Hence

$R_2 = [2\sqrt{(600 \times 150)}]/[6.324 - 0.158] = 97.3 \, \Omega$   (*Ans.*)

$R_1 = 600[(6.324 + 0.158)/(6.324 - 0.158)] - 97.3$
$= 533.5 \, \Omega$   (*Ans.*)

$R_3 = 150[(6.324 + 0.158)/(6.324 - 0.158)] - 97.3$
$= 60.4 \, \Omega$   (*Ans.*)

When a voltage ratio is specified it will generally prove to be better to use a $\pi$-network. The same design equations can be used if conductances are used in place of resistances as shown by Fig. 2.32.

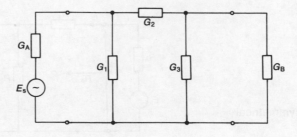

**Fig. 2.32** Non-symmetrical
π-attenuator

### Electronic attenuators

When variable attenuation is wanted a number of passive attenuator sections can be connected in cascade and switched into, and out of, circuit as required using a ladder network as in Fig. 2.27. Alternatively, an electronic attenuator can be employed. Figure 2.33 shows the circuit of a transistor attenuator that can be switched by a control voltage to provide either one of two values of attenuation. Resistors $R_4$ and $R_5$ are the series resistors of a T-attenuator. The shunt resistor consists of $R_3$ connected in series with the parallel combination of $R_2$ and the output resistance of transistor $T_1$. $C_1$ is a d.c. blocking component. When the control voltage is at 0 V $T_1$ is turned OFF and the shunt resistance of the T-attenuator is $R_3 + R_2$; when the control voltage is at $V$ volts $T_1$ is turned ON and the shunt resistance is then just $R_3$. Two, or more, such stages can be connected in cascade to obtain a switched attenuator.

A number of switched attenuator ICs can be employed to provide variable attenuation, for example the Mitel MT 8804A analogue switch array and the Harris HA 2400 four-channel op-amp. When the MT 8804A is connected as a programmable attenuator attenuation is provided, in 3 dB steps, by an input resistor chain and attenuation is provided, in 20 dB steps, by output op-amp circuitry. Selection of the input attenuation is accomplished by using the address pins $A_0$, $A_1$ and $A_2$. These give $2^3 = 8$ different combinations and each one

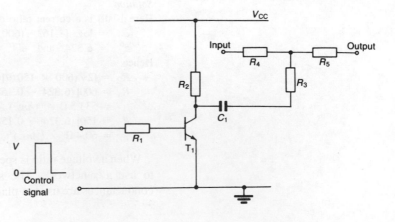

**Fig. 2.33** Transistor attenuator

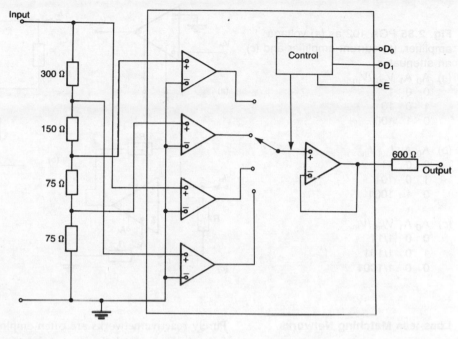

**Fig. 2.34** Programmable attenuator

enables a different input line $L_1$ through to $L_7$. The use of each successive input pin gives an extra 3 dB loss. The choice of the output decibel range, varying from $+20$ to $-40$ dB, is accomplished by applying the appropriate signal to the data inputs $D_0$ through to $D_3$. Thus, if an attenuation of 32 dB is wanted, the control signal should be $A_0, A_1, A_2 = 001$ and $D_0, D_1, D_2, D_3 = 0010$.

The HA 2400 IC can provide a programmable attenuation between 600 $\Omega$ source and load impedances and its basic block diagram is shown by Fig. 2.34. The IC has four input op-amps, any one of which may be selected by the digital signal applied to the $D_0 D_1$ input pins, provided the enable pin is at the binary 1 logic level. The output signal of the selected op-amp is applied to the output op-amp which is connected as a voltage follower. The input signal is divided by either 1, 2, 4 or 8 to give either 0, 6, 12 or 18 dB attenuation. If, for example, the control signal $D_0 D_1$ is 01 then input op-amp 3 is selected and then the division ratio is $150/(450 + 150) = 1/4$ giving 12 dB loss. The analogue device AD 7110 is an audio attenuator whose loss can be controlled, over a 0–88.5 dB range, by an input digital word.

Some programmable op-amps, such as the PGA 102, may be programmed to operate as either a current or a voltage amplifier, or as an attenuator. The device has two control terminals $A_0$ and $A_1$ to which a two-bit digital word is applied to program the circuit. Figure 2.35 shows the device operated as (*a*) a voltage amplifier; (*b*) a current amplifier; (*c*) an attenuator.

**Fig. 2.35** PGA 102 as (a) voltage amplifier, (b) current amplifier and (c) an attenuator

(a)  $A_0$  $A_1$  $V_{out}/V_{in}$
    0   0   1
    1   0   10
    0   1   100

(b)  $A_0$  $A_1$  $I_{out}/I_{in}$
    0   0   11
    1   0   101
    0   1   1001

(c)  $A_0$  $A_1$  $V_{out}/V_{in}$
    0   0   1/11
    1   0   1/101
    0   0   1/1001

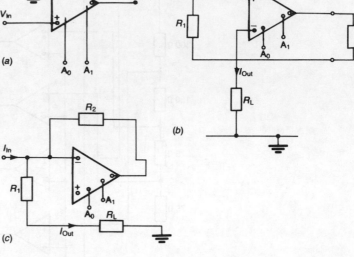

### Loss-less Matching Networks

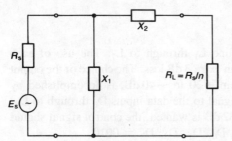

**Fig. 2.36** Loss-less matching network

Purely reactive networks are often employed, particularly at radio frequencies, to match the input and output resistances of a transistor to the source and load resistances respectively. Figure 2.36 shows a two-reactance network which is able to match a load resistance $R_L$ to a source resistance $R_s$ at one particular frequency $f$. From Fig. 2.36,

$$R_s = \frac{jX_1(jX_2 + R_L)}{[R_L + j(X_1 + X_2)]}$$

$$R_sR_L + jR_s(X_1 + X_2) = -X_1X_2 + jX_1R_L$$

Equating the real parts gives

$$R_sR_L = -X_1X_2 \quad \text{so that} \quad X_2 = -R_sR_L/X_1 \quad (2.50)$$

Equating the j parts gives

$$R_s(X_1 + X_2) = X_1R_L$$

and substituting for $X_2$ from equation (2.50) gives

$$R_sX_1 - R_s^2R_L/X_1 = X_1R_L$$
$$X_1^2(R_s - R_L) = R_s^2R_L$$
$$X_1 = \pm R_s \sqrt{[R_L/(R_s - R_L)]} = R_s/\sqrt{(n-1)} \quad (2.51)$$

Hence

$$X_2 = -R_sR_L/[\pm R_s \sqrt{R_L/(R_s - R_L)}]$$
$$= -\sqrt{R_L(R_s - R_L)} = -R_s\sqrt{(n-1)}/n \quad (2.52)$$

Note that at the frequency $f$ at which equations (2.51) and (2.52) apply $jX_1jX_2 = R_sR_L$, and this means that the two reactances must be of

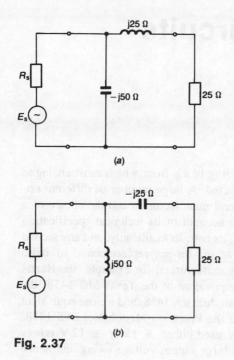

(a)

(b)

**Fig. 2.37**

the opposite sign. If $X_1$ is capacitive then $X_2$ is inductive and the matching network will also act as a low-pass filter; conversely, if $X_1$ is inductive and $X_2$ is capacitive the network will also act as a high-pass filter. The $Q$ factor of the circuit is $Q = \sqrt{(n-1)}$.

*Example 2.9*

The input resistance of a transistor radio frequency (r.f.) amplifier is 25 $\Omega$. Design a matching network that will make the input resistance 50 $\Omega$ at 40 MHz.

*Solution*

From equation (2.51), $X_1 = 50/\sqrt{[(50/25)-1]} = 50\,\Omega$, and from equation (2.52) $X_2 = 50/(2-1)/2 = 25\,\Omega$. The two possible networks are shown in Fig. 2.37(a) and (b); in Fig. 2.37(a) $C_1 = 80$ pF and $L_1 = 100$ nH and in Fig. 2.37(b) $L_1 = 200$ nH and $C_1 = 40$ pF.

If the source and/or the load impedance has a reactive component it may be possible to use a series-connected reactance of the opposite sign to cancel out the reactive component. This reactance will then become a part of either $X_1$ or $X_2$. When this procedure is found not to work an alternative method is called for, that is based upon equivalent series and parallel impedances.

The impedance of a resistor $R_s$ in series with a reactance $X_s$ is

$$Z_s = R_s + jX_s \qquad 1/Z_s = 1/(R_s + jX_s) = (R_s - jX_s)/(R_s^2 + X_s^2)$$

The equivalent parallel resistance $R_p = (R_s^2 + X_s^2)/R_s$, and the equivalent parallel reactance is $X_p = (R_s^2 + X_s^2)/X_s$. Conversely, the impedance $Z_p$ of a resistance $R_p$ and a reactance $X_p$ in parallel is

$$Z_p = jR_pX_p/(R_p + jX_p)$$

Hence the equivalent series resistance is $R_s = R_pX_p^2/(R_p^2 + X_p^2)$ and the equivalent series reactance is $X_s = R_p^2X_p/(R_p^2 + X_p^2)$.

Suppose that the load in Example 2.9 has a series inductive reactance component of j125 $\Omega$. The equivalent parallel resistance $R_p$ is

$$(25^2 + 125^2)/25 = 650\,\Omega$$

which is much bigger than the required value of 50 $\Omega$. Now

$$50 = (25^2 + X_s^2)/25 \quad \text{or} \quad X_s = \sqrt{[(25 \times 50) - 25^2]} = 25\,\Omega$$

This means that a capacitive reactance of $125 - 25 = -j100\,\Omega$ should be connected in series with the load. Then the input resistance $= R_p = 50\,\Omega$ in parallel with

$$X_p = j(25^2 + 25^2)/25 = j50\,\Omega$$

and this reactance can be cancelled out by a capacitive reactance of $-j50\,\Omega$ connected in parallel. The required matching network is shown by Fig. 2.38.

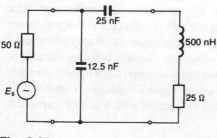

**Fig. 2.38**

# 3 Op-amp Circuits

The op-amp is the basic building block from which most analogue electronic circuits are constructed. A large number of different op-amps are available from several manufacturers and the choice of a particular device should take account of its technical specification (obtained from the data sheet), its cost, its availability, and any second sourcing. Most op-amps have pin-for-pin replacements, or close equivalents, made by other manufacturers; for example, the Harris HA 5102 is the pin-for-pin equivalent of the Texas MC 1458, the Motorola MC 1458 and the Fairchild $\mu$A 1458, and a close equivalent to the Signetics SE 5532 and the Precision Monolithics SSS 1458.

Op-amps have traditionally used either $\pm$ 15 or $\pm$ 12 V power supplies in order to obtain a large output voltage swing and hence minimize the effects of noise, output offset voltage and temperature drift. Modern designs of op-amp have greatly improved characteristics that often allow reduced power supply voltages, such as $\pm$ 5 V or just + 5 V, to be employed. If a single polarity 5 V power supply is used a common supply line can be employed for both the analogue circuitry and any on-board digital circuitry. Other advantages that arise from the use of a lower supply voltage include: (a) a higher unity gain bandwidth; (b) a larger current can be taken from the power supply without excessive power dissipation. The latter makes more current available to charge internal capacitances and/or any load capacitance and this, in turn, increases both the slew rate and the full-power bandwidth of the op-amp. Alternatively, the internal power dissipation for a given bandwidth may be reduced.

Decoupling capacitors should be fitted to the power supply lines to maintain the impedance of the supply at as low a figure as possible. This will prevent power supply ripple and any other noise voltages from reaching the op-amp's $\pm V_{CC}$ terminals. The power supply line itself will possess both inductance and resistance and this may make its impedance quite high at higher frequencies and it may often prove necessary to decouple each IC separately. Usually tantalum electrolytic capacitors are employed for this purpose with (since these have some inherent self-inductance) a lower-valued capacitor connected in parallel, e.g. a 10 $\mu$F capacitor in parallel with a 0.1 $\mu$F capacitor.

## Op-amp Circuit Design

The design of any circuit that uses an op-amp should follow a number of successive steps:

(a) The specification of the wanted circuit should be written down.

(b) The specification should be used to decide upon a suitable op-amp for the circuit.

(c) The voltage gain per stage must be smaller than the unity gain bandwidth divided by the wanted upper 3 dB frequency $f_3$. If it should prove necessary to employ more than one stage of amplification the overall upper 3 dB frequency $f_o$ is given by

$$f_o = f_3 \sqrt{(2^{1/n} - 1)} \qquad (3.1)$$

where n is the number of stages.

(d) Decide on whether the inverting or the non-inverting configuration is to be used; this is always a choice for an amplifier design but it is not for many other circuits, such as waveform generators.

(e) Calculate values for the capacitors and resistors in the chosen circuit. Resistor values should not be too high in order to reduce the effects of both noise and bias/input offset currents, and so that stray capacitances do not have too much effect upon the bandwidth of the circuit. The ratings of the components must not be exceeded and to increase reliability it is desirable that they are derated.

(f) Determine the expected output offset voltage and, if necessary, use a nulling circuit. Details of these circuits are given in the data sheet for each op-amp.

(g) If a dual-polarity power supply is not available either design a single-polarity bias circuit, or use an op-amp that will operate from a single polarity supply, e.g. the LM 124.

(h) Ensure that the designed circuit will work correctly using preferred value components and any op amp of the specified type. It should not be necessary to have to carefully select a particular value of capacitance or resistance or an individual op-amp.

The use of an internally compensated op-amp simplifies the design of a circuit but the slew rate/bandwidth of the device is then fixed. If an uncompensated, or partly compensated, op-amp is used a higher slew rate and a wider bandwidth may be obtained at the expense of requiring one, or more, external components. The data sheet of an uncompensated op-amp will give details of appropriate frequency compensation circuits.

Most op-amps have an output resistance that is somewhere in the region of $10-100\,\Omega$ and a maximum output current capability of about 20 mA. The combination of output resistance and maximum output current sets a minimum limit to the permissible load resistance. If a larger current is taken from an op-amp the output signal waveform will be distorted because of internal current limiting and/or the device may suffer damage. If this is a potential problem there are three possible solutions:

(a) Connect a current-limiting resistor in series with the output terminal of the op-amp as shown by Fig. 3.1. If, for example,

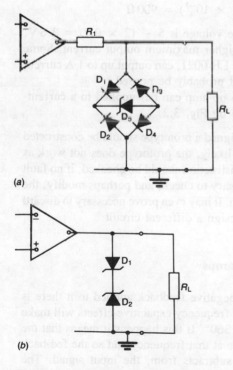

**Fig. 3.1** Two methods of limiting the output voltage/current of an op-amp

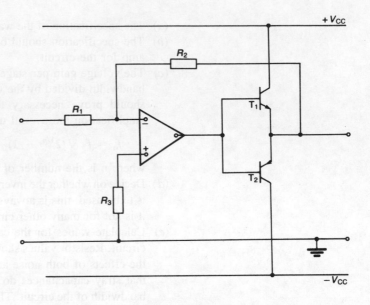

**Fig. 3.2** Current booster

the output voltage of the circuit is limited by the diode network to $\pm 5\,\text{V}$, the output saturation voltage $V_O(\text{sat}) = \pm 14\,\text{V}$, and the maximum output current is $10\,\text{mA}$, then

$$R_1 = (14 - 5)/(10 \times 10^{-3}) = 900\,\Omega$$

The required zener diode voltage is $5 - (2 \times 0.7) = 3.6\,\text{V}$.

(b) Use an op-amp with a higher maximum output current. Some types of op-amp, e.g. the LH 0021, can output up to 1 A current although a heat sink will probably be needed.

(c) The output current of an op-amp can be supplied to a current-booster circuit as shown by Fig. 3.2.

Once a circuit has been designed a prototype should be constructed and tested. If, as is only too likely, the prototype does not work as expected then, firstly, the construction should be checked; if no fault is found then it will be necessary to check, and perhaps modify, the calculations made in the design. It may even prove necessary to discard the original circuit and to design a different circuit.

### Using Uncompensated Op-amps

Whenever an amplifier has negative feedback applied to it there is always a chance that at some frequency capacitive effects will make the loop phase shift equal to $360°$. If this happens it means that the feedback has become positive at that frequency and so the fed-back signal adds to, rather than subtracts from, the input signal. The amplifier will still be stable provided the loop gain at this frequency is less than unity, but if the loop gain is unity, or greater, the amplifier will be unstable and will tend to break into oscillation at that frequency. To prevent this from happening an op-amp must be *frequency compensated*. Frequency compensation is a means by which an op-amp is given both an adequate *gain margin* and an adequate *phase margin*. The gain margin is the number of decibels by which the gain of the op-amp could be increased before the circuit would become unstable; if, for example, at the frequency at which the loop phase shift was 360° the loop gain was 0.2 then the gain margin is $20 \log_{10} 0.2 = 14$ dB. The phase margin is a measure of the loop phase shift at the frequency at which the loop gain is unity; it is the extra phase shift required to make the op-amp unstable. The most common phase margin is 45°.

Most op-amps are internally compensated to ensure that they remain stable for any value of closed-loop gain. The price paid by the user for not having to bother with instability problems is a relatively poor open-loop bandwidth, for example, the internally compensated 741 has a frequency response that starts to roll off at just 5 Hz. If a wide bandwidth is required an uncompensated, or partly compensated, op-amp must be employed. This will allow any desired phase margin to be employed. In most cases 45° is chosen since it gives the best compromise between bandwidth and the avoidance of a peak in the gain–frequency characteristic.

A number of frequency compensation techniques are available but details are usually given in the data sheet of the op-amp. One, or more, circuits are normally given, together with graphs which allow the user to choose the component values that are appropriate for the wanted closed-loop gain. If operating speed and wide bandwidth are not prime considerations, a greater stability margin can be obtained by increasing the recommended size of the compensation capacitor, but smaller values should not be used. Two commonly employed uncompensated op-amps are the LM 301 and the NE 531.

Partly compensated op-amps are provided with enough internal frequency compensation to make them stable for all closed-loop gains higher than some specified figure. If a lower closed-loop gain is required then some external frequency compensation must be added. Two examples of this type of op-amp are the LF 357 and the NE 5534. Other op-amps, such as the LM 318, are internally compensated for all values of closed-loop gain, including unity, but some external compensation can be added to the circuit to obtain an improved performance.

**Amplifiers**

An amplifier may be required to amplify all signals from 0 Hz to some maximum frequency — known as a d.c. amplifier — or be required to amplify all signals that lie within a specified bandwidth — known as an a.c. amplifier. The bandwidth is normally specified by quoting the wanted lower, and upper, 3 dB frequencies of the amplifier's gain–frequency characteristic.

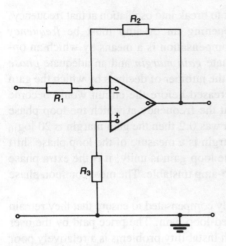

**Fig. 3.3** Inverting amplifier

## d.c. Amplifier

A d.c. amplifier may employ either the inverting or the non-inverting configuration. Particular attention needs to be paid to the reduction of both the bias current and the input offset voltage, and their drift due to temperature variations, and the op-amp ought to have a high CMRR and PSRR since all these will produce an output offset error. The bias current flows in the source resistance to produce a voltage that will be amplified to produce an offset error and this means that the source resistance should be as low as possible. The bandwidth of the op-amp is (probably) of little consequence and the slew rate only matters if the output voltage is large. Neither of the inputs of the non-inverting circuit are at earth potential but are at the signal level; hence the op-amp should have a common-mode voltage range at least equal to the maximum input signal voltage.

Figure 3.3 shows the basic circuit of an inverting amplifier. The voltage gain of the circuit is

$$A_{v(f)} = -R_2/R_1 \qquad (3.2)$$

the input resistance is equal to the input resistor $R_1$, and the closed-loop bandwidth is

$$\text{Bandwidth} = (\text{unity gain bandwidth})/(1 + A_{v(f)}) \qquad (3.3)$$

where $A_{v(f)}$ is the voltage gain as a ratio and not in decibels. Note that the closed-loop gain is $1 + R_2/R_1$ and hence not quite equal to the inverting signal gain.

The output offset voltage is given by

$$(1 + R_2/R_1)[V_{OS} + (R_1R_2/(R_1 + R_2) - R_3)I_B$$
$$+ (R_1R_2/(R_1 + R_2) + R_3)I_{OS}/2] \qquad (3.4)$$

and the gain error is

$$(1 + R_2R_1)(1/A_v) \qquad (3.5)$$

In order to minimize the output offset voltage caused by the bias currents $I_B^+$ and $I_B^-$ the resistor $R_3$ connected between the non-inverting terminal and earth should be equal to $R_1R_2/(R_1 + R_2)$, where $R_1$ includes any source resistance. If perfect balance is achieved the offset voltage will then be equal to the closed-loop gain times $I_{OS}R_2$. Bias current drift is proportional to temperature for a bipolar op-amp and doubles with each 10 °C increase in temperature for a JFET. There will be an output offset voltage due to the input offset voltage $V_{OS}$ and the input offset current $I_{OS}$. Many op-amps are provided with terminals to which a nulling circuit can be connected and adjusted to minimize the effect of $V_{OS}$. Two examples of such nulling circuits are shown by Figs 3.4(a) and (b) for the 741 and LM 301 op-amps respectively. Drift due to the input offset voltage is 3.4 $\mu$V/°C for each 1 mV of $V_{OS}$. If nulling pins are not provided, the adjustable d.c. voltage required to cancel the offset must be applied

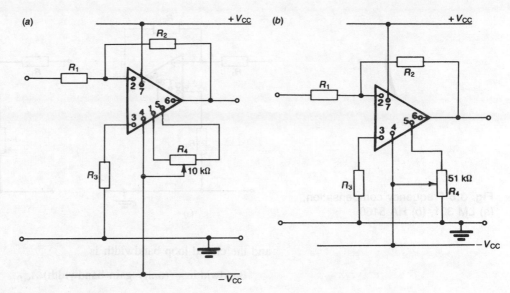

**Fig. 3.4** Two nulling circuits: (*a*) 741, (*b*) LM 301

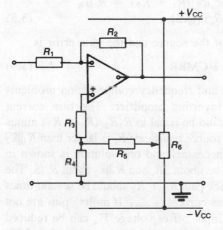

**Fig. 3.5** Nulling an op-amp without nulling terminals

to the input terminal that is not used as the signal input. For an inverting amplifier it is only necessary to connect the bias current compensating resistor $R_3$ to an adjustable d.c. voltage, instead of to earth, as shown by Fig. 3.5.

To ensure that the effect of the bias current is small the input current $I_{IN}$ should be $n$ ($n$ at least 10) times larger than the maximum bias current $I_{B(max)}$ obtained from the data sheet. If there is a known, fixed, value of input voltage $V_{IN}$ then $R_1 = V_{IN}/nI_{B(max)}$. The value of $R_2$ can then be obtained from the required voltage gain. Also the product $I_{OS}R_2$ should be smaller than $V_{IN}/10$. Since $R_2$ is effectively connected in parallel with the output resistance $R_{out}$ of the op-amp it should be at least 10 times larger than $R_{out}$, but this requirement rarely presents any difficulty. If an uncompensated op-amp is employed it will also be necessary to add frequency-compensation circuitry; details of this can be obtained from the data sheet of the op-amp. Two examples are shown by Figs 3.6(*a*) and (*b*) for the LM 301 and the HA 5160 respectively.

The nulling procedure is accurate at only one temperature and for just a short time. Both the offset current and the offset voltage will change with time because of small variations in the parameters of the internal transistors. The offset values will also vary with temperature changes. Drift due to temperature changes can be minimized by keeping the ambient temperature as constant as possible and by using a low-drift type of op-amp.

The basic circuit of a non-inverting d.c. amplifier is shown by Fig. 3.7(*a*). The voltage gain of the circuit is

$$A_{v(f)} = 1 + R_2/R_1 \tag{3.6}$$

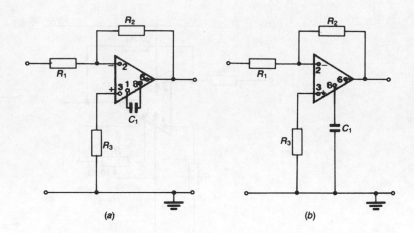

**Fig. 3.6** Frequency compensation; (a) LM 301, (b) HA 5160

and the closed-loop bandwidth is

$$\text{Bandwidth} = (\text{unity gain bandwidth})/A_{v(f)} \quad (3.7)$$

The input resistance of the amplifier is very high (and not equal to $R_3$). The output offset voltage is given by equation (3.8), i.e.

$$(1 + R_2/R_1)[V_{OS} + (R_1R_2/(R_1 + R_2) - R_s)I_B$$
$$+ (R_1R_2/(R_1+R_2) + R_s)I_{OS}/2] \quad (3.8)$$

where $R_s$ is the resistance of the source and the gain error is

$$(1 + R_2/R_1)(1/A_v) + 1/\text{CMRR} \quad (3.9)$$

The output offset voltage and frequency compensation problems are the same as for the inverting amplifier. The bias current compensation resistor $R_3$ should be equal to $R_1R_2/(R_1 + R_2)$ minus the source resistance. If the source resistance $R_s$ is larger than $R_1\|R_2$ a different method will be necessary and one solution is shown in Fig. 3.7(b). Now $R_3$ should be about $3R_s$ and $R_1\|R_2$ about $R_s/3$. The current flowing in the feedback path $R_1 + R_2$ should be several times larger than the maximum bias current $I_{B(max)}$. If nulling pins are not provided the effects of the input offset voltage $V_{os}$ can be reduced by injecting a d.c. voltage into the inverting terminal (see Fig. 3.8).

### Example 3.1

Calculate the output voltage of the amplifier given in Fig. 3.7(a) if a 741 op-amp, with worst-case values $V_{OS} = 5\,\text{mV}$, $I_{OS} = 200\,\text{nA}$, and CMRR $= 70\,\text{dB}$, is used together with $R_1 = 91\,\Omega$ and $R_2 = 9.1\,\text{k}\Omega$. The input signal voltage is $0.1\,\text{V}$ and the common mode input voltage is $100\,\mu\text{V}$.

### Solution

$A_{v(f)} = 1 + 9100/91 = 101$. Without error, $V_{OUT} = 101 \times 0.1 = 10.1\,\text{V}$. With the offset errors the output voltage is

$$V_{OUT} = 101[0.1 + (5 \times 10^{-3}) + (200 \times 10^{-9} \times 9100)$$
$$+ (100 \times 10^{-6})/3162]$$
$$= 10.789\,\text{V}. \quad (Ans.)$$

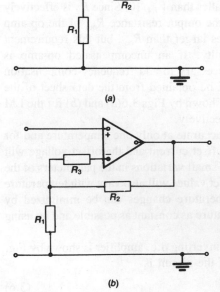

**Fig. 3.7** Non-inverting circuit

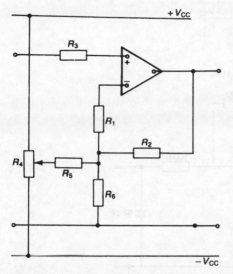

**Fig. 3.8** Nulling a non-inverting op-amp

## Example 3.2

The d.c. amplifier shown in Fig. 3.5 is to have a voltage gain of 15 and an input resistance of 22 kΩ. Use a 301 op-amp and ± 15 V power supplies.

*Solution*

Input resistance = 22 kΩ = $R_1$. Voltage gain = 15 = 1 + $R_2/R_1$, so

$$R_2 = 14R_1 = 308 \text{ k}\Omega = 309 \text{ k}\Omega$$

nearest preferred value, or 300 kΩ in series with 8.2 kΩ.
 Then $R_3 = (22 \times 308)/330 = 20.5$ kΩ.
 From the 301 data sheet $V_{OS(max)} = 7.5$ mV and $I_{OS(max)} = 50$ nA and hence the maximum output offset error is

$$7.5 \text{ mV} + (50 \times 10^{-9} \times 20.5 \times 10^3) = 8.53 \text{ mV}.$$

To cancel out this voltage the external nulling circuit must introduce a slightly larger, say ± 12 mV, compensating d.c. voltage. $R_6$ should be a variable resistor of fairly high value; 100 kΩ is suitable. Also

$$12 \text{ mV} = 15R_4(R_4 + R_5)$$
$$15/(12 \times 10^{-3}) = (R_4 + R_5)/R_4 \quad \text{or} \quad R_5 = 1249R_4$$

If $R_4$ is chosen to be 100 Ω then $R_5 = 124.9$ kΩ = 120 kΩ preferred value.
 From the data sheet single-pole frequency compensation requires a capacitor $C_1$ connected between pins 1 and 8 whose value is $30R_1/(R_1 + R_2)$ pF = $(30 \times 22)/331 = 2$ pF.

## Example 3.3

Design an amplifier to have a voltage gain of 60 dB, a closed-loop bandwidth of 0–20 kHz, and an input resistance of 10 kΩ.

*Solution*

The required amplifier could be constructed using the op-amp in either its inverting or its non-inverting configuration. $A_{(f)} = 60$ dB = 1000.

For the inverting circuit, input resistance = $R_1 = 10$ kΩ, and feedback resistor $R_2 = 1000 \times 10$ kΩ = 10 MΩ. Try the 741 op amp. $V_{OS(max)} = 5$ mV and hence, without nulling the output offset voltage would be 0.5 V. This could be considerably reduced by nulling but it does indicate that the 741 is not a good choice for a high-gain d.c. amplifier. It would be better to use a low $V_{OS}$ type of op-amp. Also, the closed-loop bandwidth would be only 1 MHz/100 = 10 kHz which is less than the wanted 20 kHz. The chosen op-amp ought to have a unity-gain bandwidth of at least 20 MHz. Suitable op-amps would be the HA 2541 and the LH 0032. The LH 0032 has $V_{OS} = 2$ mV, $I_{B(max)} = 10$ pA, $A_v = 70$ dB, slew rate = 500 V/μs and a unity-gain bandwidth of 70 MHz; the device does not need external frequency compensation if the closed-loop gain is 100 or more. Figure 3.9 shows this op-amp connected to have a voltage gain of 1000. If the upper 3 dB frequency is to be no greater than 20 kHz a 100 Ω resistor can be connected in series with the output terminal and a shunt capacitor connected between the resistor and earth. The op-amp will then need frequency compensation and this consists of a capacitor connected between terminals 2 and 3 whose value, obtained from the data sheet, depends upon the shunt capacitance; this is equal to

$$1/(2\pi \times 100 \times 20 \times 10^3) = 80 \text{ pF} = 82 \text{ pF preferred value.}$$

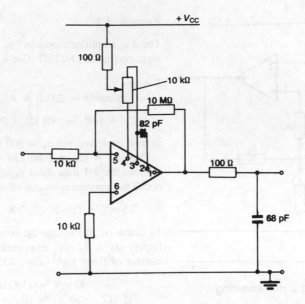

**Fig. 3.9** LH 0032 connected to give
a gain of 1000

If the non-inverting circuit is chosen the problems will be similar except
that the wanted input resistance does not specify the values of $R_1$ and $R_2$.

It will often prove to be cheaper and better to design such a high-gain
amplifier using two general-purpose op-amps connected in cascade. Each op-
amp can then provide a gain of 30 dB or 31.62. Then, if $R_1 = 10 \, \text{k}\Omega$, $R_2$
$= 316.2 \, \text{k}\Omega$, nearest preferred value 316 k$\Omega$. The voltage gain is then 316/10
$= 31.6$ and the 741 op-amp has a closed-loop bandwidth of 1 MHz/31.6 $=$
31.646 kHz. The overall bandwidth would then be

$$31.646 \, \sqrt{(2^{1/2} - 1)} = 20.37 \, \text{kHz}$$

which is only just adequate. Suppose the LM 318 is chosen instead; this
internally compensated op-amp has a unity-gain bandwidth of 15 MHz, a slew
rate of 50 V/$\mu$s, $I_{\text{B(max)}} = 500 \, \text{nA}$, $V_{\text{OS(max)}} = 10 \, \text{mV}$ and $I_{\text{OS(max)}} =$
200 nA. The bias current compensation resistor $R_3 = 10 \, \text{k}\Omega \| 316 \, \text{k}\Omega =$
10 k$\Omega$ nearest preferred value. The circuit of the amplifier is given in Fig. 3.10.

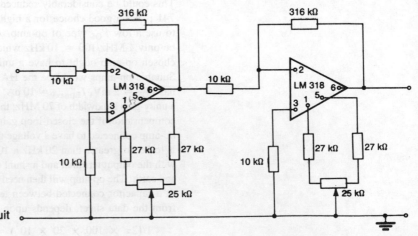

**Fig. 3.10** Two-stage op-amp circuit
has a gain of 1000

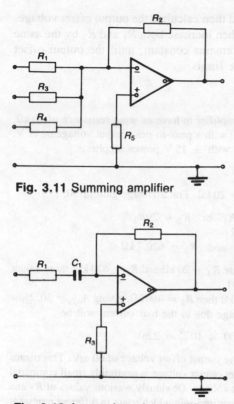

**Fig. 3.11** Summing amplifier

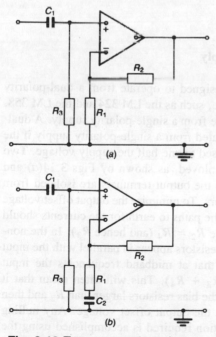

**Fig. 3.12** An a.c. inverting amplifier

When a summing amplifier is designed (see Fig. 3.11), the bias current compensating resistor $R_5$ should be equal to $1/(1/R_1 + 1/R_4 + 1/R_3)$.

### a.c. Amplifier

An a.c. amplifier is required to amplify signals from some low frequency (not 0 Hz) to some high frequency; the bandwidth is usually specified by the upper 3 dB frequency. Unless severe, input bias, current offset and drift errors have little effect since they merely shift the mean output voltage from 0 V and in so doing slightly reduce the peak output voltage that can be obtained without the op-amp going into saturation. In most cases the bandwidth is set by the unity-gain bandwidth of the op-amp; if, however, the feedback resistor is of very high value and/or the load is capacitive then the upper 3 dB frequency will be set by external capacitance(s). If the output voltage is large the slew rate of the op-amp will be of importance.

An a.c. amplifier has its input and output terminals coupled by a capacitor to the source or load resistance as shown by Fig. 3.12. In the inverting amplifier there is no d.c. path through the input resistor $R_1$ and so the bias current compensating resistor $R_3$ (if used) should be made equal to the feedback resistor $R_2$. The voltage gain of the amplifier is

$$A_{v(f)} = -Z_2/Z_1 = -j\omega C_1 R_2/(1 + j\omega C_1 R_1) \tag{3.10}$$

At zero frequency $A_{v(f)} = 0$ and at high frequencies where $R_1 \gg 1/\omega C_1$, $A_{v(f)} = -R_2/R_1$. The gain of the amplifier will have fallen by 3 dB from its low-frequency value at frequency $f_3 = 1/2\pi R_1 C_1$. The bias current compensating resistor should be equal to $R_2$, and not $R_1 \| R_2$, since $C_1$ acts as a d.c. block.

The non-inverting circuit will not have a d.c. path from the non-inverting terminal to earth and so a resistor $R_3$ must be connected between this terminal and earth as shown by Fig. 3.13(a). This means that the input resistance of the circuit is equal to $R_3$. For the optimum compensation for input bias current effects $R_3$ should be equal to $R_1R_2/(R_1 + R_2)$ but this may make the input resistance of the amplifier too low and often a compromise is necessary. The voltage gain of the amplifier is

$$A_{v(f)} = [1/(1 + 1/j\omega C_1 R_3)][1 + R_2/R_1] \tag{3.11}$$

with a 3 dB frequency of $f_3 = 1/2\pi C_1 R_3$.

Sometimes a capacitor $C_2$ is connected in series with the compensating resistor $R_3$ as shown by Fig. 3.13(b). The voltage gain of this circuit is

$$A_{v(f)} = [1/(1 + 1/j\omega C_1 R_1)][(1 + j\omega(R_2 + R_3))/(1 + j\omega C_2 R_1)] \tag{3.12}$$

The design procedure is: choose $R_3$ to give the wanted input resistance for the circuit and then calculate the output offset voltage. If this should be too large then increase both $R_1$ and $R_2$ by the same factor, so that their ratio remains constant, until the output offset voltage is within acceptable limits.

### Example 3.4

Design a non-inverting a.c. amplifier to have an input resistance of 20 kΩ, and a voltage gain of about 30 with a peak-to-peak output voltage of 12 V. If possible use a 741 op-amp with ± 15 V power supplies.

### Solution

Referring to Fig. 3.13, $R_3 = 20$ kΩ. For a voltage gain of 30,

$$30 = 1 + R_2/R_1 \quad \text{or} \quad R_2 = 29R_1.$$
$$20 \times 10^3 = 29R_1^2/30R_1$$
$$R_1 = 20.69 \text{ kΩ} \quad \text{and} \quad R_2 = 620.7 \text{ kΩ}$$

The nearest preferred values are $R_1 = 20$ kΩ and $R_2 = 620$ kΩ; these values give a voltage gain $A_{v(f)}$ of 31.

If $R_1$ is reduced to, say, 16 kΩ then $R_2 = 480$ kΩ, giving $A_{v(f)} = 30$. Now the negative input offset voltage due to the bias current will be

$$(20 - 16) \times 10^3 \times 500 \times 10^{-9} = 2 \text{ mV}$$

and this will produce a negative output offset voltage of 60 mV. This means that the change in the quiescent output voltage is negligibly small compared with the expected peak output voltage. Obviously, various values of $R_1$ and $R_2$ can be used to obtain the required gain which result in different quiescent output voltages.

### Single-polarity Power Supply

Most op-amps have been designed to operate from a dual-polarity power supply, although some, such as the LM 324 and the LM 358, have been designed to operate from a single-polarity supply. A dual-polarity op-amp can be operated from a single-polarity supply if the non-inverting terminal is biased to one-half the supply voltage. Two equal-value resistors are employed, as shown by Figs 3.14(a) and (b). The inverting input and the output terminals are isolated from earth by the coupling capacitors. To minimize the output offset voltage caused by the bias currents the paths to earth for the currents should be of equal resistance. Hence $R_2 = R_4$ (and hence $R_3$). In the non-inverting amplifier the bias resistors appear in parallel with the input terminals of the circuit so that at midband frequencies the input resistance is $R_{in} = R_3R_4/(R_3 + R_4)$. This will often mean that it becomes necessary to make the bias resistors larger than $R_2$ and then there will be an increase in the output offset voltage. Any nulling and/or frequency compensation required is accomplished using the information given in the op-amp's data sheet.

Any noise present in the power supply line will appear at the non-

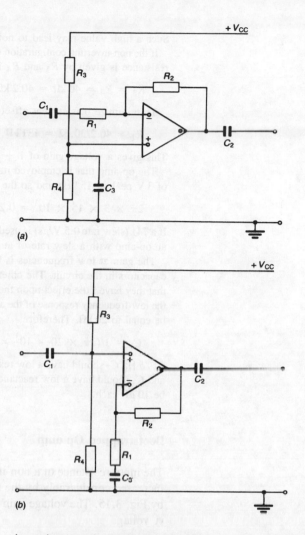

**Fig. 3.14** An a.c. amplifier using a single-polarity power supply

inverting terminal and be amplified. This unwanted effect can be eliminated in the inverting circuit by connecting a decoupling capacitor $C_3$ across $R_3$ as shown in Fig. 3.14. This is not possible in the non-inverting circuit because a decoupling capacitor would short-circuit the amplifier's input terminal.

### Example 3.5

Design an amplifier to have a voltage gain of $30 \pm 1$ dB over the frequency band 25 Hz to 15 kHz with an input resistance of about 20 kΩ at midband. The circuit is to operate from a single-polarity 9 V power supply and to produce a maximum output voltage of 3 V peak with little distortion.

### Solution

30 dB = 31.62 voltage ratio. If an inverting amplifier is chosen the input resistor must be of 20 kΩ resistance and hence the feedback resistor will be

$$31.62 \times 20 = 632.4 \text{ k}\Omega = 634 \text{ k}\Omega \text{ nearest preferred value}$$

Such a high value may lead to noise and offset voltage problems.

If the non-inverting configuration (Fig. 3.13($b$)) is chosen the required input resistance is given by $R_3$ and $R_4$ in parallel. Hence

$$R_3 = R_4 = 40 \text{ k}\Omega = 40.2 \text{ k}\Omega \text{ nearest preferred value.}$$

To minimize bias current offset effects choose $R_2 = 40.2 \text{ k}\Omega$ and then

$$R_1 = 40.2/30.62 = 1312 \,\Omega = 1300 \,\Omega \text{ nearest preferred value.}$$

This gives a voltage gain of $1 + 40.2/1.3 = 31.92 = 30.1$ dB.

The op-amp that is employed must be able to produce an output voltage of 3 V peak at 15 kHz and so the minimum acceptable slew rate is

$$2\pi \times 3 \times 15 \times 10^3 = 0.28 \text{ V/}\mu\text{s.}$$

If a 741 (slew rate 0.5 V/$\mu$s) is used some distortion may be evident, so select an op-amp with a slew rate of at least $5 \times 0.28 = 1.4$ V/$\mu$s.

The gain at low frequencies is best determined by just one of the three capacitors in the circuit. The other two capacitors should have values such that they have little effect upon the response at 25 Hz. If $C_1$ is to determine the low-frequency response of the amplifier then at 25 Hz its reactance should be equal to 20 k$\Omega$. Therefore,

$$C_1 = 1/(2\pi \times 20 \times 10^3 \times 25) = 0.32 \,\mu\text{F}$$

At 25 Hz $C_2$ should have a low reactance compared with the load resistance and $C_3$ should have a low reactance compared to $R_4$; suitable values would be 10 $\mu$F each.

### Bootstrapped Op-amp

The input resistance of a non-inverting op-amp a.c. amplifier can be increased considerably by the use of the *bootstrap* technique shown by Fig. 3.15. The voltage gain of the amplifier is $A_{v(f)} = 1 + R_2/R_1$. A voltage

$$V_{\text{out}}R_1/(R_1 + R_2) = V_{\text{out}}/A_{v(f)}$$

is fed back, via $C_2$, to the junction of resistors $R_3$ and $R_4$. This means that the a.c. voltages at either side of $R_3$ are very nearly the same, hence only a very small a.c. current flows in $R_3$ and so its a.c. resistance is very high. Voltage followers are often bootstrapped in this way and Figs 3.16($a$) and ($b$) show the circuit arrangements for operation for both dual-polarity and single-polarity power supplies.

### Current Amplifier

An op-amp may also be connected to act as a current amplifier. Figures 3.17($a$) and ($b$) show, respectively, the inverting and the non-inverting versions of the circuit.

For the inverting circuit,

$$V^- - V_{\text{out}} = I_{\text{in}}R_2 \tag{3.13}$$

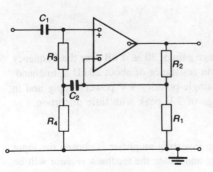

**Fig. 3.15** Bootstrapped op-amp

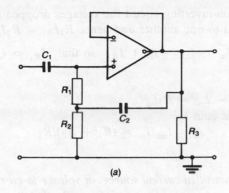

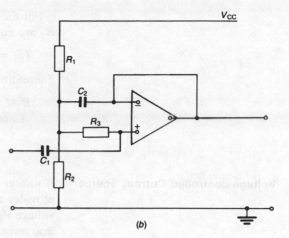

**Fig. 3.16** Bootstrapped voltage followers

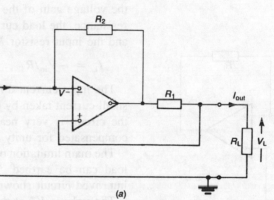

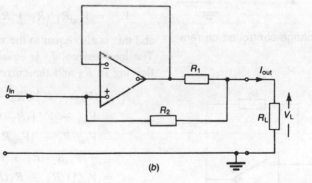

**Fig. 3.17** Inverting and non-inverting current amplifiers

$$V_{\text{out}} - V_{\text{L}} = I_{\text{out}} R_1 \qquad (3.14)$$

Since the open-loop gain of the op-amp is high $V^- = V^+ = V_{\text{L}}$ and hence equation (3.13) can be written as $V_{\text{L}} - V_{\text{out}} = I_{\text{in}} R_2$, or $V_{\text{L}} = I_{\text{in}} R_2 + V_{\text{out}}$. Substituting into equation (3.14) gives

$$V_{\text{out}} - I_{\text{in}} R_2 - V_{\text{out}} = I_{\text{out}} R_1$$
$$\text{Current gain } A_{\text{i}} = I_{\text{out}}/I_{\text{in}} = -R_2/R_1 \qquad (3.15)$$

Note that if $R_1 = R_2$ the circuit acts as a *current mirror*.

For the non-inverting circuit the voltages dropped across $R_1$ and $R_2$ are equal to one another and hence $R_1 I_{R_1} = R_2 I_{R_2}$.

$$I_{R_2} = I_{in} \qquad I_{out} = I_{in} + I_{R_1} \quad \text{so that} \quad I_{R_1} \simeq I_{out} - I_{in}.$$

Therefore

$$R_1(I_{out} - I_{in}) = R_2 I_{in}$$

Current gain

$$A_i = I_{out}/I_{in} = (R_1 + R_2)/R_1 \qquad (3.16)$$

### Voltage-controlled Current Source

A *voltage-controlled current source*, or *voltage-to-current converter*, provides a current to a load resistance $R_L$ that is set by an input voltage $V_{in}$. The simplest circuit is shown by Fig. 3.18 in which the load resistance is placed in the feedback path of the op-amp. Although the voltage gain of the circuit will vary with the value of the load resistance, the load current is determined only by the input voltage and the input resistor $R_1$, thus

$$I_L = -V_{in}/R_1 \qquad (3.17)$$

The input current differs from the load current only by the very small current taken by the op-amp itself. Hence the current gain of the circuit is very nearly unity. The op-amp must be frequency compensated for unity voltage gain.

The main limitation of this simple circuit is that neither side of the load can be earthed. The disadvantage can be overcome if the improved circuit shown in Fig. 3.19 is employed.

The voltage $V^-$ at the inverting terminal is

$$V^- = V_{out}R_1/(R_1 + R_2) + V_{in}R_2/(R_1 + R_2)$$

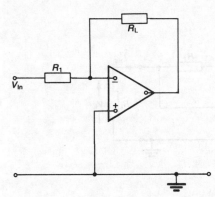

**Fig. 3.18** Voltage-controlled current source

and this is also equal to the voltage $V^+$ at the non-inverting terminal. The load current $I_L$ is equal to the difference between the current flowing in $R_4$ and the current in $R_3$, hence

$$\begin{aligned} I_L &= (V_{out} - V^+)/R_4 - V^+/R_3 = V_{out} - V^+(1/R_4 + 1/R_3) \\ &= V_{out} - V^+[(R_3 + R_4)/R_3R_4] \\ &= V_{out}/R_4 - [V_{out}R_1/(R_1 + R_2) \\ &\quad + V_{in}R_2/(R_1 + R_2)][(R_3 + R_4)/R_3R_4)] \\ &= V_{out}[1/R_4 - R_1(R_3 + R_4)/(R_1 + R_2)R_3R_4] \\ &\quad - V_{in}[R_2(R_3 + R_4)/(R_1 + R_2)R_3R_4] \end{aligned}$$

For the load current to be independent of the output voltage $V_{out}$ and hence of the value of the load resistance $R_L$

$$1/R_4 = R_1(R_3 + R_4)/(R_1 + R_2)R_3R_4 \quad \text{or} \quad R_1/R_2 = R_3/R_4$$

Then

$$\begin{aligned} I_L &= -V_{in}R_2(R_3 + R_4)/(R_1 + R_2)R_3R_4 \\ &= -V_{in}/R_3 \times R_2/(R_1 + R_2) \times (R_3 + R_4)/R_4 \\ &= -V_{in}/R_3 \times 1/(1 + R_1/R_2) \times (1 + R_3/R_4) \end{aligned}$$

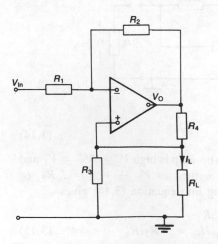

**Fig. 3.19** Improved voltage-controlled current source

or

$$I_L = -V_{in}/R_3 \qquad (3.18)$$

But $R_3 = R_1R_4/R_2$, so that

$$I_L = -V_{in}R_2/R_1R_4 \qquad (3.19)$$

When the load is a capacitor a constant current will be supplied to it so that it will charge at a constant rate. This is the basis of some types of waveform generator and analogue-to-digital converter (ADC).

### Integrators and Differentiators

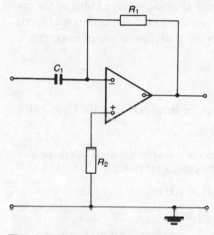

**Fig. 3.20** Basic differentiator

The basic circuit of an op-amp differentiator is shown by Fig. 3.20. The output voltage of the circuit is

$$V_{out} = R_1C_1 \; dV_{in}/dt \qquad (3.20)$$

Resistor $R_2$ is sometimes provided to minimize bias current offset problems. The choice of op-amp is based upon the same considerations as for an amplifier and the op-amp should be frequency compensated for unity voltage gain. The values of $R_1$ and $C_1$ should be chosen so that their product is small compared with the periodic time of the highest frequency to be differentiated. A standard capacitance value is then chosen and the value of $R_1$ calculated. This basic circuit is very susceptible to noise and is also prone to instability; better circuits are shown by Figs 3.21(a) and (b). The additional components, $R_3$ in Fig. 3.21(a) and $R_3C_2$ in Fig. 3.21(b), reduce the gain of the circuit at higher frequencies. In Fig. 3.21(a), for example, the maximum high frequency gain is $R_1/R_3$. The values of $R_3$ and $C_2$ are chosen so that each $CR$ circuit produces 6 dB/octave roll-off (this is a total of 12 dB/octave for Fig. 3.21(b)) to reduce the effects of noise voltages.

Figure 3.22 shows the basic circuit of an op-amp integrator. The output voltage of the circuit is

$$V_{out} = (1/C_1R_1) \int_0^t V_{in} \; dt \qquad (3.21)$$

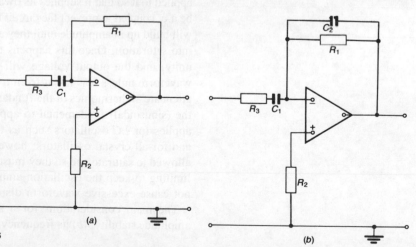

**Fig. 3.21** Improved differentiators

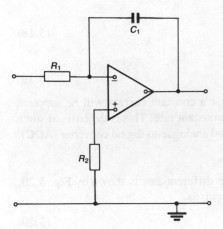

**Fig. 3.22** Basic integrator

The resistor $R_2$ is employed to reduce the output offset voltage due to the bias currents and should, if possible, be of the same value as $R_1$. The op-amp should be frequency compensated for unity gain. At low frequencies the voltage gain of the circuit is very high and even a small offset current or voltage will cause the op-amp to saturate. To prevent such 'latch-up' occurring it is common practice to connect a feedback resistor $R_3$ in parallel with $C_1$; $R_3$ should be about 10 times larger than $R_1$. When $R_3$ is fitted the circuit will be unable to retain a steady output voltage when an input d.c. signal is removed since the charge stored in $C_1$ will leak away via $R_3$. The capacitive coupling between the input and output terminals allows any transients that occur at the output to appear at the input. This means that either the common-mode or the differential-mode voltage limits of the op-amp could be exceeded. The common-mode and differential-mode voltage ratings of the chosen op-amp are therefore of importance.

### Example 3.6

Design an integrator to operate over the frequency band 100 Hz to 5 kHz.

### Solution
The time constant of an op-amp integrator should be about one-half the periodic time of the highest frequency to be integrated. Hence

$$R_1C_1 = 0.5 \times [1/(5 \times 10^3)] = 100 \, \mu s.$$

A suitable value for $C_1$ is 47 nF and then

$$R_1 = (100 \times 10^{-6})/(47 \times 10^{-9}) = 2128 \, \Omega = 2100 \, \Omega \text{ preferred value.}$$

Then

$$R_3 = 21 \, k\Omega \quad \text{and} \quad R_2 = (2.1 \times 21)/23.1 = 1.91 \, k\Omega$$

assuming that the source resistance is small.

## Sinusoidal Oscillators

Essentially, an oscillator is an amplifier that has positive feedback applied to it so that it supplies its own input signal. The feedback must be a.c. coupled to prevent the circuit from latching up. The oscillations will build up in amplitude until they are clipped by the op-amp driving into saturation. Once this happens the loop gain will be reduced to unity, and the output voltage will be stable but of non-sinusoidal waveform unless it is taken from a high-$Q$ tuned circuit which will attenuate all harmonics of the fundamental frequency and allow only the (sinusoidal) fundamental to appear at the output terminal. This applies for $LC$ oscillators such as the Hartley and Colpitts circuits and for all crystal oscillators; however, $CR$ oscillators must not be allowed to saturate and so they must employ some kind of amplitude limiting to keep the oscillation amplitude at a lower level that does not cause excessive waveform distortion.

The main considerations for a sinusoidal oscillator are: (*a*) its amplitude stability; (*b*) its frequency stability; (*c*) its frequency range;

(*d*) its distortion level. The phase-shift *CR* oscillator can operate over a frequency band of about 10 Hz to 1 MHz with an amplitude stability of about 3% and about 2% distortion. The circuit is easily tunable over a 2:1 frequency ratio and is quick to settle, and so it is often employed for cheap, moderate-performance applications. The Wien bridge *CR* oscillator will operate from about 1 Hz to about 1 MHz with an amplitude stability of about 1% and about 0.01% distortion. The Wien bridge circuit is employed for high-performance applications even though it takes some time to settle after either its amplitude or its frequency is changed. The circuit suffers from the disadvantage that tuning requires the use of dual variable resistors with good tracking and, if a wide frequency range is wanted, switched capacitors. Both the Colpitts and the Hartley oscillators can work at frequencies in the range 1 kHz to 10 MHz with about 3% amplitude stability and about 2% distortion. Both circuits are relatively difficult to tune over a wide frequency band. Crystal oscillators can operate from about 30 kHz upwards with 1% amplitude stability and 0.1% distortion. The frequency stability of a crystal oscillator is very high and only very limited tuning is possible.

The op-amp that is employed in an oscillator circuit must have a sufficiently high unity-gain bandwidth to allow the necessary gain to be provided at the highest frequency to be generated. It is usual to choose an op-amp whose unity-gain bandwidth is at least 10 times the required maximum oscillation frequency. The slew rate of the op-amp must be high enough to allow the wanted output voltage $V_{\text{out}}$ to be obtained at the highest oscillation frequency $f_{\text{max}}$ without slew rate induced distortion. The minimum slew rate is equal to $2\pi f_{\text{max}} V_{\text{out}}$ but a safety factor, of typically 5, should be allowed. The output voltage need not be large since it can always be applied to a buffer amplifier to have its amplitude increased. The op-amp is operated in its non-inverting configuration and hence a fraction of the output voltage forms a common-mode signal at the input. As the gain of the op-amp is reduced the common-mode signal increases and so does distortion. The common-mode voltage at the input terminal should be restricted to a couple of volts at most.

### Phase-shift Oscillator

The circuit of a phase-shift oscillator is shown in Fig. 3.23. The circuit oscillates at a frequency $f_o$ given by

$$f_o = 1/(2\pi \sqrt{6} \, CR) \tag{3.22}$$

where $C = C_1 = C_2 = C_3$ and $R = R_1 = R_3 = R_4$, and must have a minimum voltage gain of $-R_2/R_1 = -29$.

It is usual to make the voltage gain slightly larger than 29, say 31, to ensure that the circuit oscillates. The maximum output voltage is approximately $V_{\text{O(SAT)}} = V_{\text{CC}} - 1$ volts but this is limited to some

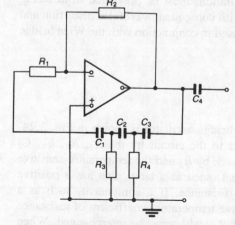

**Fig. 3.23** Phase-shift oscillator

lower figure by connecting two diodes back to back across the feedback resistor $R_2$, so that the voltage at the inverting terminal must be equal to $V_{out}/31$. The input current to the op-amp should be several times $n$ larger than $I_{B(max)}$ and hence $R_1 = V_{out}/(31\,n\,I_{B(max)})$.

### Example 3.7

Design a phase-shift oscillator to operate at 5 kHz using a $\pm$ 12 V power supply.

*Solution*
If the output voltage goes to saturation of $\pm$ 11 V the input voltage is 11/31 = 0.355 V. The op-amp should have a minimum slew rate of

$$2\pi \times 5000 \times 11 = 0.35 \text{ V}/\mu s$$

and so a device with a slew rate of at least 1.75 V/$\mu$s should be selected. The 741S is suitable. For this op-amp $I_{B(max)} = 500$ nA so that $I_{in}$ must be at least 50 $\mu$A. Then

$R_1 = 0.355/(50 \times 10^{-6}) = 7.1 \text{ k}\Omega = 7.15 \text{ k}\Omega$ nearest preferred value
$R_2 = 31 \times 7.1 = 220.1 \text{ k}\Omega = 220 \text{ k}\Omega$ preferred value

These values give a voltage gain of 220/7.1 = 31.

$$R_3 = R_4 = R_1 = 7.1 \text{ k}\Omega$$
and
$$C_1 = C_2 = C_3 = 1/(2\pi \times \sqrt{6} \times 7100 \times 5000)$$
$$= 1.83 \text{ nF} = 2.2 \text{ nF preferred value}$$

This gives a frequency of 4159 Hz.
Try $C = 22$ nF, then

$R = 1/(2\pi \times \sqrt{6} \times 22 \times 10^{-9} \times 5000) = 591 \, \Omega$
$= 590 \, \Omega$ preferred value
$R_2 = 31 \times 590 = 18.29 \text{ k}\Omega = 18.2 \text{ k}\Omega$ preferred value

This gives a voltage gain of 18 200/590 = 30.85 and a frequency of 5006 Hz.

The amplitude of the oscillations must be prevented from taking the op-amp into saturation with consequent waveform distortion and ways of doing this are discussed in conjunction with the Wien bridge oscillator.

### Wien Bridge Oscillator

The basic circuit of a Wien bridge oscillator is shown in Fig. 3.24. Positive feedback is applied to the circuit by the $R_2$, $R_3$, $C_1$, $C_2$ network, and negative feedback by $R_1$ and the temperature-sensitive element L. L is a component, such as a lamp, that has a positive temperature coefficient of resistance. If a component, such as a thermistor, that has a negative temperature coefficient of resistance is used then the positions of $R_1$ and L must be interchanged. When power is first applied to the circuit the power dissipated in L is small

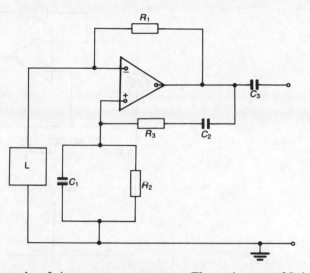

**Fig. 3.24** Wien bridge oscillator

and so L is at room temperature. The resistance of L is at its room temperature value $R_0$ and the gain of the op-amp is $-R_2/R_0$ and this is large enough to allow the amplitude of the oscillation to build up. As the oscillation amplitude increases the current flowing in L increases also and the temperature of L goes up; the resistance of L increases and this reduces the gain of the circuit. The amplitude of the oscillations stabilize at the level at which the contribution of the positive feedback is exactly equal to the contribution of the negative feedback. The correct operation of the circuit depends upon the resistance of L not changing significantly during each cycle of oscillation. This is not a problem at frequencies greater than about 100 Hz but it is at lower frequencies.

The frequency of oscillation is

$$f_0 = 1/2\pi RC \tag{3.23}$$

where $R = R_2 = R_3$ and $C = C_1 = C_2$, and the minimum voltage gain is 3.

### Example 3.8

Design a Wien bridge oscillator to operate over the frequency range 20 Hz to 20 kHz. Use (a) a lamp and (b) a thermistor to limit the oscillation amplitude. Employ an LF 356 BiFET op-amp and $\pm 9$ V power supplies.

### Solution

The LF 356 has a slew rate of 5 V/$\mu$s, an open-loop voltage gain of 103 dB, a unity-gain bandwidth of 18 MHz, and $V_{OS} = 0.5$ mV, $I_{B(max)} = 30$ pA. $I_B$ is so small that it can be ignored in the selection of the resistances from each op-amp input terminal to earth. This will allow a lamp to be used as the amplitude limiter.

Divide the frequency range into the decades 20–200 Hz, 200 Hz to 2 kHz, and 2–20 kHz. Suitable decades of capacitance that can be switched into circuit

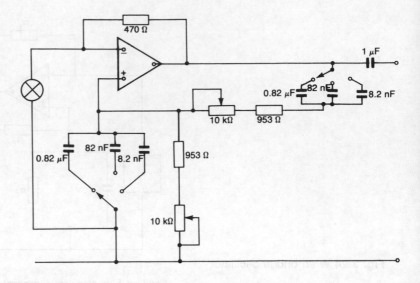

**Fig. 3.25**

for each sub-band must then be selected; suitable values are 0.68 $\mu$F, 68 nF and 6.8 nF. Then at 20 Hz,

$$R_{(max)} = 1/(2\pi \times 0.68 \times 10^{-6} \times 20) = 11.7 \text{ k}\Omega$$

and at 200 Hz

$$R_{(min)} = 11.7/10 = 1.17 \text{ k}\Omega$$

This resistance range can be obtained using a 1000 $\Omega$ resistor in series with a 22 k$\Omega$ potentiometer. A more delicate control of the oscillation frequency would be possible if the capacitor values were increased to 0.82 $\mu$F, 82 nF and 8.2 nF when $R = 9705$ $\Omega$ at 20 Hz and 970 $\Omega$ at 200 Hz. This resistance range could be obtained using a 953 $\Omega$ resistor in series with a 10 k$\Omega$ potentiometer.

(a) The gain of the circuit is set by a lamp. Any 12 V lamp that takes a current of less than 50 mA can be used; the resistance of the lamp is $12/(50 \times 10^{-3}) = 240$ $\Omega$ . So choose $R_1$ to be either a 470 $\Omega$ resistor or a 1 k$\Omega$ potentiometer. The circuit of the oscillator is shown by Fig. 3.25. The op-amp used must have a slew rate of about

$$5 \times (2\pi \times 20 \times 10^3 \times 8) = 5 \text{ V/}\mu\text{s}.$$

(b) A thermistor is used to control the oscillation amplitude. Thermistors are available that have resistances, at 25 °C, of 470 $\Omega$, 1 k$\Omega$, 4.7 k$\Omega$, 10 k$\Omega$, etc. Choose a 10 k$\Omega$ device, then $R_1$ (now in the position of L in Fig. 3.24) must be a 4.99 k$\Omega$ resistor (then gain = 1 + 10/4.99 = 3). The thermistor will have a voltage of (2/3) $\times$ 8 = 5.33 V across and so it must have a rated power dissipation of at least

$$(8 - 5.33)/(4.99 \times 10^3) \times 5.33 = 2.85 \text{ mW}$$

Such a device is easily obtained.

As an alternative to the use of a temperature sensitive element a pair of diodes can be employed to regulate the amplitude of the oscillations. The diodes are used to obtain a non-linear feedback

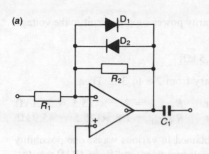

(a)

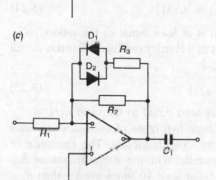

(b)

(c)

**Fig. 3.26** Regulating the output voltage of an *RC* oscillator

characteristic that will limit the amplitude without introducing too much distortion. Several different arrangements are possible and Fig. 3.26 shows three of them. Initially, in Fig. 3.26(a) the two zener diodes are non-conducting and the gain of the circuit is $1 + R_2/R_1$. When the oscillation amplitude has built up the diodes conduct on alternate half-cycles to reduce the feedback resistance and so the gain falls. The arrangements of Figs 3.26(b) and (c) operate in a similar fashion except that when the diodes conduct the gain falls to a specified lower value.

### Example 3.9

Obtain values for the components in the circuit shown in Fig. 3.27 if the frequency of oscillation is to be 1 kHz and the single polarity supply is 9 V.

### Solution

Both the inverting and the non-inverting terminals of the op-amp must be held at + 4.5 V by the equal-value resistors $R_2$ and $R_3$, and $R_4$ and $R_5$, respectively.

Select $C_1 = C_2 = 10$ nF, then

$$R_1 = R_2 \| R_3 = 1/(2\pi \times 10 \times 10^{-9} \times 1000)$$
$$= 15.9 \text{ k}\Omega = 15.8 \text{ k}\Omega \text{ preferred value}$$

Then $R_2 = R_3 = 2 \times 15.8 = 31.6$ kΩ.

The choice of $R_4 = R_5 = 33$ kΩ will minimize bias current offset effects but increasing their value will reduce the current taken from the supply — possibly a battery. Choose $R_4 = R_5 = 51$ kΩ. These two resistors in parallel

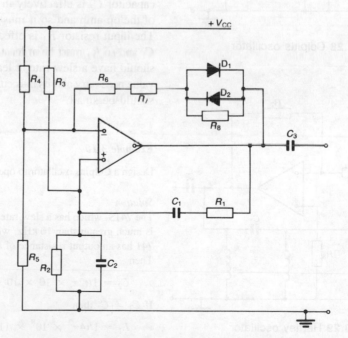

**Fig. 3.27**

act as the input resistor in a dual-polarity power supply circuit so the voltage gain is

$$[1 + (\text{feedback resistance})/25.5 \text{ k}\Omega]$$

Design for the voltage gain to vary from 2.8 to 3.2. Then,

$$3.2 = 1 + R_{f(max)}/25.5 \quad \text{or} \quad R_{f(max)} = 2.2 \times 25.5 = 56.1 \text{ k}\Omega$$
$$2.8 = 1 + R_{f(min)}/25.5 \quad \text{or} \quad R_{f(min)} = 1.8 \times 25.5 = 45.9 \text{ k}\Omega$$

This range of resistances can be obtained in various ways, one possibility is $R_6 = 3 \text{ k}\Omega$ resistor, $R_7 = 47 \text{ k}\Omega$ potentiometer, and $R_8 = 10 \text{ k}\Omega$ resistor. $C_3$ can be any value that will have a low reactance at 1 kHz, say 10 $\mu$F and the op-amp can be a 741.

## Colpitts and Hartley Oscillators

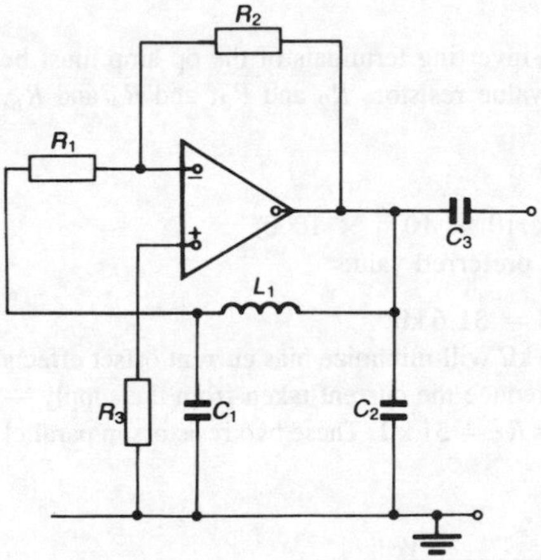

The circuit of a Colpitts oscillator is shown in Fig. 3.28. The frequency of oscillation $f_o$ is given by

$$f_o = 1/[2\pi \sqrt{[L_1 C_1 C_2/(C_1 + C_2)]}] \tag{3.24}$$

provided that the voltage gain is at least equal to the ratio $C_1/C_2$.

Figure 3.29 shows the circuit of a Hartley oscillator. For this circuit the frequency of oscillation $f_o$ is

$$f_o = 1/[2\pi \sqrt{[C_1(L_1 + L_2)]}] \tag{3.25}$$

provided the voltage gain is at least equal to the ratio $L_1/L_2$.

The design procedures for these two types of oscillator are more or less the same. Consider the Colpitts circuit. The reactance of capacitor $C_2$ is effectively in parallel with the output resistance $R_{out}$ of the op-amp and so it must be at least 10 times greater than $R_{out}$. The input resistor $R_1$ is effectively in parallel with the reactance of $C_1$ and so $R_1$ must be at least 10 times greater than $X_{C_1}$. The op-amp should have a slew rate at least five times larger than the minimum value necessary and its bias current and input offset current and voltage should be small.

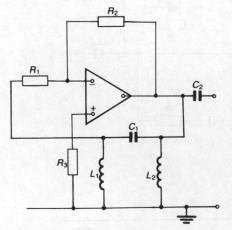

**Fig. 3.28** Colpitts oscillator

**Fig. 3.29** Hartley oscillator

### *Example 3.10*

Design a Colpitts oscillator to operate at 10 kHz from a $\pm$ 12 V power supply.

*Solution*
The 741S, which has a slew rate of 20 V/$\mu$s and a full-power bandwidth that is much greater than 10 kHz, will be a suitable choice for the op-amp. The 741 has an output resistance of about 70 $\Omega$ and so $X_{C_2} > 700 \Omega$, say 800 $\Omega$. Then

$$C_2 = 1/(2\pi \times 10 \times 10^3 \times 800) = 22 \text{ nF preferred value}$$

If $C_1 = C_2$ then

$$L_1 = 1/(4\pi^2 \times 10^8 \times 11 \times 10^{-9}) = 23 \text{ mH}$$

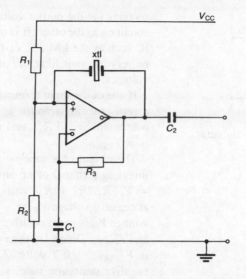

$V_{CC}$

**Fig. 3.30** Crystal oscillator

$R_1$ should be chosen to be several times larger than $X_{C_1}$, say 20 kΩ . The minimum voltage gain is unity so select $R_2$ to be 22 kΩ.

The circuit of an op-amp crystal oscillator is shown in Fig. 3.30.

## Voltage Comparators

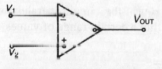

**Fig. 3.31** Voltage comparator

A *voltage comparator* is a circuit whose output is always at either its positive saturation voltage $V_{O(SAT)}^{+}$ or its negative saturation voltage $V_{O(SAT)}^{-}$, depending upon which of its two input terminals is the more positive. The basic voltage comparator circuit consists merely of an op-amp with separate input voltages applied to its inverting and non-inverting input terminals, as shown by Fig. 3.31. The output voltage of the comparator is equal to

$$V_{OUT} = (V_1 - V_2)A_v \qquad (3.26)$$

When $V_1 > V_2$ the output voltage is $V_{O(SAT)}^{-}$ and when $V_2 > V_1$ the output voltage is $V_{O(SAT)}^{+}$. Because the open-loop voltage gain is very high the difference between the two input voltages need be only very small to cause the op-amp to go into its output saturation state.

If both the input voltages are variables the circuit is known as a *differential comparator* but if either one is held at a constant value it is known as a *threshold detector* or a *level detector*.

The important parameters of a voltage comparator are: (*a*) its *resolution*, which is the smallest difference between $V_1$ and $V_2$ to which the circuit is able to respond; (*b*) its switching speed; (*c*) the accuracy and stability of the two threshold voltages; (*d*) negligible jitter; (*e*) its output current. An op-amp employed for this purpose must have a maximum differential input voltage that is greater than the maximum difference that can occur between $V_1$ and $V_2$, low bias and input offset currents, and low input offset voltage. The slew rate of the op-amp must be high since it will take $2V_{O(SAT)}/$(slew rate)

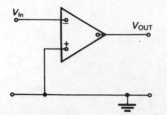

**Fig. 3.32** Zero-crossing detector

seconds for the output voltage to switch from one output saturated condition to the other. It is often best to employ a voltage comparator IC such as the LM 311 or the LM 339, since these are designed to be very fast switching and often have TTL/CMOS compatible output voltages.

If one of the input terminals is connected to earth, as in Fig. 3.32, a *zero-crossing detector* is obtained. Whenever $V_{in}$ is positive the output will be at $V_{O(sat)}^-$ and whenever the input is negative the output is at $V_{O(sat)}^+$.

The circuit of a *window comparator* is shown in Fig. 3.33. The inverting terminal of op-amp A is held at a positive voltage of $V_A = V_{CC}R_1/(R_1 + R_3)$ volts, and the non-inverting terminal is held at negative voltage $V_B = -V_{CC}R_2/(R_2 + R_4)$ volts. When the input voltage $V_{in}$ is more positive than $V_A$ the output of op-amp A will go to $V_{O(SAT)}^+$. Diode $D_1$ then conducts and so the output voltage $V_{out}$ is $V_{O(SAT)}^+ - 0.7$ volts. At the same time op-amp B goes to its negative saturation state, since $V_{in}$ is more positive than $V_B$ but its output voltage turns diode $D_2$ OFF. Conversely, when $V_{in}$ becomes more negative than $V_B$ the output voltage of op-amp B switches to its positive saturation voltage $V_{O(SAT)}^+$, while op-amp A's output switches to $V_{O(SAT)}^-$. Now $D_1$ is OFF and blocks the output of op-amp A so that the output voltage of the circuit is again equal to $V_{O(SAT)}^+ - 0.7$ volts.

Whenever the input voltage is somewhere in between the threshold voltages $V_A$ and $V_B$ the output voltage of the comparator is 0 V because both diodes are turned OFF. Hence the circuit operates to indicate when the input voltage lies within a certain range of values, i.e within a 'window'.

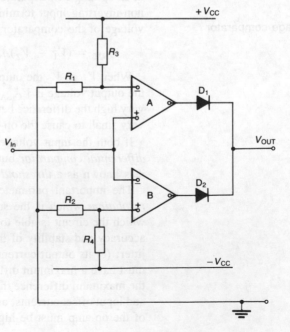

**Fig. 3.33** Window comparator

*Example 3.11*

Design a window comparator to operate from a ± 15 V power supply which uses a 10 mA LED as an indicator. The threshold voltages are to be + 2.8 and −1.5 V.

*Solution*

Here $2.8 = 15R_1/(R_1 + R_3)$, $R_3 = 4.357R_1$. Choose $R_1 = 11$ kΩ and then

$R_3 = 47.93$ kΩ $= 47.5$ kΩ preferred value

Also, $-1.5 = -15R_2/(R_2 + R_4)$, $R_4 = 9R_2$. Choose $R_2 = 11$ kΩ then

$R_4 = 99$ kΩ $= 100$ kΩ preferred value

The diodes $D_1$ and $D_2$ can be any signal diode and the LM 311 voltage comparator can be used.

The maximum output voltage of each comparator can be set to any desired value less than 14 V by the use of an external pull-up resistor $R_5$ that is connected between pin 7 and the $V_{CC}$ line. For TTL compatibility connect $R_5$ to + 5 V and pin 1 to earth; the + 5 V is obtained by a potential divider connected between the + 15 V line and earth. A suitable value for $R_5$ is 10 kΩ. The LED requires a forward bias of about 1.7 V to glow and therefore a current limiting resistor of

$$(5 - 1.7)/(10 \times 10^{-3}) = 330 \, \Omega$$

should be connected in series with the LED. The circuit is shown in Fig. 3.34.

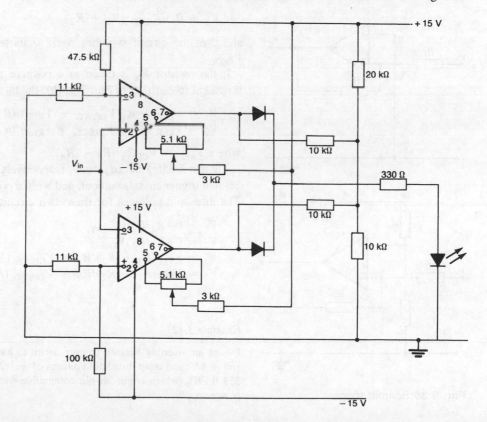

**Fig. 3.34**

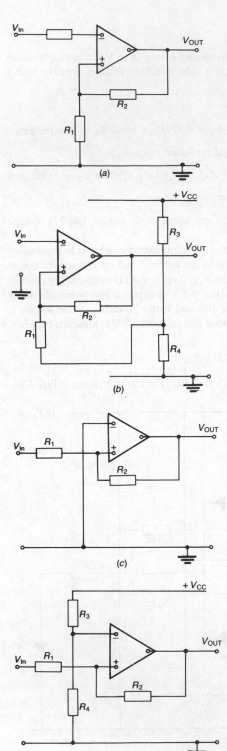

**Fig. 3.35** Schmitt trigger

## Schmitt Trigger

The *Schmitt trigger* is a kind of voltage comparator which is often employed to convert an input signal of any waveform into a rectangular output waveform. The output voltage of the circuit can only be at either one of two possible values, namely $\pm V_{OUT}$. The output voltage will be high when the input voltage is more positive than a specified upper threshold voltage $V_1$ and it will remain at this value until such time as the input voltage becomes less positive than the lower threshold voltage $V_2$. If this point is reached the circuit will rapidly change states so that its output voltage becomes low. The difference between the two threshold voltages is known as the hysteresis of the circuit. Typically, the hysteresis is about 1 V.

An op-amp Schmitt trigger may be either inverting or non-inverting and may, or may not, include a reference voltage. Figure 3.35 shows the circuit of four different op-amp Schmitt trigger circuits. Figure 3.35($a$) is the basic Schmitt trigger; the output voltage switches to its negative saturation voltage $V_{O(SAT)}^-$ whenever the input voltage $V_{in}$ is more positive than

$$V_1 = R_1 V_{O(SAT)}^+ / (R_1 + R_2) \tag{3.27}$$

and will remain in that state until the input voltage becomes more negative than

$$V_2 = R_1 V_{O(SAT)}^- / (R_1 + R_2) \tag{3.28}$$

and then the circuit switches back to its positive saturation state $V_{O(SAT)}^+$.

If the resistor $R_1$ is taken to a positive reference voltage $V_{REF}$ instead of to earth, as in Fig. 3.35($b$), the threshold voltages become

$$V_1 = V_{REF} + R_1 (V_{O(SAT)}^+ - V_{REF}) / (R_1 + R_2) \tag{3.29}$$
$$V_2 = V_{REF} - R_1 (V_{O(SAT)}^- + V_{REF}) / (R_1 + R_2) \tag{3.30}$$

where $V_{REF} = V_{CC} R_4 / (R_3 + R_4)$.

Figures 3.35($c$) and ($d$) show, respectively, non-inverting op-amp Schmitt trigger circuits without, and with, a positive reference voltage. The threshold voltages for these two circuits are:

$$V_1 = (-R_1/R_2) V_{O(SAT)}^+ \tag{3.31}$$
$$V_2 = (-R_1/R_2) V_{O(SAT)}^- \tag{3.32}$$
$$V_1 = V_{REF}(R_1 + R_2)/R_1 + V_{O(SAT)}^+ (R_1/R_2) \tag{3.33}$$
$$V_2 = V_{REF}(R_1 + R_2)/R_2 - V_{O(SAT)}^- (R_1/R_2) \tag{3.34}$$

*Example 3.12*

Design an inverting Schmitt trigger circuit to have output voltages of 0 V and + 5 V, and input threshold voltages of + 1.2 and + 1.8 V. Use an LF 353 BiFET (which is pin-for-pin compatible with the 741) and a ± 12 V power supply.

*Solution*

For both the threshold voltages to be positive a reference voltage must be applied to the circuit. From equation (3.29)

$$V_1 = 1.8 = V_{REF} + [R_1/(R_1 + R_2)][11 - V_{REF}] \tag{3.35}$$
$$V_2 = 1.2 = V_{REF} - [R_1/(R_1 + R_2)][-11 + V_{REF}] \tag{3.36}$$

Subtracting equation (3.36) from equation (3.35) gives

$$0.6 = (22R_1)/(R_1 + R_2) - (2R_1V_{ref})/(R_1 + R_2)$$

and adding the two equations gives

$$3 = 2V_{REF} \quad \text{so} \quad V_{REF} = 1.5 \text{ V}.$$

Hence

$$0.6 = (22 - 3)[R_1/(R_1 + R_2)] = (19R_1)/(R_1 + R_2)$$
$$0.6 R_2 = 18.4R_1 \quad \text{and} \quad R_2 = 30.67R_1.$$

Choose $R_1 = 1.2$ k$\Omega$, then $R_2 = 36.8$ k$\Omega$ = 36.5 k$\Omega$ preferred value.

$$1.5 = 15R_4/(R_3 + R_4) \quad \text{or} \quad R_3 = 9R_4.$$

Choose $R_4 = 12$ k$\Omega$, then $R_3 = 108$ k$\Omega$ = 107 k$\Omega$ preferred value.

If an OTA, such as the CA 3080, is employed as a Schmitt trigger the hysteresis of the circuit can be programmed by the control current. The circuit of a 3080 Schmitt trigger is shown by Fig. 3.36. When the output voltage $V_{OUT}$ of the circuit is high it is equal to the product $I_cR_2$. Whenever the input voltage becomes more positive than $I_cR_2$ the output voltage switches rapidly to its negative saturation value which is equal to $-I_cR_2$. The output voltage will then remain in this state until such time as the input voltage becomes more negative than $-I_cR_2$ when $V_{OUT}$ rapidly switches back to its original value of $+I_cR_2$.

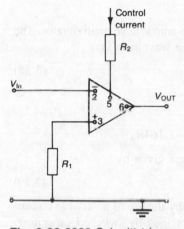

**Fig. 3.36** 3080 Schmitt trigger

## Monostable Multivibrators

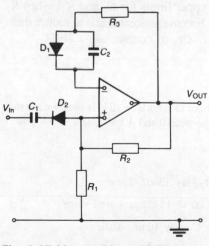

**Fig. 3.37** Monostable multivibrator

The circuit of an op-amp monostable multivibrator is shown in Fig. 3.37. When the circuit is in its stable state its output voltage is at its positive saturation value $V_{O(SAT)}^+$. When a negative voltage pulse is applied to the input terminal of the circuit the circuit rapidly switches to its negative saturation state when the output voltage is $V_{O(SAT)}^-$. Here it will remain for a time $T$ and then it will switch back again to its stable positive saturation state. The time for which the output negative voltage pulse lasts is given by

$$T = C_2R_3 \log_e[1/(1 - \beta)] \tag{3.37}$$

where $\beta = R_1/(R_1 + R_2)$.

One of the main uses of the monostable circuit is as a pulse shaper that is able to receive distorted and/or attenuated input pulses and then output good rectangular pulses.

The op-amp employed in a monostable circuit must have a minimum slew rate of (peak-to-peak output voltage)/(risetime + falltime). Hence, if the output pulse is to be ± 6 V with maximum rise- and fall-times of 20 $\mu$s, the minimum slew rate would be $12/(40 \times 10^{-6})$

$= 30 \, \text{V}/\mu\text{s}$. The op-amp must be able to withstand a large differential input voltage without an increase in its input current. Some FET types conduct an increased current when the input differential voltage exceeds a few hundred millivolts because their input diode protection network conducts. If the output frequency is higher than about 5 kHz a general-purpose op-amp will not be able to recover from saturation rapidly enough. The 741, for example, is unable to recover from saturation in less than about 12 $\mu$s. The best performance is obtained if a voltage comparator IC is used. At the lower frequencies, when the timing resistor(s) are of large value, the input bias current can cause variations in the duty cycle. To avoid this the chosen op-amp should have a bias current that is several times smaller than the charging current that flows in the timing resistor(s).

## Astable Multivibrators

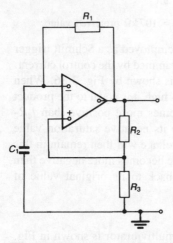

**Fig. 3.38** Astable multivibrator

Figure 3.38 shows the circuit of an op-amp astable multivibrator. The periodic time $T$ of the output voltage waveform is

$$T = 2C_1 R_1 \log_e [1 + 2R_3/R_2] \tag{3.38}$$

If $\log_e [1 + 2R_3/R_2] = 1$, then

$$e^1 = e = 1 + 2R_3/R_2$$
$$e - 1 = 2R_3/R_2 \quad \text{and} \quad R_2 = 1.164R_3$$

The frequency of oscillation $f_o$ is then given by

$$f_o = 1/(2C_1 R_1) \tag{3.39}$$

The frequency can be made variable by the use of a variable resistor for $R_1$ and decades of frequency obtained by switched capacitors. The minimum slew rate for the op-amp is determined in the same way as for the monostable circuit. If the rise- and fall-times are not specified then the slew rate should be at least 10 times greater than the full-power bandwidth for the same frequency and voltage. The lowest frequency is set by the upper limits for $C_1$ and $R_1$; when $R_1$ is large bias current effects may become noticeable and to reduce these make $R_1 = R_2 \| R_3$ if possible. Or, of course, use a JFET.

### Example 3.13

Design an op-amp astable multivibrator to generate a 10 kHz square wave that varies (*a*) between $\pm 5 \, \text{V}$ and (*b*) between 0 and $+5 \, \text{V}$. Use a $\pm 9 \, \text{V}$ power supply.

### Solution
First select a suitable value for $R_3$, say 13 k$\Omega$. Then

$$R_2 = 1.164 \times 13 = 15.13 \, \text{k}\Omega = 15 \, \text{k}\Omega \text{ preferred value}$$

Next select a suitable value for $C_1$, say 10 nF. Then

$$R_1 = 1/(2 \times 10 \times 10^3 \times 10 \times 10^{-9}) = 5 \, \text{k}\Omega$$
$$= 4.99 \, \text{k}\Omega \text{ preferred value}$$

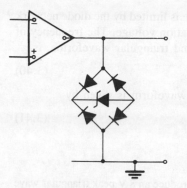

**Fig. 3.39** Limiting the output voltage of a multivibrator

(Then $f_0 = 10.02$ kHz.) If the op-amp chosen is a 741S, $I_{B(max)} = 500$ nA and so the minimum current flowing in $R_2$ should be 5 $\mu$A. The voltage at the non-inverting terminal will be

$$5 \times 13/(13 + 15) = 2.32 \text{ V}$$

and hence the minimum allowable value for the feedback resistor $R_2$ is $(5 - 2.32)/(5 \times 10^{-6}) = 536$ k$\Omega$

The paths to earth for the bias current are 4.99 k$\Omega$ and 5||13 = 7 k$\Omega$ which are not very different and so there will be little output offset voltage due to this effect.

The output voltage will switch between $\pm V_{O(SAT)}$ and so it must be limited to the required $+5$ V and 0 and 5 V. This may be achieved by connecting:

(a) diode network between the output of the circuit and earth (see Fig. 3.39). Each of the signal diodes drops 0.7 V when conducting and so the required zener voltage is $5 - 1.4 = 3.6$ V. It may be necessary to connect a current-limiting resistor in series with the output terminal of the op-amp.

(b) a 5.1 V zener diode across the output terminal of the circuit.

## Triangle-wave Generator

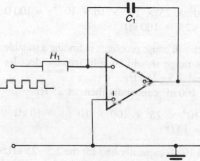

A triangular waveform is generated by applying first a positive d.c. voltage and then a negative d.c. voltage to the input terminal of an integrator (Fig. 3.40(a)). In turn, the alternately negative and positive d.c. voltages can be obtained by applying the triangular wave to a Schmitt trigger and the complete circuit is shown by Fig. 3.40(b). The square wave output of op-amp A is applied to the inverting terminal of op-amp B and integrated so that positive pulses produce a negative-going ramp voltage and negative pulses produce a positive-going ramp voltage at the output of the circuit. The combined ramp voltages make up a triangular wave and this is fed back to the input of op-amp A, which is connected as a Schmitt trigger, to be converted into a square wave.

*(a)*

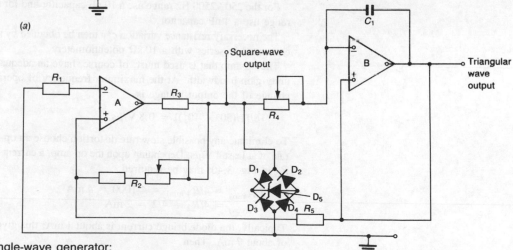

**Fig. 3.40** Triangle-wave generator; (a) basic circuit, (b) complete circuit

*(b)*

The output voltage of the first stage is limited by the diode network to some value $V_D$ less than the saturation voltage. The frequency of oscillation $f_0$, for both the square and triangular waveforms, is

$$f_0 = R_2/(4R_4R_5C_1) \qquad (3.40)$$

The peak voltage of the triangular waveform is given by

$$V_T = V_D R_5/R_2 \qquad (3.41)$$

### Example 3.14

Design a triangular-wave generator to produce an 8 V peak triangular wave and a 4 V peak square wave at a frequency that is continuously variable from 25 Hz to 25 kHz.

### Solution

The 4 V peak square wave can be obtained by connecting a diode network between the square-wave output terminal and earth. The required zener voltage is $4 - 1.4 = 2.6$ V but a 2.7 V device will have to be used.

$$8 = 4R_5/R_2 \qquad R_2 = (4/8)R_5 = 0.5R_5$$

Choose $R_5 = 4\,k\Omega$, then $R_2 = 2\,k\Omega$.
$R_1 = R_2 \| R_5 = 1.33\,k\Omega = 1.3\,k\Omega$ preferred value. Choose $C_1 = 10\,nF$, then

$$R_{4(min)} = 2000/(4 \times 20 \times 10^3 \times 25 \times 10^3 \times 10 \times 10^{-9}) = 100\,\Omega$$
$$R_{4(max)} = 100 \times (25 \times 10^3)/25 = 100\,k\Omega$$

This wide range of resistance values will cause problems in finding a suitable potentiometer and so the frequency range should be split into the decades 25–250 Hz, 250 Hz to 2.5 kHz, and 2.5–25 kHz.
For the 25–250 Hz range use a 100 nF capacitor; then, at 25 Hz,

$$R_{4(max)} = 2000/(4 \times 20 \times 10^3 \times 25 \times 100 \times 10^{-9}) = 10\,k\Omega$$
$$R_{4(min)} = 10\,k\Omega \times 25/250 = 1\,k\Omega$$

For the 250–2500 Hz range use a 10 nF capacitor and for the 2.5–25 kHz range use a 1 nF capacitor.
The necessary resistance variation can then be obtained by an 820 $\Omega$ resistor connected in series with a 10 k$\Omega$ potentiometer.
The op-amp that is used must, of course, have an adequate slew rate and unity-gain bandwidth. At the maximum frequency of operation the rate of change of the output voltage is

$$16/[1/(50 \times 10^3)] = 0.8\,V/\mu s$$

To eliminate any possible slew rate distortion choose an op-amp with a slew rate of at least 4 V/$\mu$s. Depending upon the op-amp, a current-limiting resistor, $R_3$ in Fig. 3.40, may be required.

$$I_{R3(max)} = 4/R_{4(min)} = 4/1000 = 4\,mA$$
$$I_{R2(max)} = 4/R_2 = 4/2 = 2\,mA$$

Typically, the diode bridge current is about 1 mA; this gives a total current of about 7 mA. Then

$$R_3 = (11 - 4)/(7 \times 10^{-3}) = 1000\,\Omega$$

Choose a higher preferred value of, say, 1500 Ω. If the square-wave output of the circuit is employed and supplies a current this must be taken into account.

## Ramp Generators

A ramp, or sawtooth, generator is used to generate a sawtooth waveform. The voltage ramps positively (or negatively) from 0 V until it reaches a predetermined maximum value and then it rapidly 'flies back' to 0 V ready to begin another ramp. There are two main types of ramp generator in common use, namely the Miller circuit and the bootstrap circuit.

The circuit of an op-amp Miller ramp generator is shown in Fig. 3.41 and it can be seen to consist of two op-amps A and B that are connected as an integrator and a voltage comparator, respectively. When a negative d.c. voltage $V_{IN}$ is applied to the input terminal of the circuit it will be integrated by op-amp A to produce a positive-going ramp voltage of $V_{IN}t/R_1C_1$ volts at the non-inverting terminal of op-amp B. As long as this ramp voltage is less positive than the reference voltage, $V_{REF} = V_{CC}R_4/(R_3 + R_4)$, at the inverting terminal of op-amp B, the output of op-amp B will remain at its negative saturation voltage of $V_{O(SAT)}^-$. Both the diode $D_1$ and the bipolar transistor $T_1$ are then OFF. Immediately the ramp voltage becomes more positive then $V_{REF}$ the comparator switches states and its output voltage becomes $V_{O(SAT)}^+$. Diode $D_1$ then turns ON and this, in turn, turns $T_1$ ON. An effective short-circuit is now placed across capacitor $C_1$ and so it rapidly discharges, causing the ramp voltage to fly back to 0 V. This then causes the comparator to switch back to its negative saturation state turning $D_1$ and $T_1$ OFF. If the d.c. input voltage has been maintained another ramp will now start and a sawtooth waveform is generated.

The frequency $f_0$ of the sawtooth waveform is

$$f_o = V_{IN}/V_{REF}R_1C_1 \qquad (3.42)$$

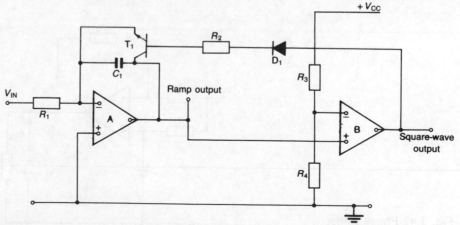

**Fig. 3.41** Miller ramp generator

The frequency of the waveform is controlled by the input voltage $V_{IN}$ and its amplitude is controlled by the reference voltage $V_{REF}$. $D_1$ and $T_1$ should be fast switching devices and the op-amps must have a sufficiently high slew rate; it will probably be best if op-amp B is actually an IC voltage comparator. Op-amp A should have a low bias current and offset voltage and a JFET is often employed.

*Example 3.15*

Design a sawtooth generator to have a peak output voltage of 6 V at a frequency of 3 kHz. Use a supply voltage of ± 12 V and the LF 411 JFET.

*Solution*
$V_{REF} = 6 V = 12 R_4/(R_3 + R_4)$, or $R_3 = R_4$. A suitable value would be 20 kΩ.

If the input voltage is chosen to be 5 V then the product

$$R_1 C_1 = 5/(6 \times 3000) = 278 \ \mu s$$

If $C_1$ is chosen to be 10 nF then $R_1 = 2780 \ \Omega = 2800 \ \Omega$ preferred value.

The action of the voltage comparator stage can be speeded up by the connection of a capacitor $C_2$ between its non-inverting and output terminals. The charge stored in $C_2$ prevents the comparator switching immediately the voltage in $C_1$ falls below $V_{REF}$ and this allows time for $C_1$ to completely discharge before the next ramp commences.

**Op-amp Phase Shifters**

Sometimes there is a need for a circuit that can shift the phase of a sinusoidal signal without changing its amplitude. Figure 3.42 shows the circuit of an op-amp phase shifter. To obtain a voltage gain of unity $R_1 = R_2$ and then phase lag $\theta$ introduced by the circuit is

$$\theta = 2 \tan^{-1}(2\pi f R_3 C_1) \tag{3.43}$$

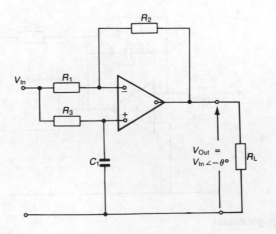

**Fig. 3.42** Phase shifter

*Example 3.16*

Design the phase shift circuit to give a phase shift of 60° at a frequency of 2 kHz.

*Solution*

Choose $R_1 = R_2 = 68$ k$\Omega$. Choose $C_1 = 10$ nF. Then

$$R_3 = (\tan 30°)/(2\pi \times 2000 \times 10 \times 10^{-9}) = 4594 \, \Omega$$
$$= 4640 \, \Omega \text{ preferred value.}$$

## Thermistors in Op-amp Circuits

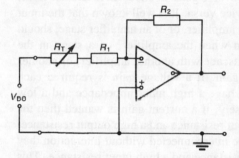

**Fig. 3.43** Linearizing a thermistor

If a thermistor is connected in series with a fixed resistor $R_1$ and the inverting terminal of an op-amp it can be used to convert, firstly, changes in temperature to changes in resistance, and then, secondly, changes in resistance into changes in voltage. The basic circuit is shown by Fig. 3.43. The voltage gain of the circuit is $A_{v(f)} = -R_2/(R_1 + R_T)$. The non-linear changes in the resistance of the thermistor as the temperature changes are linearized by the series resistance $R_1$ and hence very nearly linear changes in the output voltage are obtained. The resistance of $R_1$ must be several times larger than $R_T$ over the whole of the temperature range to be measured. The applied voltage $V_{DC}$ must be supplied by a constant voltage source with good temperature stability and the op-amp should be one having low drift.

# 4 Interfacing

To obtain the best performance from two circuits that are different from one another in respect of their logic family, or their d.c. voltage levels, or their impedances, etc. care must be taken with the way the circuits are *interfaced*. Interfacing also means the hardware and/or software that is needed to convert signals in one form to another form, e.g. analogue to digital or vice versa. It is well known that the input and output impedances of an amplifier, or of an amplifier stage, should be taken into consideration when the amplifier, or a stage in the amplifier, is connected in cascade with another amplifier, or another stage, to obtain a greater gain. If a voltage gain is required each amplifier, or stage, should have a high input impedance and a low output impedance. Conversely, if a current gain is wanted then the requirement is for a low input resistance and a high output resistance. In general, for devices to be interconnected without interaction they should have a low output resistance and a high input resistance. This is true for most digital devices.

Perhaps the simplest example of an interface is the transformer which can also provide an increase in voltage equal to its turns ratio. For maximum possible power to be transferred from a source, of resistance $R_s$, to a load, of resistance $R_L$, the turns ratio $n$ of the *matching* transformer must be equal to $n = \sqrt{(R_s/R_L)}$.

## Transducer Interfacing

Transducers are employed to change a physical parameter, such as the flow or level of a liquid, or the position of a mechanical item, or gas or fluid pressure, or, most commonly, temperature, into an equivalent electrical signal. In most cases the electrical signal that is generated is of analogue form. If the analogue signal is to be processed by a digital computer or a microprocessor, it must first be converted into the equivalent digital signal. This conversion is carried out by applying the analogue signal to an analogue-to-digital converter (ADC).

Temperature is probably the most commonly controlled physical parameter and so it will be used to provide an example of a sensor interface. Some kind of temperature sensor, such as a thermocouple or a temperature-sensitive IC like the LM 135 (which gives a linear output voltage of 10 mV/°C over the temperature range 0−100 °C), must be employed. The sensor can be placed in one arm of a

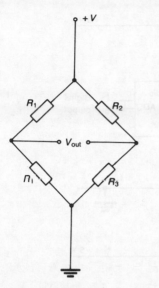

**Fig. 4.1** Temperature sensor $R_T$ in a Wheatstone bridge

Wheatstone bridge as shown by Fig. 4.1. The output voltage $V_{out}$ of the bridge is

$$V_{out} = V[R_T/(R_1 + R_T) - R_3/(R_2 + R_3)] \qquad (4.1)$$

The output voltage will be equal to zero when $R_1 R_3 = R_2 R_T$ and this fact can be used as the basis of a null measurement. In practice, however, it is more common to use the bridge offset voltage as a measure of temperature.

If $R_1 = R_2 = R_3 = R_T = R$ at the calibration temperature and the fractional deviation in $R_T$ at any other temperatures is $x$, then equation (4.1) can be rewritten as

$$V_{out} = V[R(1 + x)]/[R + R(1 + x)] - V/2$$
$$= (V/2)[x/(2 + x)] = (V/4)[x/(1 + x/2)] \qquad (4.2)$$

The offset voltage can be applied to an op-amp circuit for amplification and one typical circuit is shown in Fig. 4.2. Resistors $R_4$ and $R_5$ provide a means of nulling out any component tolerances. From Fig. 4.2,

$$V_T = VR_3/(R_2 + R_3)$$

and

$$V_{out} = [(V_T/R_T) - (V - V_T)/R_1]R_6 + V_T \qquad (4.3)$$

The output voltage must be applied to some kind of processing circuit; if this is to be either a digital computer or a microprocessor then an ADC will be required.

Usually, the output voltage of a temperature sensor is only slowly changing and this means that the analogue signal can be applied directly to an ADC. Figure 4.3 shows how an IC temperature sensor, the AD 590, can be interfaced to an ADC. The output signal of the IC is actually a current and this is converted into a voltage by passing the current through resistors $R_1$ and $R_2$. A voltage reference IC, the AD 580, is used to produce a constant reference voltage that is used

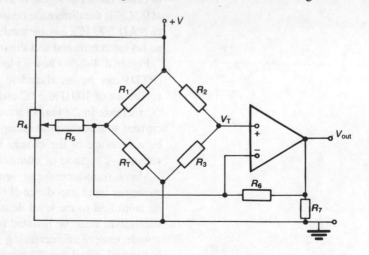

**Fig. 4.2** Measurement of the offset voltage

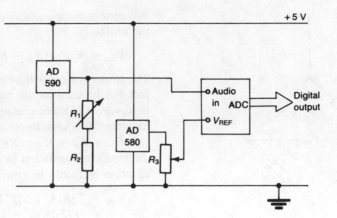

**Fig. 4.3** Use of the AD 590
temperature sensor

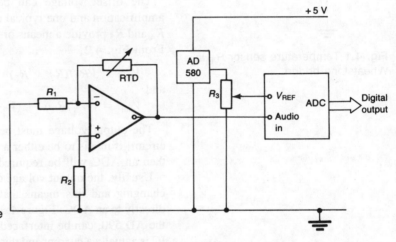

**Fig. 4.4** Use of a platinum resistance
detector to measure temperature

to offset the input signal to allow the circuit to be calibrated (usually
at 0 °C). If the difference between two temperatures is to be measured
two AD 590 ICs can be used; one of them is then connected to the
audio input terminal and the other to the $V_{REF}$ terminal of the ADC.

Figure 4.4 shows how a platinum resistance–temperature detector
(RTD) can be interfaced to an ADC. Typically, an RTD has a
resistance of 100 Ω at 0 °C and this resistance increases by 0.4 Ω per
°C increase in the temperature. If the RTD is supplied by a constant
current source then the change in the voltage dropped across it will
be a measure of the change in the temperature of the device. An
alternative method of connecting the RTD is shown by Fig. 3.43.

Many transducers and sensors must apply their signal to the
particular input impedance of the processing system, their signals must
be amplified to the level demanded by the ADC, and their input to
the system must be isolated from any noise sources. There is such
a wide variety of interfacing problems that very often a number of
individual signal conditioning boards are used to handle impedance

matching and amplification, and the outputs of these boards are applied, via a multiplexer, to the ADC.

## Interfacing between Analogue and Digital Circuitry

The basic block diagram of a signal processing circuit is shown in Fig. 4.5. The analogue input signal is first applied to a band-limiting filter whose main purpose is to prevent *aliasing* taking place. The band-limited signal is then passed on to the ADC and here it is converted into the equivalent digital signal. The digitalized signal is processed by the digital signal processor, which is very often either a digital computer or a microprocessor. Alternatively, it may be an application specific microprocessor known as a *digital signal processor* (DSP). A DSP differs from an ordinary microprocessor in two ways: firstly, it can execute most instructions in one clock cycle and secondly, it can execute a complete multiply and accumulate instruction in one cycle. The processed data is then sent to the digital-to-analogue converter (DAC) which produces a quantized version of the required analogue output signal. The ADC operates by taking samples of the analogue input voltage at a *sampling frequency* $f_s$ hertz. The analogue signal must be sampled at a minimum rate of twice the highest frequency contained in the signal if the digital signal is later to be recovered without error. If the sampling frequency should be less than twice the highest audio frequency an unwanted effect known as aliasing occurs. This sampling frequency limit only applies if the analogue signal is to be reconstituted later on; if it is not, a lower sampling rate will often suffice.

Most audio applications can ignore frequencies above about 20 kHz, while instrumentation systems may need to deal with specific frequencies, such as 25 Hz, 15 and 30 kHz. Other systems, particularly those with temperature sensor inputs, may only contain low frequencies and the bandwidth of the input filter may be as low as 25 Hz in order to minimize noise problems.

Conversion of an analogue signal into the corresponding digital signal is accomplished by means of a process known as *quantization*. An analogue signal with a peak voltage of $V_m$ can be quantized into $V_m/2^n$ different values, where $n$ is the number of bits employed by the converter. Each of the steps can then be represented by a binary number. This is shown by Fig. 4.6(a). Clearly each step represents a range of analogue voltages and so can only be an approximation to the correct value. The greater the number of steps, and hence the greater the number of bits used, the smaller will be each step. The difference between the analogue signal and its quantized approximation

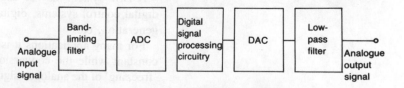

**Fig. 4.5** Signal processing

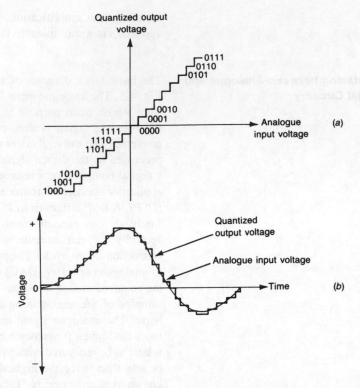

**Fig. 4.6** (a) The quantization process; (b) quantized output voltage

produces an error in the reconstituted signal that is known as the *quantization error* or *noise*. Quantization error is illustrated by Fig. 4.6(b). The quantized signal is then passed through a low-pass filter that removes all the unwanted high-frequency components that the quantization process has introduced and so the filter's output is a purely analogue signal.

The dynamic range of an ADC is ideally given by

$$\text{Ideal dynamic range} = 6.02n + 1.76 \text{ dB} \tag{4.4}$$

where $n$ is the number of bits used in the ADC. Various errors in the conversion process result in the worst case dynamic range being somewhat worse than this and it is given by equation (4.5).

$$\text{Worst case dynamic range} = 6.02n - 4.24 \text{ dB} \tag{4.5}$$

When a change in the amplitude of the analogue input signal is smaller than the resolution of the ADC that signal will not be detected and converted. For example, a 12-bit ADC will not detect any signal that is smaller than $1/2^{12} = 1/4096$ times the FSR.

A DSP system may be used for many varied purposes including digital control systems, digital recording, digital filters, waveform generation, etc.

For many applications it is necessary to hold the analogue signal constant while the conversion process is being carried out. This 'freezing' of the analogue signal is the function of a *sample-and-hold amplifier* which, when used, is connected between the band-limiting filter and the ADC. In most cases the ADC and the sample-and-hold amplifier are operated at the same frequency.

### Analogue-to-digital Converter

There are a number of different types of ADC that are available in IC form and it is necessary to have some idea of their relative merits so that the right device can be chosen for each application.

The simplest technique that is employed for analogue-to-digital conversion is *single-slope* or *ramp conversion*. A converter of this type works by comparing the analogue voltage with an internally generated ramp voltage while a counter notes the number of clock pulses that occur before the two voltages are equal to one another. The ramp ADC has a relatively long conversion time of some milliseconds. The *dual-slope* or *integrating* converter operates using a similar principle but it is more accurate. The conversion is again slow, being in the millisecond range. Both of these types of converters are used for low-frequency data loggers and digital voltmeters.

*Tracking converters* are faster to operate, typically less than 1 ms, but the time required before an input signal is acquired could be as much as $2^n$, where $n$ is the number of bits used. The tracking converter is best suited to the continuous monitoring of an analogue signal and the conversion of this into a digital code sequence. The conversion time of a *successive approximation converter* may be just a few microseconds and it is independent of the amplitude of the analogue voltage. The fastest type of converter is the *flash converter* and its conversion time may well be less than 1 $\mu$s. Unfortunately, flash converters are expensive and in an attempt to offer some of the speed benefits of this type of converter at a lower cost *half-flash converters* have been developed. A half-flash converter has the number of internal voltage comparators reduced by a factor of eight, this reduces the cost considerably and still gives a faster conversion than a successive approximation converter can manage. Devices of this kind include the ADC 0820 and the MAX 158.

The majority of the IC converters offered by various manufacturers are of either the successive approximation and/or of the flash/half-flash types. The cost of a successive approximation converter increases rapidly with increase in the number of bits used for each conversion. Typically, the cost of a 10-bit device is about six times that of an 8-bit device, while a 14-bit device may be more than ten times as expensive.

In the choice of an ADC for a particular application the following points should be considered:

(a) The input and output requirements such as current and/or voltage ranges, impedances, logic levels and data rates.

(b) The required accuracy; this involves the resolution determined by the number of bits and the linearity.

(c) The tolerable errors such as allowable non-linearity and missing codes.

(d) The conversion speed required.

(e) Is the converter to output its data to a microprocessor? Some converters have been designed to be microprocessor compatible. Short-form data should be consulted to pick some suitable devices and then a final choice can be made after a scrutiny of their data sheets.

### Example 4.1

An ADC has a maximum dynamic range of 60 dB, and it is preceded by a Butterworth low-pass filter that has a 3 dB frequency of 1 kHz and 25 dB loss at 3 kHz. It is to be used to convert an analogue signal in the frequency band 0−3000 Hz. It is required that the aliasing error is less than 3 dB. Calculate (a) the number of bits the ADC should have, (b) the order of filter required, and (c) the sampling frequency.

### Solution

(a) At 1 kHz the filter has 3 dB loss and therefore

$$63 = 6.02n - 4.24 \quad \text{or} \quad n = (63 + 4.24)/6.02 = 11.2$$

Hence, a 12-bit ADC is required. (Ans.)

(b) The attenuation/frequency characteristic of a Butterworth low-pass filter is given by

$$A = [1 + (f/f_c)^{2n}]^{1/2}$$

where $f$ is any frequency, $f_c$ is the frequency at which the loss is 3 dB, and $n$ is the order of the filter.

$$25 \text{ dB} = 17.78 \text{ voltage ratio} = [1 + (3)^{2n}]^{1/2}$$
$$316 = 1 + 3^{2n} \quad 2n \log_{10} 3 = \log_{10} 315 = 2.5$$
$$n = 2.5/(2 \times 0.477) = 2.6$$

Therefore, a third-order filter is necessary. (Ans.)

(c) For an attenuation of 63 dB = 1413 the frequency must be $f'$.

$$1413 = [1 + (f'/1)^6]^{1/2} \quad 1413^2 \simeq (f')^6$$
$$6.3 = 6 \log_{10} f' \quad \text{or} \quad f' = \text{alog}_{10}(6.3/6) = 11.2 \text{ kHz}$$

The maximum signal frequency is 3 kHz = $f_s$ − 11.2 kHz. Hence the minimum sampling frequency $f_s$ = 14.2 kHz. (Ans.)

### Input Impedance

The impedance of the source presented to an ADC will have an effect upon its operation. Each of the three most common types of ADC has different input loading requirements which ought to be satisfied if the best accuracy is to be obtained from the device.

### (i) Integrating

The integrating converter is fairly tolerant of the signal source impedance since the audio input terminal is usually connected to the input of a unity gain buffer amplifier which has a very high input resistance. Only if the source resistance is itself also of high resistance will a problem arise.

### (ii) Successive approximation

The input resistance of a successive approximation converter varies from a few kilohms to about 100 k$\Omega$, e.g. 2.5 k$\Omega$ for the 12-bit MAX 162 and 100 k$\Omega$ for the 8-bit ZN 447. Suppose that the peak analogue voltage is 1 V, then the input voltage $V_{in}$ to the MAX 162 converter is

$$V_{in} = (1 \times 2500)/(R_s + 2500)$$

The LSB represents a voltage of $1/2^{12} = 1/4096$ V $= 244.1\ \mu$V and for 0.5 LSB error (i.e. 122 $\mu$V),

$$V_{in} = 1\ V - 122\ \mu V = 0.999878\ V = 2500/(R_s + 2500)$$

or

$$R_s = 0.3\ \Omega$$

Such a low source resistance can only be obtained if an input buffer amplifier is employed. Obviously, the op-amp used must have both an adequate bandwidth and an adequate slew rate to handle the analogue signal without the introduction of distortion.

### (iii) Flash and Half-flash Converter

The input impedance of a flash converter is mainly capacitive and this capacitance must be charged each time a conversion is carried out. For the delay caused by the charging time to cause no more than 0.5 LSB error the source resistance should be less than 500 $\Omega$. If the source resistance is greater than this then a buffer amplifier will be necessary.

## Sample-and-hold Amplifier

While the conversion of the analogue signal into digital form is proceeding it is essential that the signal remains constant within $\pm$ 0.5 LSB. If the change in the input analogue signal is larger than this then a *sample-and-hold* amplifier will be necessary at the ADC's input to keep the input voltage constant during the conversion process. The maximum frequency $f_{(max)}$ that may be converted without the use of a sample-and-hold amplifier connected in front of the converter can be calculated. If the input analogue signal is sinusoidal,

$$v = V_m \sin 2\pi ft = 2^n \sin 2\pi ft$$
$$dv/dt = \text{slew rate} = 2\pi f 2^n \cos 2\pi ft$$

Maximum slew rate $= dv/dt = 2\pi f_{(max)} 2^n$

$dv = 2\pi f_{(max)} 2^n dt = 2\pi f_{(max)} 2^n T_c,$

or, where $T_c$ is the conversion time

$$f_{(max)} = dv/(2\pi 2^n T_c) \tag{4.6}$$

For 0.5 LSB error,

$$f_{(max)} = 1/(2\pi 2^n T_c) \tag{4.7}$$

*Example 4.2*

A 10-bit, 15 $\mu$s conversion time, successive approximation ADC is used in a system. Calculate the frequency above which a sample-and-hold amplifier will have to be used.

*Solution*
From equation (4.7)

$$f_{(max)} = 1/(2\pi \times 1024 \times 15 \times 10^{-6}) = 10.4 \, \text{Hz} \quad (Ans.)$$

There is another reason for using a sample-and-hold amplifier. An ADC places transient loads on the signal source during each bit of a conversion. This may well cause transient errors in the applied analogue signal producing both conversion noise and non-linearity. The effect becomes steadily worse as the number of bits used by the ADC increases whatever the signal amplitude.

The demands made upon the sample-and-hold amplifier vary according to whether the signal is continuous or is a train of pulses. The main criterion for a sample-and-hold amplifier used to convert pulsed information is its acquisition time. This (see Fig. 4.7(*a*)) is the sum of the time taken for the switch to close, the time taken for the sample capacitor to charge to within an acceptable error band, and the settling time after any overshoot. The acquisition time of the chosen sample-and-hold amplifier should be less than the duration of the shortest input pulse.

Once a signal has been acquired the amplifier is switched into its hold mode and then the voltage across the sample capacitor appears at the output terminals. The buffer amplifier has a settling time before its output is settled within an acceptable error band, and after this time the conversion of the input analogue signal to digital form may commence. This is shown by Fig. 4.7(*b*). The charge stored in the sample capacitor gradually discharges through the input resistance of the buffer amplifier and the ADC must be fast enough to have completed the conversion process before the sample-and-hold amplifier's output voltage droops out of the error band.

The conversion of a continuous analogue signal demands consideration of droop rate, aperture delay and jitter. Such a signal will be tracked continuously by the sample-and-hold amplifier, and

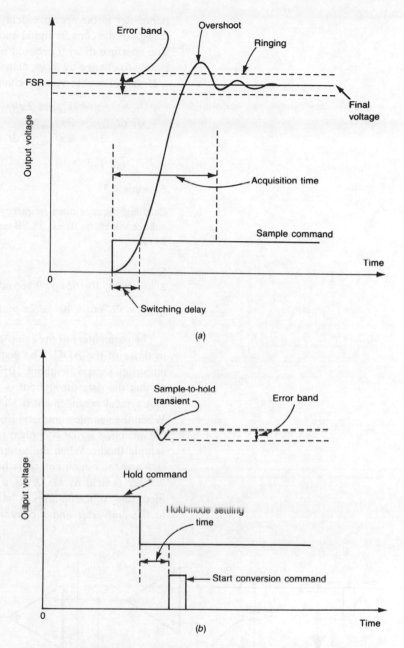

**Fig. 4.7** Sample-and-hold amplifier: (a) acquisition time; (b) settling time

some manufacturers describe one optimized for such a purpose as being a *track-and-hold* amplifier. A track-and-hold amplifier will remain in the sample mode for relatively long periods of time and so its acquisition time is of little importance. The most important parameter is the delay that occurs when the circuit is switched from track mode to hold mode. The sampling of the analogue signal can be carried out either on a regular basis, or whenever a command is received from an external source. In the latter case the most important parameter is the aperture delay; this is the time delay between the receipt of the convert signal and the circuit going into its hold mode. The aperture delay $T_d$ should not be so long that the signal voltage is able to change by more than 0.5 LSB of the conversion accuracy.

If the allowable voltage change is $dv$ and the input signal is

$$v = V_m \sin 2\pi ft \quad \text{then} \quad dv/dt = 2\pi f V_m \cos 2\pi ft$$
$$dv/dt_{(max)} = 2\pi f_{(max)} V$$
$$dv = 2\pi f_{(max)} V \, dt = 2\pi f_{(max)} V \, T_d \tag{4.8}$$

*Example 4.3*

Calculate the maximum frequency for a 10-bit conversion of a 10 V peak voltage waveform if $\pm 0.5$ LSB accuracy is required. The aperture delay is 24 ns.

*Solution*
$2^{10} = 1024$, $10/1024 = 9.766$ mV. From equation (4.8),

$$f_{(max)} = (9.766 \times 10^{-3})/(2\pi \times 10 \times 24 \times 10^{-9}) = 6476 \, \text{Hz} \quad (Ans.)$$

The parameters of the sample-and-hold amplifier must be matched to those of the ADC. The acquisition time of the sample-and-hold amplifier should be about 10% of the converter's conversion time so that the data throughput is only slowed down by about 10%.

A typical arrangement is shown, in simplified form, in Fig. 4.8. When the sample command line is taken HIGH FET $T_1$ turns ON and the analogue signal is applied to the sample capacitor $C_1$. This is the sample mode. When the sample command line goes LOW $T_1$ turns OFF and the circuit enters its hold mode in which the sampled input voltage is held by $C_1$. $R_1$ is a bias component and $R_2$ protects the input stage of the op-amp. A 'start conversion' command is then given to the converter and a conversion commences. At the end of the

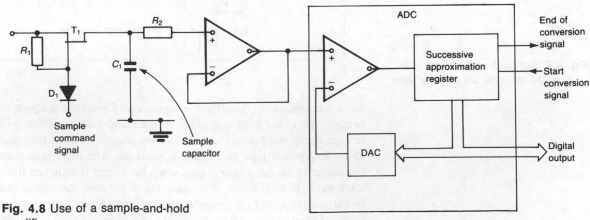

**Fig. 4.8** Use of a sample-and-hold amplifier

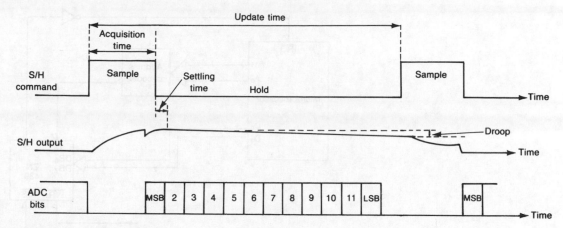

**Fig. 4.9** Timing diagram for Fig. 4.8

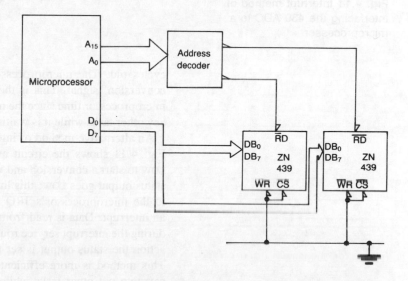

**Fig. 4.10** Interfacing the ZN 439
ADC to a microprocessor

conversion an 'end-of-conversion' signal is sent to the control logic.
The timing diagram for the system is shown in Fig. 4.9.

The data sheet for the Plessey ZN 439 8-bit ADC is shown in Fig.
1.13. The data sheet includes diagrams which show how the ADC
can be used for basic, unipolar or bipolar operation. Figure 4.10 shows
how the ZN 439 can be directly interfaced to a microprocessor. The
converter is able to read valid data at any time no matter what its
conversion status happens to be, and it can be located anywhere in
the memory map of the microprocessor. The decode enable line is
connected to the $\overline{\text{RD}}$ input of the ZN 439. Whenever the microprocessor reads data from this address the $\overline{\text{RD}}$ line is taken LOW
and the ZN 439 then places valid data on to the bus. The $\overline{\text{WR}}$ and
$\overline{\text{CS}}$ lines are connected to earth and so they are always LOW; the
data sheet for the device shows that this makes it continuously cycle.
This means that any data read by the microprocessor is at most eight

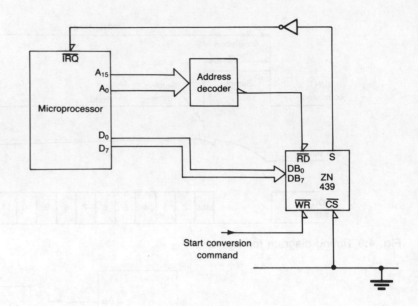

**Fig. 4.11** Interrupt method of interfacing the 439 ADC to a microprocessor

cycles old. The microprocessor continually polls for the 'end-of-conversion' signal. This is the simplest method, but it does waste microprocessor time since the microprocessor cannot be dealing with any other tasks while it is waiting for the 'end-of-conversion' signal.

An alternative method of interfacing makes use of interrupts, and Fig. 4.11 shows the circuit arrangement. The $\overline{WR}$ input is taken LOW to start a conversion and when the conversion is completed the status output goes LOW; this line is connected via a buffer amplifier to the microprocessor's $\overline{IRQ}$ input and hence the LOW generates an interrupt. Data is read from the converter to the microprocessor during the interrupt service routine. At the completion of the reading action the status output is set HIGH and this removes the interrupt. This method is more efficient because the microprocessor can be carrying out other tasks while waiting for the 'end-of-conversion' signal. If a very fast converter is employed the conversion time may be the less than the time required for the microprocessor to service the interrupt.

### Digital-to-analogue Converters

The function of a DAC is to convert digital data into the equivalent analogue voltage. Figure 1.14 shows the data sheet of the Plessey ZN 558 DAC. Details are given on how to employ the device for both unipolar and bipolar operation. The converter can easily be interfaced to a microprocessor and Fig. 4.12 shows one method that may be employed. The analogue output voltage of a DAC is usually applied to a buffer amplifier and some converters may require that this is a current-to-voltage converting amplifier.

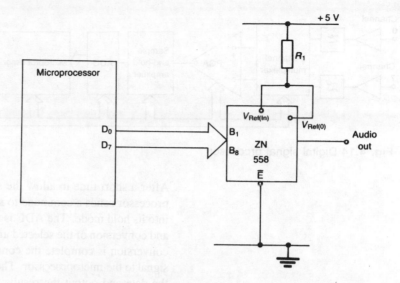

**Fig. 4.12** Interfacing a
microprocessor to the ZN 558 DAC

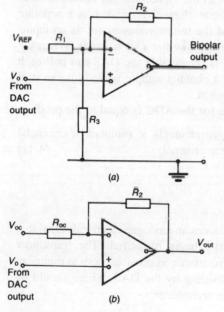

**Fig. 4.13** (a) Making a DAC output
voltage bipolar; (b) equivalent circuit
of (a)

The analogue output voltage of a DAC is normally unipolar, i.e.
of single polarity (usually positive). Some applications may require
that the output voltage is bipolar, i.e. both positive and negative. This
may be accomplished by adding a negative offset voltage, equal to
$V_{REF}/2$, to the output so that when the DAC output is equal to
$V_{REF}/2$ the output of the circuit is at 0 V. One commonly employed
arrangement is shown by Fig. 4.13(a).

Applying Thevenin's theorem to the left of the inverting terminal
gives

$$V_{oc} = V_{REF} R_3/(R_1 + R_3) \quad \text{and} \quad R_{oc} = R_1 R_3/(R_1 + R_3)$$

The Thevenin equivalent circuit is shown by Fig. 4.13(b).
From Fig. 4.13

$$V_{out} = V_0[1 + (R_2/R_{oc})] - V_{oc}R_2/R_{oc}$$
$$= V_0[1 + R_2/R_{oc}] - V_{REF} R_2/R_1 \quad (4.9)$$

For $V_0 = V_{REF}/2$, $V_{out}$ is to be equal to 0 V. Hence,

$$(V_{REF}/2)[1 + R_1R_2R_3/(R_1 + R_3)] = V_{REF} R_2R_1$$

or

$$R_1 = R_2R_3/(R_2 + R_3) \quad (4.10)$$

## Digital Signal Processing

The basic block diagram of a DSP is shown by Fig. 4.14. The blocks
up to the ADC are known as the *digital acquisition system* (DAS).
An 8-channel multiplexer is shown but 4- and 16-channel versions
are also available. When the signal present in any of the eight analogue
channels is to be converted into digital form the microprocessor sends
the address of the wanted channel to the multiplexer. The signal in
the selected channel is then passed to the sample-and-hold amplifier.

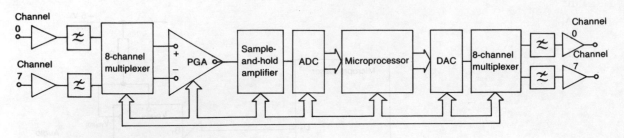

**Fig. 4.14** Digital signal processor

After a short time to allow the sampled voltage to settle, the microprocessor sends a command to switch the sample-and-hold amplifier into its hold mode. The ADC is then sent the 'start conversion' signal and conversion of the selected analogue signal commences. When the conversion is complete the converter sends the 'end-of-conversion' signal to the microprocessor. The microprocessor can then (a) process the data and output the results to the DAC, and (b) access another analogue channel. Each analogue channel is treated as a separate address in the memory map of the microprocessor (or as an input/output device if a microprocessor like the Z80 is used). When the channel data being monitored is predictable the DAS can poll each channel in turn and whenever a channel sample is ready it can start the analogue-to-digital conversion.

The minimum sampling rate for the ADC is equal to the product

$$\text{(highest bandwidth analogue channel)} \times \text{(number of channels)}$$
$$\times \text{(number of samples per channel)} \qquad (4.11)$$

### Transducer

The transducer, or sensor, produces an analogue electrical signal that represents the physical quantity being measured. The transducer should have as low an internal resistance as possible, both to minimize internal noise and to reduce loading by the DAS. There should be some means of calibrating the transducer.

### Transducer-to-DAS Connection

If the transducer output voltage is at a relatively high level, say 0.1–10 V, and a common-mode voltage is unlikely to be picked up, then a single conductor can be used for each channel with a common return line. The maximum distance for the connection is then limited to about 10 m. When the transducer voltage is at a low level and/or common-mode voltages are anticipated, a differential input connection must be employed, and usually this consists of two screened and twisted conductors. Single-ended, and differential, signal paths are

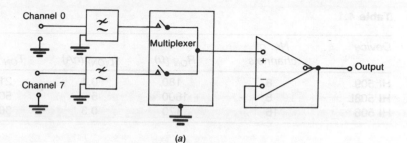

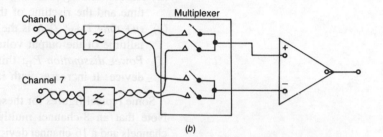

**Fig. 4.15** (a) Single-ended and (b) differential transducer-to-DAS connections

shown by Figs 4.15(a) and (b). When the transducer voltage is at a low level it may be necessary to amplify each channel separately before it arrives at the multiplexer. If the twisted conductors are screened it may not be necessary to use a filter in each channel but when one is required it must be a *CR* filter and not an active circuit, since the latter will often have a large d.c. offset voltage.

Some transducers, particularly piezo-electric types, have a very high source impedance. Such transducers require that a high-impedance amplifier is used to buffer the connection between the transducer and the DAS. An input amplifier is also desirable if the input signal is smaller than about 1 V since most ADCs are designed to work best with a higher level signal. A low-level signal will be converted with both a poor resolution and poor accuracy.

**Analogue Multiplexer**

Most analogue multiplexers use CMOS switches. The most important parameters quoted in the manufacturer's data sheets are:

(a) *Analogue signal range*: this is the range of analogue input voltages that can be accurately switched.

(b) *ON resistance $R_{ON}$*: this is the resistance of a channel path through the multiplexer when it has been turned ON. Ideally $R_{ON}$ is zero.

(c) *Leakage current $I_{D(ON)}$*: when a channel is ON a small leakage current flows through it that creates an offset voltage equal to $I_{D(ON)}R_{ON}$. This leakage current is quoted at 25 °C but it doubles with every 10 °C increase in temperature.

**Table 4.1**

| Device | No. of channels | $R_{ON}$ ($\Omega$) | $I_{D(ON)}$ (nA) | $T_{ON}$ (ns) | $T_{OFF}$ (ns) | $P_D$ (mW) |
|---|---|---|---|---|---|---|
| HI 509 | 8 | 180 | 0.1 | 210 | 180 | 23 |
| HI 508L | 8 | 1000 | 5 | 500 | 500 | 40 |
| HI 506 | 16 | 170 | 0.3 | 300 | 300 | 30 |

(d) *Access time* $T_{ON}$: the access time is the sum of the logic delay time and the risetime of the output voltage.

(e) *OFF time* $T_{OFF}$: $T_{OFF}$ is the sum of the logic delay time and the falltime of the output voltage.

(f) *Power dissipation* $P_D$: this is the power dissipated within the device. It increases with increase in the switching frequency.

Some typical figures for these parameters are given by Table 4.1. Note that an 8-channel multiplexer can provide four differential channels and a 16-channel device can give eight differential channels.

## Programmable-gain Instrumentation Amplifier

The amplifier that precedes the sample-and-hold amplifier is usually an instrumentation amplifier and very often it is one whose gain can be varied by a command given by the microprocessor. It is desirable that the gain of the amplifier is variable, because, otherwise, the MSBs of the converted digital output voltage will become idle, one after the other, if the analogue input voltage decreases in amplitude — and this means that the resolution of the system, referred to the input, is degraded.

Consider, as an example, a 12-bit ADC. The full 12-bit resolution only applies for input voltages greater than, or equal to, one-half the full scale. If the input voltage should only be, say, 0.06 times full scale it receives only 8 bits digitalization. To fully utilize the converter the gain of the instrumentation amplifier should be varied before each conversion to ensure that

$$\text{(full scale)}/2 < V_{in} < \text{(full scale)}$$

The gain of the programmable-gain amplifier (PGA) is recorded by the microprocessor and is taken into account when it processes the digitalized signal. Typically, a PGA might have programmable gains of 1, 2, 4 and 8. The PGA also buffers the output of the multiplexer, provides a single-ended source for the sample-and-hold amplifier, and provides rejection of common-mode signals.

The DAS need not be interfaced to either a digital computer or a microprocessor. Figure 4.16 shows a DAS that outputs its digital data to a storage register. The outputs of the binary counter are applied to both the 8-channel multiplexer to select the wanted channel (000

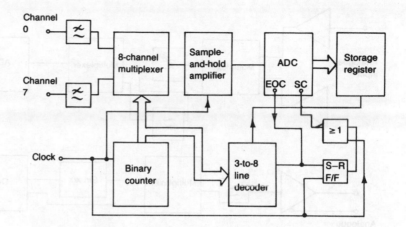

**Fig. 4.16** DAS to register system

selects channel 0, 111 selects channel 7), and to the 3-to-8 line decoder. One of the analogue channels is selected at the leading edge of a clock pulse and at the same time the sample-and-hold amplifier is placed into its sample mode. At the trailing edge of the clock pulse the sample-and-hold amplifier is switched into its hold mode and the ADC commences a conversion. When the conversion is complete the 'end-of-conversion' terminal goes low and this enables the decoder to send a signal to the storage register telling it to store the converted digital signal. The set−reset (S−R) flip-flop introduces a short delay before the next conversion can commence. A better, but more expensive, arrangement is to connect a sample-and-hold amplifier in each channel before the multiplexer.

It is possible to obtain the analogue multiplexer, the PGA, and the sample-and-hold amplifier in one package, a typical IC being the Harris HI 5900/1 analogue data acquisition signal processor.

Now, ICs are available that include all the circuitry required to interface an analogue system to a DSP. The block diagram of the Texas TLC 32046 is shown by Fig. 4.17. Two analogue signals are applied to a multiplexer whose output is connected to a low-pass filter of 7.6 kHz cut-off frequency. The output of the low-pass filter is applied to another multiplexer and to a high-pass filter of cut-off frequency 300 Hz. The multiplexer can be programmed to select the output of either the low-pass filter or of the high-pass filter; if the latter is selected then a band-pass characteristic of 300−7600 Hz is obtained. The output of the second multiplexer is applied to a 14-bit ADC which can operate at any frequency up to 25 kHz. In the other direction of transmission the DAC output is applied, either directly or via a (sin $x$)/$x$ correction filter, to a multiplexer whose output is applied to a low-pass filter.

## Interfacing between Logic Families

Some parts of a digital system may be required to operate at much faster speed than other parts and it may then be necessary to employ

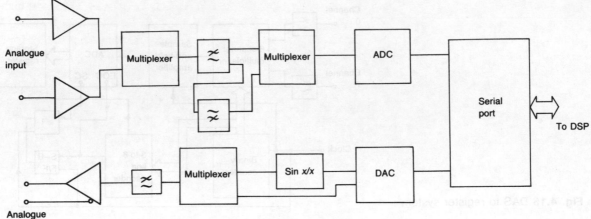

**Fig. 4.17** Block diagram of the TLC 32046 (*courtesy of Texas Instruments*)

devices from two different logic technologies. Each of the digital logic technologies has its own unique current and voltage specifications with variations between their various families. Within a logic family, e.g. LS TTL, interfacing is not necessary unless the fan-out of a device is to be exceeded. A logic interface must ensure that:

(a) the driving IC is able to source, or sink, enough current to meet the total requirements of all the driven ICs, and

(b) the high and low output voltages of the driving IC are within the specified range of high and low input voltages for the driven ICs.

The problems associated with logic interfacing fall into two categories:

(a) between logic families that operate at the same power supply voltage and so do not need logic level conversion;

(b) between logic families that operate from different power supply voltages and hence do require a level translation to convert logic-level voltages from one value to another.

### Low-power Schottky to High-speed CMOS

Devices in the high-speed CMOS (HC) family can operate from power supply voltages in the range 2−6 V but for most applications a common 5 V supply will probably be used.

### (a) Same Power Supply Voltage

#### LS driving HC
The low-level output voltage of the LS family is compatible with the HC low-level input voltage. The high-level LS output voltage is

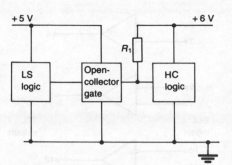

**Fig. 4.18** LS to HC interface

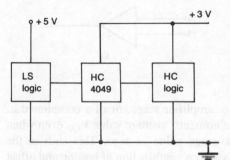

**Fig. 4.19** LS to HC interface

specified to pull up to a minimum of 2.7 V but the HC minimum high-level input voltage is 3.2 V (for $V_{SS} = 5$ V). If the fan-out of the LS device is only one its high-level output voltage will probably reach 3.2 V but this is not guaranteed by the specification. To ensure high-level compatibility a $2-6.2$ kΩ pull-up resistor must be connected between the output pin of the LS device and the 5 V supply line. Alternatively, a TTL-to-CMOS logic converter such as the 40104 can be employed.

A better solution, of course, is to employ high-speed CMOS devices that have TTL compatible inputs, i.e. HCT devices, but, if the LS device is of the open-collector type, a pull-up resistor will be required in any case.

### HC Driving LS

When an HC device drives an LS device both the low-level and the high-level voltages are compatible but the HC device can only have a fan-out of one. This is because the low-level output current of an HC device is 0.4 mA minimum to 1 mA typical while the LS device will need a minimum input current of 0.36 mA. If the fan-out is to be two, or more, then a CMOS buffer, such as the 4010, must be employed.

### (b) Different Power Supply Voltages

If, in order to increase the speed of operation, it is desired to operate the HC device at a higher supply voltage than 5 V then logic-level translation will be necessary. This can be accomplished either by using a CMOS open-drain gate or an LS open-collector gate. Figure 4.18 shows how the TTL gate is employed for this purpose.

When the HC logic is to be operated at a lower voltage than 5 V a logic-level down converter IC, such as the HC 4049/50, must be employed. Figure 4.19 shows how the converter is connected to interface between the two logic families. When the direction of signal flow is from HC to LS, level translation will be required when the HC power supply voltage is greater than 5 V but not if it is less than 5 V.

### Emitter-coupled Logic to Low-power Schottky

The emitter-coupled logic (ECL) family uses a power supply voltage of $-5.2$ V and the LS family uses $+5$ V. Interfacing the two families requires the use of level translation. Different level translators are needed for each direction of transmission and Fig. 4.20($a$) shows how the MC 10124 quad TTL-to-ECL translator is used to interface TTL devices to ECL devices. Figure 4.20($b$) shows how the 40125 quad ECL-to-TTL translator is connected.

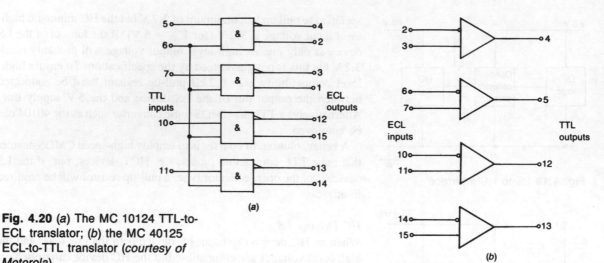

**Fig. 4.20** (a) The MC 10124 TTL-to-ECL translator; (b) the MC 40125 ECL-to-TTL translator (*courtesy of Motorola*)

## Level Shifting

The output voltage of a d.c. amplifier stage, or of a complete d.c. amplifier, may often have a non-zero average value $V_{QO}$ even when the input voltage has an average value of zero. This shift in the quiescent value may be caused by a combination of biasing and offset effects. When stages are connected in cascade without the use of coupling capacitors it may often be necessary to shift the quiescent output voltage $V_{QO}$ of a stage nearer to 0 V before it is applied to the input of the next stage. Otherwise the next stage may be driven into saturation.

A *level shifter* is an amplifier, often with unity gain, that adds, or subtracts, a known d.c. voltage to or from the input of a stage/amplifier to compensate for d.c. offset effects. A level-shifter stage, for example, is included in the internal circuitry of most op-amps. The level shifter should have a high input resistance and a low output resistance and very often an emitter follower can be employed. If the output voltage $V_{QO}$ is taken from the emitter itself the change in the d.c. level would be equal to $V_{QO} - V_{QI} = 0.7$ V but usually this shift will be too small. To increase the d.c. voltage shift the circuit given by Fig. 4.21(a) can be used and this gives a shift of $(0.7 + I_C R_1)$ volts. The disadvantage of the circuit is fairly obvious — the signal voltage is attenuated by the ratio $R_2/(R_1 + R_2)$. This difficulty can be overcome by making $R_2$ a very high value resistance, i.e. by replacing $R_2$ by a current source. This is shown by Fig. 4.20(b). The current source can be obtained in a number of ways, one of which is used in the circuit given in Fig. 4.21(c).

If the d.c. resistance of the previous stage is $R_0$ and the current flowing into the level shifter is $I_B$ then

$$V_{QI} = I_B R_0 + V_{BE} + I_C R_1 + V_{QO}$$

and

$$V_{QO} = V_{QI} - (I_C R_0)/h_{FE} - V_{BE} - I_C R_1 \qquad (4.12)$$

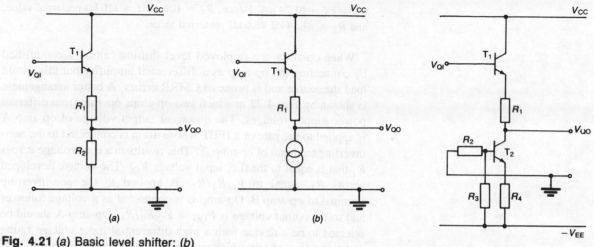

**Fig. 4.21** (a) Basic level shifter; (b) use of a current source instead of $R_2$; (c) practical circuit

By suitable choice of the value of $R_1$ the quiescent output voltage can be shifted downwards by the required amount. If it is required to shift the quiescent voltage upwards then p-n-p transistors should be used instead of n-p-n.

### Example 4.4

Two d.c. voltage amplifier stages are to be connected in cascade. When the input signal is at 0 V the d.c. output voltage of the first stage is 3.5 V across a 1200 Ω resistance. To prevent the second stage going into saturation a level shifter is to be connected between the two stages and it is to shift the d.c. voltage down to 1 V. Design the level shifter. The power supply voltage is ± 12 V, the transistors have $h_{FE} = 80$ and $V_{BE} = 0.7$ V, and the constant current is to be 1 mA.

*Solution*

$$1 = 3.5 - (1 \times 10^{-3} \times 1200)/80 - 0.7 - (1 \times 10^{-3})R_1$$

or

$$R_1 = 1785 \, \Omega$$

The nearest preferred value is 1800 Ω.

To design the constant current source set the operating point of $T_2$ in the middle of the load line. Then,

$$V_{CE} = 6 \text{ V} \qquad R_4 = 6/(1 \times 10^{-3}) = 6 \text{ k}\Omega$$
$$= 6.04 \text{ k}\Omega \text{ preferred value}$$
$$R_2 \| R_3 < 0.1 \, h_{FE}R_4 < 0.1 \times 80 \times 6040 = 48.32 \text{ k}\Omega$$
$$V_{R2} = V_{R4} + 0.7 = 6.7 \text{ V}$$
$$V_{R3} = 12 - 6.7 = 5.3 \text{ V}$$

Hence, $R_2 = 6.7/I$ and $R_3 = 5.3/I$, where $I$ is the current flowing through the two resistors assuming that the base current is negligibly small.

$$[(6.7/I) \times (5.3/I)]/[(6.7/I) + (5.3/I)] = 48.32 \times 10^3$$

Solving $I = 61.24 \,\mu A$. Hence, $R_1 = 109.4 \,k\Omega = 110 \,k\Omega$ preferred value, and $R_2 = 84.54 \,k\Omega$ 86.6 $k\Omega$ preferred value.

When op-amps are employed level shifting can be accomplished by connecting the op-amp as a differential amplifier but this would load the source and is prone to CMRR errors. A better arrangement is shown by Fig. 4.22 in which two op-amps operate from different power supply voltages. The quiescent output voltage of op-amp A is applied to the gate of a JFET whose drain is connected to the non-inverting terminal of op-amp A. This results in a d.c. voltage across $R_1$ that is equal to the d.c. input voltage $V_{QI}$. The voltage developed across $R_2$, equal to $V_{QI}R_2/R_1$, is applied to the non-inverting terminal of op-amp B. Op-amp B is connected as a voltage follower and so the output voltage is $V_{QO} = V_{QI}R_2/R_1$. Op-amp A should be selected to be a device with a high differential input voltage rating and low value of input bias current and offset voltage.

Sometimes a level shifter that is able to convert a 0 V referenced signal into an identical positive-voltage referenced signal is wanted. One method of accomplishing this is shown by Fig. 4.23. The voltage gain of the circuit is $R_3/R_4$ and the d.c. output reference voltage is the supply voltage $V_{CC}$. If a positive-referenced voltage is to be

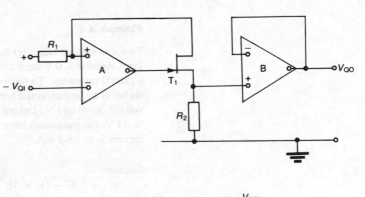

**Fig. 4.22** Op-amp level shifter.

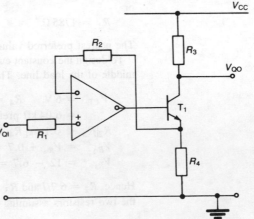

**Fig. 4.23** Converting a zero-referenced signal to a positive-referenced signal.

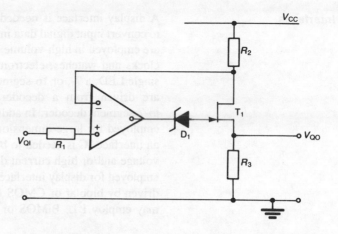

**Fig. 4.24** Converting a positive-referenced signal to a zero-referenced signal

shifted to become zero-referenced then the circuit given in Fig. 4.24 can be employed. The voltage gain of the circuit is $R_2/R_3$. Typically, $R_1 = 47\,\text{k}\Omega$, $R_2 = 1\text{–}10\,\text{k}\Omega$, $R_3 = 1\,\text{k}\Omega$, and the op-amp is a CMOS device such as the LF 351.

## Op-amp to Logic Interface

When there is a requirement to interface op-amp circuitry with logic circuitry it will simplify design if an op-amp manufactured to operate from a single-polarity power supply is employed. The TL 090 series of BiFET devices can operate from a single-polarity power supply voltage of 3–36 V and hence 5 V would usually be chosen. The interface will need an output load resistor to ensure that the output voltage of the op-amp is able to fall below the maximum input voltage that will be recognized as logic 0 by the driven logic device. The slew rate of the op amp must be high enough to avoid logic errors caused by a too low risetime for the logic input signal. The minimum required slew rate is much higher (typically 70 V/$\mu$s) than most op-amps are able to provide and so a comparator IC, such as the LM 311, or a Schmitt trigger, such as the LS 0014, is often employed. The use of the latter device is shown by Fig. 4.25.

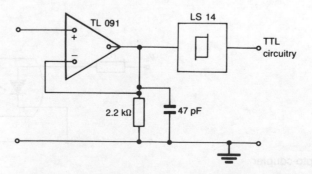

**Fig. 4.25** Op-amp to logic interface

## Display Interfaces

A display interface is needed to provide readable information, i.e. to convert input digital data into an analogue visual display. Displays are employed in high-volume consumer products such as calculators, clocks and watches, electronic games, etc. They may consist of a single LED, a 7-, or 16-segment display, or a dot matrix display that are driven from a decoder/driver IC, e.g. the LS 248 BCD-to-7-segment decoder. In addition, various other display devices are employed for large-dimensioned, large-current displays. For these, an interface IC is needed to buffer the low-level logic from the high voltage and/or high current display load. Different technologies are employed for display interface devices. Low-power LED displays are driven by bipolar or CMOS logic but high-voltage/current displays may employ $I^2L$, BiMOS or CMOS/DMOS logic.

## Opto-couplers

Interference between a transmitter and a receiver in a circuit can be reduced in several different ways, e.g. the use of screened cable, but one of the best is complete electrical isolation between transmitter and receiver. The electrical isolation can be obtained by using an *opto-coupler*. Opto-couplers are also employed to interface between two devices from two different logic families.

An opto-coupler has two parts; these are an optical receiver and an optical transmitter. The optical transmitter consists of an LED that emits light energy when a current is passed through it. The optical receiver, or detector, is either a phototransistor or a photodiode that generates a current proportional to the magnitude of the light intensity incident upon it. In some cases the phototransistor is Darlington-connected. A photodiode is used when high-speed operation is required, the diode current may be amplified by an on-chip amplifier. Some other opto-couplers have a silicon—controlled rectifier or triac or Schmitt trigger output. The basic arrangement of an opto-coupler is shown in Fig. 4.26 in which a phototransistor is shown. When a voltage is applied to the input of the circuit a current flows in the LED and it emits light energy. This light energy illuminates the base

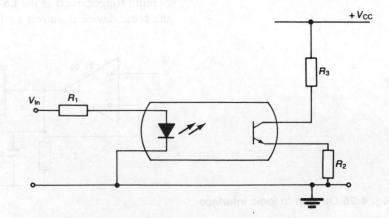

**Fig. 4.26** Opto-coupler

of the phototransistor and causes a collector current to flow that is more or less proportional to the input voltage applied across the LED. The collector current passes through the load resistor $R_2$ and develops a voltage across it.

The most important parameters of an opto-coupler are:

(a) *Input-to-output isolation voltage*: this is the maximum voltage difference that can exist between the input and output terminals of the device without causing damage.

(b) *Current transfer ratio*: the current transfer ratio $\beta$ is the ratio (output current)/(input current), expressed as a percentage. Typically $\beta$ is some 20–200%.

(c) *Switching time*: the switching time is the sum of the risetime and the falltime of the output waveform when an input pulse is applied.

(d) *Breakdown voltage*: this is the maximum collector-to-emitter voltage of the phototransistor and is typically some 30–70 V.

Figure 4.27 shows how an opto-coupler can be used as the interface between two different logic families.

The current transfer ratio $\beta$ characteristic of an opto-coupler is not linear and so some devices employ optical negative feedback to reduce distortion. An example of this is shown by Fig. 4.28.

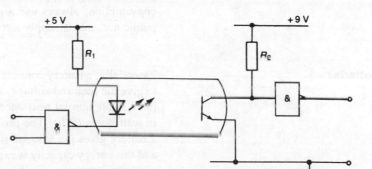

**Fig. 4.27** Opto-coupler interfaces two logic families

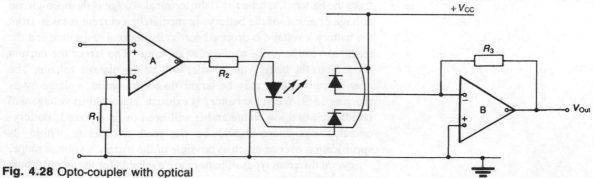

**Fig. 4.28** Opto-coupler with optical feedback

# 5 Large-signal Electronics

Although all digital and most analogue electronic circuits process low-level signals, some circuits are designed to handle high-level signals. A high-level signal may be of high current, high voltage and/or high power and the large-signal circuits include power supplies, power amplifiers, both audio-frequency and radio-frequency, and various control circuits using devices such as thyristors.

All electronic circuits need to have some kind of power supply to provide the d.c. currents and voltages needed to power the various devices in the circuit. Many portable appliances, such as domestic radio receivers whose power consumption is small, are battery powered. Larger appliances, both physically and in terms of power consumption, always use a power supply circuit that converts the public a.c. mains supply into a d.c. supply at the required voltage.

**Batteries**

Essentially, a battery consists of one or more cells connected in series to give the required voltage. A battery provides a voltage source that has a small internal resistance. The capacity of a battery is expressed in watt-hours (Wh). The product of the capacity and the voltage of a battery gives the amount of energy that is contained in the battery and this energy capacity is expressed in ampere-hours (Ah). All three of these parameters vary with both temperature and the rate of discharge. Figure 5.1 shows how the terminal voltage of some common types of battery varies with time as a constant current is taken from the battery. At time $t = 0$ the terminal voltage is the open-circuit voltage or e.m.f. of the battery. Immediately a current is taken from the battery a voltage is dropped across its internal resistance and the terminal voltage drops to the plateau value. The larger the current taken from the battery the smaller will be the plateau voltage. The open-circuit voltage may be larger than the nominal voltage by as much as 15%. When the battery is exhausted its terminal voltage will rapidly fall to a low value and it will need to be replaced. Battery-operated equipment should be designed to operate within its specification over as much as possible of the battery's voltage range.

Several different types of battery are available for use in electronic equipment and these are listed in Table 5.1.

The most commonly employed battery is the zinc carbon; it is of

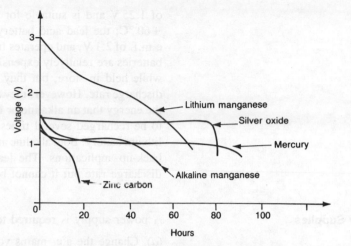

**Fig. 5.1** Variation with time of the terminal voltage of various types of battery

**Table 5.1**

| Type | Nominal voltage (V) | e.m.f. (V) | Capacity (mAh) | Operating temperature range (°C) |
|------|---------------------|------------|----------------|----------------------------------|
| Alkaline manganese | 1.5 | 1.5 | 800–15 000 | 20 to +60 |
| Lithium manganese | 3.0 | 3.5 | 50–1000 | −20 to +60 |
| Mercury | 1.35 | 1.35 | 350–14 000 | −5 to +70 |
| Silver oxide | 1.5 | 1.6 | 45–130 | −10 to +50 |
| Zinc carbon | 1.5 | 1.7 | 10–30 | −5 to +45 |
| Zinc chloride | 1.5 | 1.6 | 40–100 | −10 to +50 |

low cost, it is suitable for all non-critical applications, and it is available in a variety of sizes and shapes. Unfortunately, its capacity is poor compared with the other types of battery and its plateau voltage starts to fall off early in its life. There is also a 'heavy-duty' version of the zinc carbon cell that is able to supply a high current for a longer period of time. The zinc chloride battery is somewhat more expensive than the zinc carbon battery but it has a better performance. The alkaline manganese battery is of moderate cost and has a better performance than the zinc carbon and zinc chloride batteries as well as having a longer life. The alkaline battery is particularly suited to supplying both a continuous heavy load and also high intermittent currents. The mercury battery is suited to applications where voltage stability of the power supply is important but this type of battery is expensive. Both the mercury and the silver oxide batteries are available in button form for use in small electronic devices such as watches, pocket calculators, cameras, etc. The lithium battery is of light weight, has a very long shelf life and high energy density but it is fairly expensive. It is often used in CMOS RAM back-up systems.

In addition, two types of rechargeable battery are also employed. The nickel cadmium battery has a nominal voltage of 1.2 V, an e.m.f.

of 1.25 V and is suitable for use in temperatures of from −40 to +60 °C; the lead acid battery has a nominal voltage of 2.0 V, an e.m.f. of 2.1 V, and operates from −20 to +50 °C. Nickel cadmium batteries are relatively expensive and do not retain their charge well while held in store, but they are easy to use and can have a high discharge rate. However, they are able to store only about one-quarter the energy that an alkaline or lithium battery can store and will need to be recharged several times during the lifetime of an alkaline or lithium battery. Both alkaline and lithium batteries are used for CMOS back-up applications. The lead acid battery can also have a high discharge rate but it cannot be left in a discharged state.

## Power Supplies

A power supply is required to perform the following functions:

(a) Change the a.c. mains voltage into the required d.c. voltage.
(b) Filter the rectified voltage to remove ripple.
(c) Regulate the d.c. output voltage to keep it at a constant value as the input voltage and/or the load vary. The former is known as *line regulation* and the latter is known as *load regulation*.

Power supply circuits employ either *linear* or *switching* regulators. Figures 5.2(a) and (b) show the two most common linear power supply

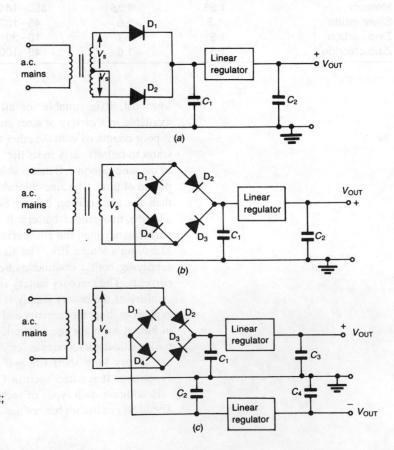

**Fig. 5.2** (a) Full-wave rectifier circuit; (b) bridge rectifier circuit; (c) dual power supply

circuits. The full-wave circuit of Fig. 5.2(a) uses a centre-tapped transformer and has only one diode at a time in series with the regulator. The diodes must therefore be able to withstand a peak inverse voltage (PIV) of twice the peak voltage $V_{s(max)}$ across each half of the secondary winding of the mains transformer. Since the mains voltage could increase by up to 10% high the PIV ratings of the diodes should be at least $2.2V_{s(max)}$ or 3.1 times the no-load d.c. output voltage. The bridge circuit shown in Fig. 5.2(b) avoids the need for a centre-tapped transformer but it has two diodes in series with the voltage regulator. The PIV ratings of the diodes must be at least $1.1V_{s(max)}$ or 1.56 times the no-load d.c. output voltage of the rectifier unit. Bridge rectifier units are available in various packages including dil. A dual power supply is frequently required and Fig. 5.2(c) shows how this is accomplished.

The linear regulator may be either a discrete circuit or an IC device. The use of an IC is simpler and uses fewer components than a discrete design and it is the common practice. The efficiency of a linear regulator is rather poor, generally in the region of 40−55%, but perhaps as low as 20% for a 5 V supply. The basic block diagram of a switching power supply is shown by Fig. 5.3. A switching regulator regulates the output voltage by varying the duty cycle of a switching transistor and this leads to much higher efficiencies. Switching power supplies usually operate *off-line*, i.e. they directly rectify and filter the a.c. mains voltage without first using a step-down mains transformer. The rectified voltage is 'chopped' by a transistor switch before further rectification and filtering take place. Several different arrangements are employed for switching power supplies, such as the *buck* and the *boost regulators*, the *flyback converter*, the *forward converter*, the *half-bridge converter* and the *full-bridge converter*.

The advantages of a linear power supply over a switching power supply are as follows:

(a) Simpler, cheaper, more reliable, and easier to fault-find.
(b) Good line regulation, typically 0.02−0.05% compared to 0.05−0.1% for a switching supply.
(c) Good load regulation, typically 0.01−0.1% compared to 0.1−1% for a switching supply.
(d) Very low output ripple and noise, typically 1−10 mV peak-to-peak compared with 25−100 mV for a switching circuit.

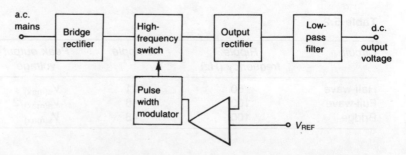

**Fig. 5.3** Switching power supply

(e) Rapid recovery from load transients, typically 50 $\mu$s compared with 300 $\mu$s for a switching power supply.

(f) A much lower level of radio-frequency interference (RFI) is generated.

The advantages of switching power supplies over linear power supplies are as follows:

(a) Since the transformer operates at a much higher frequency than 50 Hz (usually between 15 and 25 kHz), it can be smaller, lighter and much cheaper.

(b) Smoothing is much easier to accomplish because the ripple frequency is at least 15 kHz.

(c) The efficiency is much higher, typically 65−90% compared with 40−55% for a linear power supply.

(d) The input voltage range is wider; typically this is ± 10% for a linear supply and ± 20% for a switching supply.

The much lower efficiency of a linear power supply occurs because it must dissipate the difference between the unregulated input power and the regulated output power. This does not present a problem with mains-operated equipment requiring up to about 25 W power because the power losses in the regulator can then be tolerated. However, the linear power supply is more or less obsolete for the higher-power end of the market.

**Linear Power Supplies**

The rectifier unit may employ the half-wave, the full-wave or the bridge configuration and Table 5.2 lists the main parameters of these three circuits.

Suppose that the voltage $V_s$ across the secondary winding of the mains transformer is 12 V and the diodes are ideal, i.e. they have zero forward resistance.

Then the output voltage $V_{OUT}$ of the half-wave circuit is

$$V_{OUT} = (0.637/2) \sqrt{2} \times 12 = 5.4 \text{ V}$$

The full-wave circuit uses only one-half of the mains transformer secondary winding at a time so that only 6 V is applied to each diode.

**Table 5.2**

| Circuit | Ripple frequency (Hz) | % ripple | Peak output voltage | Average output voltage | Diode PIV |
|---|---|---|---|---|---|
| Half-wave | 50 | 121 | $V_{s(max)}$ | $V_{s(av)}/2$ | $V_{s(max)}$ |
| Full-wave | 100 | 48 | $V_{s(max)}/2$ | $V_{s(av)}/2$ | $2V_{s(max)}$ |
| Bridge | 100 | 48 | $V_{s(max)}$ | $V_{s(av)}$ | $V_{s(max)}$ |

The output voltage is $0.637 \times \sqrt{2} \times 6 = 5.4\,\text{V}$ (again). It can be seen that although the average output voltages are equal to one another the half-wave circuit has a peak voltage twice as large as the full-wave circuit.

The bridge circuit uses all of the secondary winding of the mains transformer. Hence the average output voltage $V_{\text{OUT}}$ is

$$V_{\text{OUT}} = 0.637 \times \sqrt{2} \times 12 = 10.8\,\text{V}$$

and this is exactly twice the average output voltage of the other two circuits.

In practice, of course, the diodes are *not* ideal and so there is a voltage drop of approximately 0.6 V across each diode when it is conducting. This means that the average output voltage for the half-wave and the full-wave circuits is $5.4 - 0.6 = 4.8\,\text{V}$, and for the bridge circuit is $10.8 - 1.2 = 9.6\,\text{V}$. The half-wave rectifier circuit is rarely employed to supply power to electronic circuits and the bridge rectifier circuit is much more commonly employed than the full-wave rectifier circuit.

*Capacitor Filter*

Usually a capacitor filter is employed to remove the ripple content of the output voltage. The d.c. output voltage $V_{\text{OUT}}$ of the rectifier circuit is then

$$V_{\text{OUT}} = V_{\text{s(max)}} - I_{\text{DC}}/4fC \tag{5.1}$$

where $V_{\text{s(max)}}$ is the peak voltage across the secondary winding, $I_{\text{DC}}$ the average load current, $f$ the frequency of a.c. mains supply and $C$ is the filter capacitance.

The percentage ripple voltage superimposed upon the d.c. output voltage is

$$\% \text{ ripple} = 100/[4\sqrt{3}\,R_L fC] \tag{5.2}$$

where $R_L = V_{\text{OUT}}/I_{\text{DC}}$ and the $\sqrt{3}$ appears because the ripple voltage is of approximately triangular shape.

The mains transformer should have the required secondary voltage and an adequate power rating. The power rating is the product of the r.m.s. values of the a.c. secondary voltage and the full-load secondary current. The reservoir, or filter, capacitance will be either an aluminium or a tantalum electrolytic type. The required capacitance is directly proportional to the load current and indirectly proportional to the desired ripple voltage. The important parameters of the capacitor are its capacitance, its rated voltage and its ripple current rating. The ripple current, which flows as the capacitor is alternately charged and discharged, is bigger than the load current by a factor of typically 1.5. It is usual to use a capacitor whose ripple current rating is approximately twice the load current. Short-form data for rectifier diodes include the following: the mean forward current $I_{\text{F(av)}}$, the surge (non-repetitive) forward current $I_{FSM}$, the $I^2t$ rating (which is used to select a suitable fuse), the repetitive peak reverse voltage $V_{RRM}$, and the mean forward voltage $V_{FM}$ for a quoted forward current. If the bridge rectifier circuit is used then a bridge rectifier unit can be employed.

### Example 5.1

Design a bridge rectifier power supply that uses a capacitor filter to provide 12 V at 0.8 A with a maximum ripple content of 1% and a voltage regulation of ± 2%.

### Solution

First select a suitable bridge rectifier unit. One such unit has $I_{FSM} = 50$ A, $I_{F(av)} = 1$ A, $V_{RRM} = 50$ V and $V_{FM} = 1$ V at 1 A.

For ± 2% voltage regulation the output voltage could rise to 12.24 V when the peak secondary voltage would be 12.24 + 1.2 = 13.44 V. Hence a 13.44/√2 = 9.5 V transformer is required. Select a 9 V transformer with a current rating in excess of 1 A.

The percentage ripple is to be 1% (or less) and $R_L = 12/0.8 = 15$ Ω. Therefore, from equation (5.2),

$$C = 100/[4\sqrt{3} \times 15 \times 50] = 19\ 245\ \mu F = 22\ 000\ \mu F \text{ nearest preferred value}$$

Using this value, equation (5.1) gives

$$V_{OUT} = 12.24 - 0.8/[4 \times 50 \times 22 \times 10^{-3}] = 12.06 \text{ V}$$

The capacitor should have a voltage rating of 16 V and a ripple current rating of 1.8 A or more. A fuse should be provided in series with the primary winding of the mains transformer to protect the circuit against damage should the output terminals be inadvertently short-circuited. An anti-surge fuse whose rating is about 1.5 times the r.m.s. primary current should be used.

There is a wide selection of IC voltage regulators offered by the various manufacturers. These devices can be classified into one of three groups: (a) low drop-out voltage (the *drop-out voltage* is the minimum difference between the input and output voltages of the regulator that will allow it to operate correctly); (b) fixed voltage; (c) adjustable voltage. The choice of a suitable voltage regulator can be made after consulting short-form data and then, if need be, the data sheets of possible choices. Table 5.3 shows the data typically given in short-form and also details of a few selected devices. The fixed voltage regulators are available in positive voltages of 5, 12, 15 and 24 V, as well as in negative voltages of the same value.

The data sheet of a voltage regulator gives details of how the device should be connected into a circuit. Figure 5.4 shows how a LM 7805 IC is connected to regulate a 7.5−35 V input voltage to a 5 V output voltage with an output current of up to 1 A. The adjustable voltage

**Table 5.3**

| Type | Output voltage (V) | Maximum input voltage (V) | Minimum input voltage (V) | Typical line regulation (mV) | Typical load regulation (mV) | Output noise (μV) | Ripple rejection (dB) | Drop-out voltage (V) |
|------|--------|--------|--------|--------|--------|--------|--------|--------|
| LM 2930T5 | 5 | 26 | 5.9 | 30 | 14 | 140 | 56 | 0.32 |
| LM 78L05 | 5 | 30 | 6.9 | 55 | 11 | 40 | 49 | |
| LM 7812 | 12 | 35 | 14.6 | 4 | 12 | 75 | 55 | |
| LM 317LZ | 1.2–37 | 40 | 4 | 0.01% | 0.1% | | 80 | |
| LM 723CN | Variable | 40 | | 0.1% | 0.6% | | 74 | |

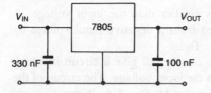

**Fig. 5.4** Use of an IC voltage regulator

regulators employ external resistors to program the wanted output voltage.

### Switching Power Supplies

Switching power supplies are increasingly employed, particularly for powers greater than about 25 W, because of their higher efficiency. A switching regulator achieves a constant d.c. output voltage by varying the duty cycle of a switching transistor under the control of a feedback circuit. To achieve precise regulation the difference between the output voltage and a reference voltage is amplified to produce an error voltage. This error voltage controls a pulse width modulator whose output is applied to the base of the switched transistor to turn it ON and OFF. The average value of the output voltage is the d.c. voltage supplied to the load. It is maintained at a constant value by keeping a constant charge in the output capacitor.

The basic circuit of a *buck*, or step-down, voltage regulator is shown by Fig. 5.5. The d.c. output voltage $V_{OUT}$ is compared with a reference voltage $V_{REF}$ by applying both voltages to a voltage comparator, and the error voltage produced drives the pulse width modulator. The transistor $T_1$ is turned ON and OFF with the duty

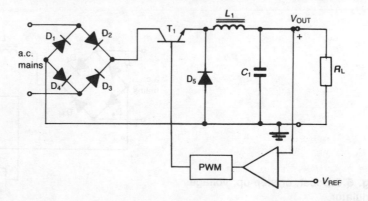

**Fig. 5.5** Buck, or step-down, voltage regulator

cycle needed to regulate $V_{OUT}$ to the required value. When $T_1$ is ON the voltage applied to the inductor is equal to the input voltage minus the output voltage. All of the current in the inductor will then flow into the load until the inductor current becomes larger than the load current. Then the excess current will flow into the output capacitor to charge it and hence the output voltage will rise. When $T_1$ is turned OFF the voltage across the inductor falls to approximately $V_{OUT}$ and the energy stored in the magnetic field of the inductor is transferred to the load. The average current flowing in the inductor is equal to the load current.

If $T_1$ is turned ON for a time $t_1$ and turned OFF for a time $t_2$ then

$$V_{OUT} = V_{IN}t_1/(t_1 + t_2) = V_{IN} \times \text{(duty cycle)} \qquad (5.3)$$

The output voltage is always smaller than the input voltage. A negative supply can be obtained from a negative input voltage by reversing the direction of diode $D_5$.

A rearrangement of the buck circuit will give a circuit in which the output voltage is bigger than the input voltage. The circuit of the *boost*, or step-up, regulator is shown by Fig. 5.6. When $T_1$ is ON the input voltage is applied across the inductor $L_1$. A current flows through the inductor and a magnetic field builds up around it. Diode $D_1$ is then reverse biased and the load current is supplied by the capacitor $C_1$. When $T_1$ is turned OFF the magnetic field collapses and the energy stored in it is transferred to the capacitor by a current that flows in the same direction as before. This current recharges the capacitor. Since the transistor switch is operated at a much faster rate than the time constant $C_1R_L$ of the output circuit a steady d.c. voltage is obtained across the load.

The output voltage of this circuit is

$$V_{OUT} = V_{IN}/(1 - \text{duty cycle}) \qquad (5.4)$$

The buck and boost voltage regulators have evolved into more complex circuits that add a transformer to the circuit to provide electrical isolation between the input and the load. These circuits are generally known as *switched-mode power supplies*.

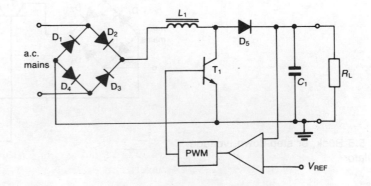

**Fig. 5.6** Boost, or step-up, voltage regulator

### Switched-mode Power Supplies

Four main versions of switched-mode power supplies (SMPS) are in common use. The *flyback converter* is employed for powers of up to 100 W, the *forward converter* for powers of 100−250 W, the *half-bridge converter* for powers of 200−400 W, and lastly the *full-bridge converter* for powers in excess of 400 W. The basic circuits of the four kinds of SMPS are shown in Fig. 5.7.

Figure 5.7(*a*) shows the basic circuit of a flyback converter. A transformer is used in the forward path to isolate the load from the input line and this transformer also transfers energy from the input to the load. When $T_1$ is ON current flows in the primary winding and energy is stored in the transformer's core. The polarity of the winding is such that the diode $D_2$ is reverse biased. When $T_1$ turns OFF the stored energy is transferred to the output capacitor; this maintains the charge stored in the capacitor and keeps the output voltage constant at the wanted value.

The d.c. output voltage $V_{OUT}$ is given by

$$V_{OUT} = nV_{IN} \times (\text{duty cycle})/(1 - \text{duty cycle}) \qquad (5.5)$$

For any chosen value of transformer turns ratio $n$ a duty cycle can be selected that will give the required value of $V_{OUT}$. The output voltage is regulated by comparing $V_{OUT}$ with a reference voltage $V_{REF}$ and using the difference, or error, voltage to control the output of a pulse width modulator. The output of the pulse width modulator drives the base of transistor $T_1$ to turn it ON and OFF for the time periods necessary to give the desired duty cycle. The output voltage may be either higher or lower than the input voltage. The output of the circuit is isolated from the input line by either a small transformer or an opto-isolator in the feedback path. The main disadvantage of this kind of converter is that the energy stored in the transformer is high and this means that the transformer must have a large core. This is the reason why the power output is limited to about 100 W.

The basic circuit of a forward converter is shown by Fig. 5.7(*b*). In this circuit power is not stored in the transformer core and so a smaller core can be used. The output voltage of the converter is

$$V_{OUT} = V_{IN}\, n \times (\text{duty cycle}) \qquad (5.6)$$

Figures 5.7(*c*) and (*d*) give the basic circuits of the half-, and full-wave bridge converters.

A variety of ICs are available from different manufacturers that provide all of the power and control functions for the buck, boost, flyback and forward switching regulators. Examples are the LM 1577-12, the LM 2575-5-0, the MC 78S40 voltage regulators, and the LT 1070 current-mode switcher. The data sheets for each of these devices give the design details. Figures 5.8(*a*) and (*b*) show how the LM 1577-12 can be connected to give a boost regulator and a flyback converter respectively.

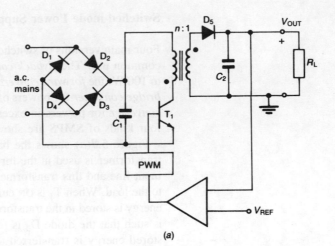

(a)

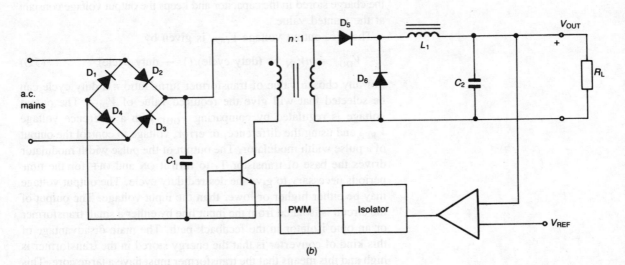

(b)

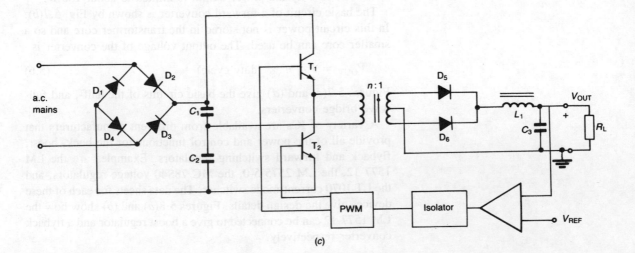

(c)

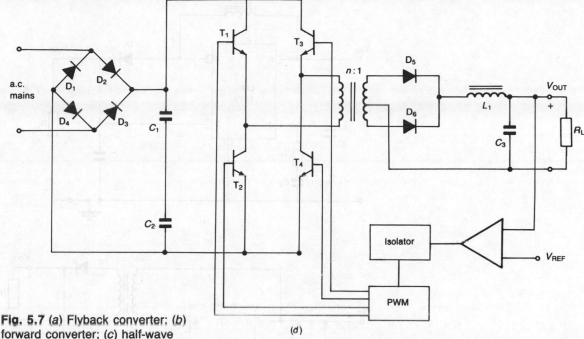

**Fig. 5.7** (*a*) Flyback converter; (*b*) forward converter; (*c*) half-wave bridge converter; (*d*) full-wave bridge converter

The relative merits of the various types of SMPS are as follows:

(*a*) *Flyback converter*:
    (i) Uses the least components.
    (ii) The transformer needs a large core.
    (iii) Has good line regulation but poor load regulation.
    (iv) Has high output ripple and noise.
    (v) The maximum efficiency is about 70%.

(*b*) *Forward converter*:
    (i) Uses a smaller transformer than the flyback converter.
    (ii) Has moderate output ripple and noise.
    (iii) Has rapid recovery from load transients.
    (iv) The maximum efficiency is about 75%.
    (v) Can be the source of high RFI.
    (vi) Has a more complex circuit than the flyback converter.

(*c*) *Half-bridge and full-bridge converters*:
    (i) The transformer requires only a small transformer core.
    (ii) Have low output ripple and noise.
    (iii) Give a rapid recovery from load transients.
    (iv) Have good line and load regulation.
    (v) Both circuits are complex.

### Current-mode Loop Control

Current-mode loop control (CML) is a method of controlling an SMPS by means of dual feedback loops. The first loop consists of the output

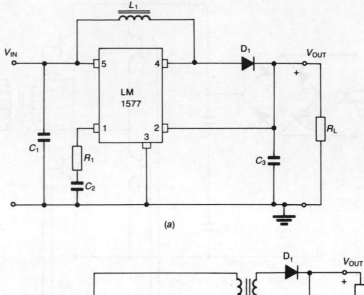

(a)

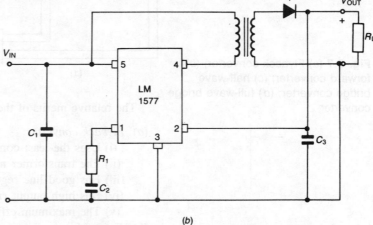

(b)

**Fig. 5.8** Use of the LM 1577-12 as
(a) a boost regulator and (b) a
flyback converter

voltage fed back to a voltage comparator where it is compared with a reference voltage $V_{REF}$. An error current proportional to the difference between the two voltages is produced. The second loop monitors the current that flows in the primary winding of the transformer by passing it through a series-connected resistor; the voltage thus developed is compared with the output of the voltage comparator. The switching transistor is turned ON by a clock pulse but is turned OFF when the primary current becomes large enough to reduce the error current to zero. The basic block diagram of a CML system is shown by Fig. 5.9; CML gives an improved transient response, current limiting, and, since the transformer core is never taken into saturation, RFI is reduced. Also CML is relatively easy to apply to the flyback and forward converters but it is more difficult to apply it to either of the bridge circuits.

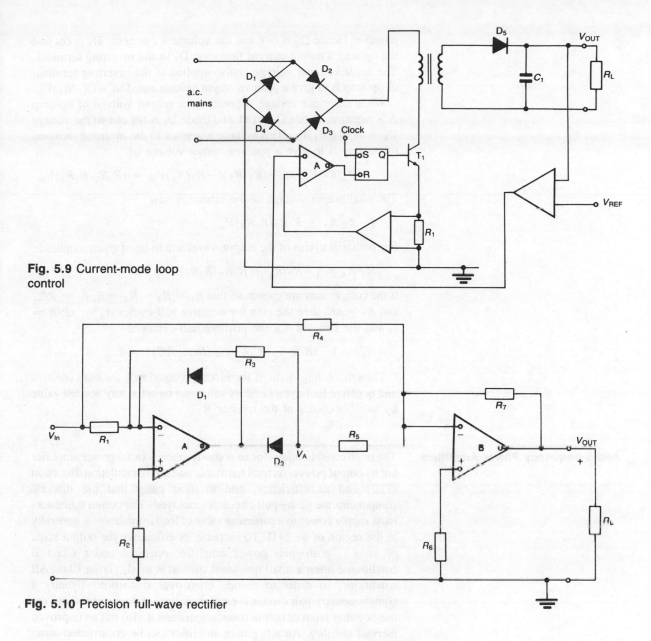

**Fig. 5.9** Current-mode loop control

**Fig. 5.10** Precision full-wave rectifier

## Precision Full-wave Rectifier

For some applications, other than power supplies, the 0.6 V voltage drop across an ON diode in a rectifier circuit cannot be tolerated. In such cases a *precision rectifier* circuit can be employed. One example of such a circuit is shown by Fig. 5.10. Op-amp A acts as an inverting half-wave rectifier and op-amp B adds the two half-sinewave outputs of op-amp A to give a full-wave rectified output voltage. When the input voltage $V_{in}$ is negative the output voltage of op-amp A is positive. Hence $D_2$ is OFF and the voltage $V_A$ is zero. $D_1$ is ON and the op-amp's output current flows via $D_1$ to the inverting terminal. The negative input voltage is also applied to the inverting terminal of op-amp B to give a positive output voltage equal to $-(R_7/R_4)V_{in}$.

When the input voltage is positive the output voltage of op-amp A is negative. Diode $D_1$ is OFF and diode $D_2$ is ON and so the voltage $V_A = -(R_3/R_1)V_{in}$. This voltage is applied to the inverting terminal of op-amp B to give a positive output voltage of

$$-(R_7/R_5)V_A = -(R_7/R_5)(-R_3/R_1)V_{in} = (R_3R_7/R_1R_5)V_{in}$$

The total output voltage of the circuit is now

$$[-R_7/R_4 + R_3R_7/R_1R_5]V_{in}$$

For both half cycles of the output waveform to be of equal amplitude

$$R_7/R_4 = [-R_7/R_4 + R_3R_7/R_1R_5]$$

If the components are chosen so that $R_1 = R_3 = R_4 = R$, $R_5 = R/2$, and $R_7 = xR$, then the gain for negative half-cycles $A_p^- = xR/R = x$, and the gain $A_p^+$ for the positive half-cycles is

$$A_p^+ = [-xR/R + (R \times xR)/(R \times R/2)] = x$$

Thus the voltage gain of the circuit is equal to $x$ for both positive and negative half-cycles and its value can be set at any wanted value by suitable choice of the resistor $R_7$.

## Audio-frequency Power Amplifiers

The main considerations for an audio-frequency (a.f.) power amplifier are its output power, its total harmonic and intermodulation distortion (TID) and its efficiency, and in most cases that use discrete components the push—pull circuit is employed. The output transistors must supply power to a particular value of load impedance — generally in the region of $8-24\ \Omega$. To increase its efficiency the output stage of an a.f. push—pull power amplifier operates under Class B conditions; often a small quiescent current is used, giving Class AB conditions, in order to reduce cross-over distortion. Usually a complementary pair circuit is employed since it does not require the use of either input or output transformers and it also has an improved thermal stability. An a.f. power amplifier can be constructed using discrete components only, using a power amplifier IC such as the LM 380, or using an op-amp followed by a power booster circuit.

Figure 5.11 shows the circuit of a typical a.f. power amplifier that is suitable for output powers of up to about 5 W. The specification for such an amplifier might include any of the following:

(a) The output power into a specified load impedance.
(b) The maximum total harmonic and intermodulation distortion.
(c) The input impedance.
(d) The bandwidth.

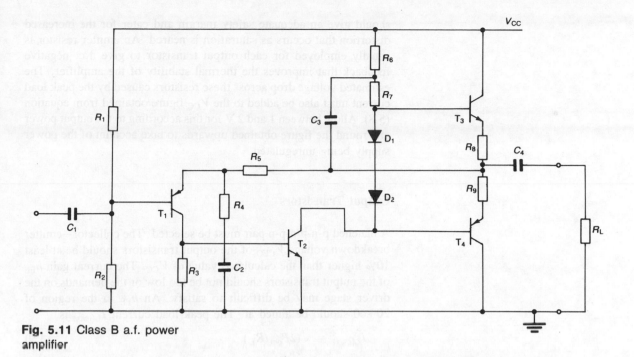

**Fig. 5.11** Class B a.f. power
amplifier

(e)  The maximum allowable hum and noise at the output.

(f)  Whether or not short-circuit protection is required.

A non-regulated power supply is normally employed whose voltage
may drop by as much as 10% when the amplifier is supplying the
full load power.

**Supply Voltage**

The first step in the design of an a.f. power amplifier is to determine
the necessary power supply voltage. To obtain the specified output
power $P_{out}$ across the specified load impedance $R_L$ demands a certain
r.m.s. signal voltage across the load. The quiescent voltage at the
junction of $R_8$ and $R_9$ is equal to $V_{CC}/2$ and hence the maximum
possible r.m.s. load voltage is $V_{CC}/(2\sqrt{2})$. This means that the
maximum output power is

$$P_{out} = V_{CC}^2/8R_L \tag{5.7}$$

Rearranging,

$$V_{CC} = \sqrt{(8\ P_{out}R_L)} \tag{5.8}$$

To this figure must be added an allowance for the collector saturation
voltage of each output transistor. This is usually taken to be about
1 V per transistor for amplifiers of up to about 5 W output power,
and 2 V per transistor for higher-power amplifiers. These figures

should give an adequate safety margin and cater for the increased distortion that occurs as saturation is neared. An emitter resistor is usually employed for each output transistor to give d.c. negative feedback that improves the thermal stability of the amplifier. The estimated voltage drop across these resistors caused by the peak load current must also be added to the $V_{CC}$ figure obtained from equation (5.8). Allow between 1 and 2 V for this according to the output power and round the figure obtained upwards to take account of the power supply being unregulated.

## Output Transistors

A matched p-n-p/n-p-n pair must be selected. The collector−emitter breakdown voltage $V_{CEO}$ of the output transistors should be at least 10% higher than the calculated value of $V_{CC}$. The current gain $h_{FE}$ of the output transistors should not be too low or the demands on the driver stage may be difficult to satisfy. An $h_{FE}$ in the region of 20−60 should be aimed at. The peak load current $I_{L(max)}$ is

$$I_{L(max)} = \sqrt{(2P_{out}/R_L)} \tag{5.9}$$

and hence the peak base current is $I_{L(max)}/h_{FE(min)}$. The maximum collector dissipation of two transistors operating under Class B conditions is

$$P_{c(max)} \simeq 0.4P_{out(max)} \tag{5.10}$$

The safe operating area (SOAR) of the proposed transistor should be checked to ensure that its limits will not be exceeded. When the Class AB operating condition is used the maximum output power that the amplifier is able to provide will be reduced and a safety margin is required. The diodes $D_1$ and $D_2$ connected between the base terminals of the output transistors are provided to slightly forward bias the transistors so that they conduct a small quiescent current. This gives Class AB operation and so reduces cross-over distortion. Alternatively, a $V_{BE}$ *multiplier* may be used (an example of the employment of this circuit is shown in the amplifier circuit given in Fig. 5.12).

## Class A Driver

Since the two output transistors are connected as emitter followers much of the voltage gain of the circuit is provided by the Class A driver stage. The collector load for $T_2$ consists of the two diodes in series with resistors $R_6$ and $R_7$ and this collector load is bootstrapped by capacitor $C_3$. When the base of $T_3$ is driven near to the saturation point of the transistor its potential approaches that of the collector, and without bootstrapping very little base current would be supplied.

The charge stored in the bootstrap capacitor $C_3$ provides the current necessary to keep $T_3$ fully driven and allows the maximum output power to be developed by the amplifier. The time constant $C_3 R_6 R_7/(R_6 + R_7)$ must be larger than $(T/2\pi)$, where $T$ is the periodic time of the lowest frequency to be amplified.

The d.c. collector current $I_{C2}$ of $T_2$ must be larger than the peak base current taken by the output transistors and given by equation (5.9). The current rating of $T_2$ should be about 10% greater than this value, the breakdown voltage $V_{CEO}$ should be the same as for the output transistors. The maximum collector dissipation of $T_2$ is $(V_{CC}/2)I_{C2}$, and the power rating should be at least 10% greater than $V_{CC}I_{C2}/2$. An audio driver type of transistor should be selected. The current gain of $T_2$ should be specified at a d.c. collector current of twice its expected collector current $I_{C2}$ because $T_2$ must be able to supply the full peak load driving current plus the Class A d.c. current. The base current of $T_2$ is equal to $I_{C2}$ divided by the $h_{FE}$ of the chosen transistor.

The total resistance of $R_6$ and $R_7$ in series is given by

$$(R_6 + R_7) = (V_{CC}/2 - 0.6)/I_{C2}.$$

Let $R_6 = R_7$ if preferred values are then obtained. The voltage gain of the driver stage is given by

$$A_v = g_m(R_6 + R_7) = 38.5 I_{C2} \times V_{CC}/2 = 19.25 V_{CC}$$

## Input Stage

The main function of the input stage is to provide the specified input impedance for the amplifier but, of course, it also provides some voltage gain. The voltage gain of the stage is $h_{FE}R_L'/R_{in}$, where $R_L' = R_3 \| h_{ie2}$ and $R_{in} = h_{FE}R_4$ since

$$R_4 \ll R_5. \quad h_{ie2} = h_{fe2} \times (26 \text{ mV}/I_{C2})$$

The collector current $I_{C1}$ of $T_1$ should be chosen to be about 10 times the base current of $T_2$ to give adequate operating point stability. Hence, $R_3 = V_{BE2}/I_{C1} = 0.7/I_{C1}$. $R_5$ must have between 1 and 2 V dropped across it by the collector current of $T_1$, $R_5 = 1/I_{C1}$. The d.c. voltage at the junction of resistors $R_8$ and $R_9$ is $V_{CC}/2$ and hence the emitter voltage of $T_1$ is equal to $(V_{CC} - 1)$ volts. The base voltage of $T_1$ should then be 0.7 V less than this value.

The input impedance of the amplifier is equal to $R_1 \| R_2 \| R_{in,T_1}$, where $R_{in,T_1}$ is the input resistance of $T_1$. $R_{in,T_1}$ is of high value and so the input resistance of the amplifier is determined solely by the bias resistors. $T_1$ should have a high current gain $h_{FE}$ so that it will need only a small base current and this will allow the bias resistors to be of high value. The closed-loop gain of the amplifier is $(R_4 + R_5)/R_4 \simeq R_5/R_4$ and this equation enables $R_4$ to be determined.

Capacitor $C_2$ is merely a d.c. blocking component and it should have a reactance that is no greater than $R_4/2$ at the lowest frequency to be amplified. The low-frequency 3 dB point of the amplifier is determined by the combined contributions of the three capacitors $C_1$, $C_2$ and $C_3$. If the hum and noise at the output of the amplifier are above the specified value they can be reduced by splitting $R_1$ into two and decoupling the junction to earth by a capacitor. The output transistors, and perhaps the driver transistors, will need to be mounted on a heat sink.

### Example 5.2

Design an audio-frequency power amplifier to produce 5 W power across an 8 Ω load resistance with an input impedance of at least 200 kΩ, a bandwidth of 20 Hz to 20 kHz, and an input sensitivity of 0.5 V.

### Solution

From equation (5.7), $V_{CC} = \sqrt{(8 \times 5 \times 8)} = 17.89$ V. Allowing 3 V for the sum of the output transistor's saturation voltages and the voltages dropped across the emitter resistors gives $V_{CC} = 20.89$ V. Adding a little extra voltage to cater for the power supply being unregulated gives $V_{CC} = 22$ V. The peak load current $I_{L(max)} = \sqrt{(2 \times 5/8)} = 1.12$ A.

$$\text{Let } V_{CE(max)} = 22 + 2.2 = 24.2 \text{ V} \simeq 25 \text{ V}$$

For the two output transistors choose the cheap and readily available BD 131/2. These transistors have $V_{CEO(max)} = 45$ V, $V_{CBO(max)} = 70$ V, $h_{FE(min)} = 20$, $P_{d(max)} = 15$ W at 60 °C, $I_{C(max)} = 3$ A and $f_t = 60$ MHz.
The two emitter resistors should each drop at least 0.5 V to ensure adequate thermal stability. The peak collector current of each transistor is 1.12 A and hence

$$R_8 = R_9 = 0.5/1.12 \simeq 0.5 \,\Omega = 0.56 \,\Omega \text{ preferred value}$$

Capacitor $C_4$ should be chosen so that it has negligible reactance at all the frequencies to be amplified; a suitable value is 1000 $\mu$F.
The d.c. voltage at the junction of $R_8$ and $R_9$ is 11 V; hence the base voltage of $T_3$ is $11 + 0.5 + 0.7 = 12.2$ V and the base voltage of $T_4$ is $11 - 0.5 - 0.7 = 9.8$ V.
The peak base current of the output transistors is $1.12/20 = 56$ mA. The collector current of $T_2$ must be larger than this, say 68 mA. Then $T_2$ should have a current rating of at least 75 mA, a breakdown voltage $V_{CEO}$ of 25 V, and a power rating of $(11 \times 0.075) + 10\% = 0.9$ W. An audio driver transistor is the BC 337 which has $I_{c(max)} = 500$ mA, $V_{CEO(max)} = 45$ V, $V_{CBO(max)} = 45$ V, $h_{FE(min)} = 100$, and $P_{d(max)}$ at 70 °C = 8 W.
The total resistance of $R_6$ and $R_7$ in series is

$$(11 - 0.6)/(68 \times 10^{-3}) \simeq 150 \,\Omega \quad \text{so} \quad R_6 = R_7 = 75 \,\Omega$$

The lowest frequency to be amplified is 20 Hz so the time constant of the collector load of $T_2$ should be larger than $1/(2\pi \times 20) \simeq 8$ ms. Hence

$$C_3 \times 75/2 = 8 \times 10^{-3} \quad \text{or} \quad C_3 = 213 \,\mu\text{F} = 220 \,\mu\text{F preferred value}$$

The voltage gain of the Class A driver stage is equal to $19 \times 22 = 418$.

The base current of $T_2$ is $68/100 = 0.68$ mA.

$$R_3 = 0.7/(0.68 \times 10^{-3}) = 1029\,\Omega = 1000\,\Omega \text{ preferred value}$$
$$R_5 = 1/(0.68 \times 10^{-3}) = 1470\,\Omega = 1500\,\Omega \text{ preferred value.}$$

The emitter voltage of $T_1$ is $V_{CC}/2 - I_{CT2}R_5 = 10$ V and so the base voltage must be 9.3 V. Hence

$$9.3 = 22\,R_2/(R_1 + R_2) \quad \text{and} \quad 200 = R_1R_2/(R_1 + R_2)$$

since the input resistance of $T_1$ is high. Suitable values are $R_1 = 470$ k$\Omega$ and $R_2 = 360$ k$\Omega$.

The closed loop gain of the amplifier is equal to $1500/R_4$ and this is to be $11/0.5 = 22$. Hence, $R_4 = 1500/22 = 68\,\Omega$. At 20 Hz, the reactance of $C_2 = 34\,\Omega - 1/(2\pi \times 20C_2)$, or $C_2 \simeq 220\,\mu$F. $C_1$ can now be chosen to set the lower 3 dB frequency of the circuit:

$$X_{C_1} = 200 \times 10^3 = 1/(2\pi \times 20C_1)$$

or

$$C_1 = 40 \text{ nF} = 47 \text{ nF preferred value}$$

The circuit of a higher power amplifier is shown by Fig. 5.12. This uses the output resistors in series with common collectors and a form

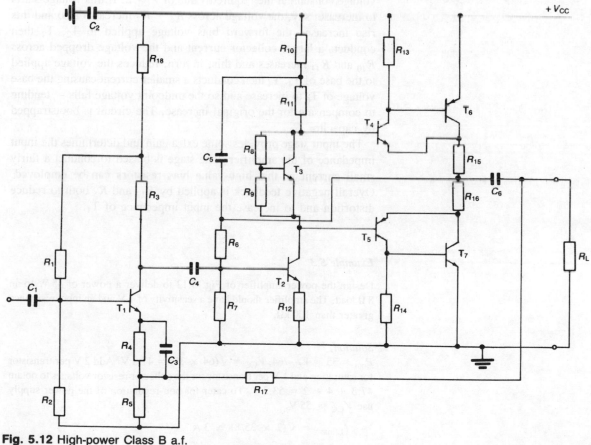

**Fig. 5.12** High-power Class B a.f. amplifier

of the Darlington connection. The output transistors $T_6$ and $T_7$, and the driver transistors $T_4$ and $T_5$, are operated under Class AB conditions. They are biased to have a selected quiescent current by $T_3$ which is connected as a $V_{BE}$ multiplier. The predriver stages, $T_1$ and $T_2$, operate with Class A bias. The small-value collector resistors of the output transistors provide negative feedback to improve linearity and to stabilize the output current. The current gains of the driver and output transistor pairs are multiplied together and hence the base drive to be supplied by $T_2$ is only a few milliamps.

When transistors $T_4$ and $T_6$ are ON, transistors $T_5$ and $T_7$ are OFF and vice versa. When a transistor pair is OFF their base-emitter junctions are reverse-biased and $R_{13}/R_{14}$ provide a path for the base charge to be rapidly removed.

The quiescent current of $T_2$ and $T_3$ must be greater than the peak base current taken by $T_4$ or $T_5$ by at least 20%. The base bias voltage of $T_2$ is obtained from the potential divider $R_6 + R_7$ that is connected between the output midpoint, which is at $V_{CC}/2$ volts, and earth. This midpoint connection applies both d.c. and a.c. negative feedback to the base of $T_2$. The d.c. feedback tends to maintain the midpoint voltage constant at the required value of $V_{CC}/2$. If this voltage starts to increase, say, the voltage across $R_6 + R_7$ increases also and this rise increases the forward bias voltage applied to $T_2$. $T_2$ then conducts a larger collector current and the voltage dropped across $R_{10}$ and $R_{11}$ increases and this, in turn, reduces the voltage applied to the base of $T_4$. $T_4$ then conducts a smaller current causing the base voltage of $T_6$ to increase and so the midpoint voltage falls — tending to compensate for the original increase. The circuit is bootstrapped by capacitor $C_5$.

The input stage provides some extra gain and determines the input impedance of the amplifier. The stage is biased to conduct a fairly small current so that high-value bias resistors can be employed. Overall negative feedback is applied by $R_{17}$ and $R_5$ both to reduce distortion and to increase the input impedance of $T_1$.

### Example 5.3

Design the power amplifier of Fig. 5.12 to deliver a power of 35 W to an 8 Ω load. The amplifier should have a sensitivity of 1 V and an input resistance greater than 100 kΩ.

### Solution

$P_{out} = 35 = V_{CC}^2/64$, $V_{CC} = \sqrt(64 \times 35) = 47.3$ V. Add 2 V per transistor for saturation and 1 V per transistor for the collector resistor voltages to obtain $47.3 + 4 + 2 = 53.2$ V. To cater for non-regulation of the power supply use $V_{CC} = 55$ V.

$$I_{L(max)} = \sqrt(2 \times 35/8) \simeq 3 \text{ A}.$$

Thus the output transistors should have the following ratings: $V_{CEO(max)} = 60$ V, $I_{C(max)} = 3.3$ A and $P_{d(max)} = 14$ W. Suitable transistors are the TIP

33A/34A which have $P_{d(max)} = 80\,\mathrm{W}$ at $25\,°\mathrm{C}$, $V_{CEO(max)} = 60\,\mathrm{V}$, $V_{CBO(max)} = 100\,\mathrm{V}$, $h_{FE(min)} = 20$ and $f_t = 3\,\mathrm{MHz}$. The output characteristics of these transistors show that a collector current of 35 mA is above the 'knee' of the curves and so this current can be chosen to be the (relatively) small quiescent current.

The collector resistors $R_{15}$ and $R_{16}$ are both equal to $0.5/3 = 0.17\,\Omega = 0.18\,\Omega$ preferred value.

The base voltage of transistor $T_6 = 55/2 + 0.5 + 0.7 = 28.7\,\mathrm{V}$ and its base current is to be $3/20 = 0.15\,\mathrm{A}$. The current in $R_{13}$ must be larger than this, say 380 mA. Then

$$R_{13} = R_{14} = (55 - 28.7)/0.38 = 69.2\,\Omega = 68\,\Omega \text{ preferred value}$$

Transistor $T_4$ must have a $V_{CEO}$ rating of at least 32 V and a peak current rating of at least 10% more than $(0.15 + 0.38)\,\mathrm{A} = 0.59\,\mathrm{A}$. The same ratings apply to $T_5$. Suitable devices are the TIP 31A/32A which has $V_{CEO(max)} = 60\,\mathrm{V}$, $V_{CBO(max)} = 60\,\mathrm{V}$, $h_{FE(min)} = 20$, and $P_{d(max)} = 0.8\,\mathrm{W}$.

The overall current gain of the $T_4/T_6$ pair is the product of their individual current gains, i.e. a minimum of 400. Hence, the base current to $T_4$ is $3/400 = 7.5\,\mathrm{mA}$. The quiescent collector current of $T_2$ must be greater than this in order to prevent clipping and making it about 15 mA provides adequate safety margin. Allow 1 V across the emitter resistor $R_{12}$, then

$$R_{12} = 1/(15 \times 10^{-3}) = 67\,\Omega = 68\,\Omega \text{ preferred value}$$

The base voltage of $T_2$ should be about 1.5 V. Then

$$1.5 = (55/2) \times R_7/(R_6 + R_7) \quad \text{or} \quad R_7/(R_6 + R_7) \simeq 0.055$$

A variety of different values could be selected; suitable values are $R_6 = 47\,\mathrm{k}\Omega$ and $R_7 = 2.7\,\mathrm{k}\Omega$.

The $V_{BE}$ multiplier should have about 0.25 times the quiescent current of $T_2$ flowing in the two bias resistors. Then (see Fig. 5.13) the voltage between the bases of $T_4$ and $T_5$ is 2.4 V. The $V_{BE}$ of $T_3$ should be about 0.6 V and so

$$R_9 = 0.6/(3.75 \times 10^{-3}) = 160\,\Omega$$

and

$$R_8 = (2.4 - 0.6)/(3.75 \times 10^{-3}) = 480\,\Omega$$

To adjust for minimum cross-over distortion $R_8$ will be a $1000\,\Omega$ potentiometer.

The output capacitor $C_6$ should be of large value so that negligible signal voltage is dropped across it at the lowest frequency to be amplified. Choose a 2000 μF capacitor.

The time constant $C_5 R_{10} R_{11}/(R_{10} + R_{11})$ must be larger than

$$1/(2\pi \times 10) = 16\,\mathrm{ms}$$

The total resistance of $R_{10}$ and $R_{11}$ in series is equal to the d.c. voltage dropped across them, divided by the quiescent current of $T_2$. The voltage drop is $55 - 28.7 = 26.3\,\mathrm{V}$ and the current is $15 - 7.5 = 7.5\,\mathrm{mA}$. Hence

$$R_{10} + R_{11} = 26.3/(7.5 \times 10^{-3}) = 3506\,\Omega$$

Let $R_{10} = 1500\,\Omega$ and $R_{11} = 1800\,\Omega$. Then

$$16 \times 10^{-3} = C_6 \times 1500 \times 1800/3300$$

or

$$C_6 = 20\,\mu\mathrm{F} = 22\,\mu\mathrm{F} \text{ preferred value.}$$

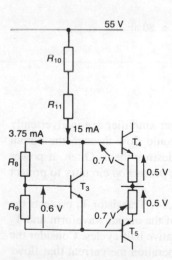

**Fig. 5.13** $V_{BE}$ multiplier

$T_2$ should have a peak current rating of at least 20 mA,

$$V_{\text{CEO(max)}} = (V_{\text{BE5}} - 1) + 10\% = (27.5 - 1.5 - 1) + 10\% \simeq 28 \text{ V}.$$

These requirements can be satisfied by many general-purpose transistors one of which is the 2N 3704. This transistor has rated values of $V_{\text{CEO(max)}}$ = 30 V, $V_{\text{CBO(max)}} = 50$ V, $I_{\text{C(max)}} = 0.8$ A, $P_{\text{d(max)}} = 360$ mW at 25 °C, $h_{\text{FE(min)}} = 100$ and $f_t = 100$ MHz.

Let the quiescent current of the input stage be 1 mA. The power supply to the stage is decoupled to reduce hum and noise, suitable values are $R_{18}$ = 33 k Ω and $C_2 = 47$ μF. The d.c. voltage drop across $R_{18}$ is then 33 V so that the collector supply voltage to $T_1$ is 22 V. The chosen transistor must have $V_{\text{CEO(max)}}$ of at least 25 V and again the 2N 3704 could be used. For an emitter voltage of about 6 V,

$$R_4 + R_5 = 6/(1 \times 10^{-3}) = 6 \text{ k}\Omega$$

If $R_4 = 4.7$ kΩ and $R_5 = 470$ Ω then $V_E = 5.17$ V and then

$$I_{\text{B,T}_1} = (1 \times 10^{-3})/100 = 10 \text{ μA}$$

Let the current in $R_2$ be 40 μA when

$$R_2 = 5.67/(40 \times 10^{-6}) = 142 \text{ k}\Omega = 150 \text{ k}\Omega \text{ preferred value}$$
$$R_1 = (22 - 5.67)/(50 \times 10^{-6}) = 327 \text{ k}\Omega = 330 \text{ k}\Omega \text{ preferred value}$$

The closed-loop gain is then $(27.5/\sqrt{2})/1 = 19.44$ and this is equal to $R_{17}/R_5$. Hence

$$R_{17} = 19.44 \times 470 = 9137 \text{ } \Omega = 10 \text{ k}\Omega \text{ preferred value.}$$

$R_{18}$ can be selected to give a suitable operating point and voltage gain for $T_1$ and a suitable value is 10 kΩ. Lastly, the values of the coupling capacitors $C_1$ and C4 must be determined. $C_4$ should have negligible reactance at the lowest frequency to be amplified and a 22 μF capacitor can be used. $C_1$ should determine the lower 3 dB frequency of the circuit, i.e. 20 Hz. The input resistance of the amplifier is

$$150 \times 330/480 = 103.135 \text{ k}\Omega$$

in parallel with the high resistance of $T_1$, say 100 kΩ. The reactance of $C_1$ at 20 Hz should be

$$X_{C_1} = 1/(2\pi \times 20 \times 100 \times 10^3) \simeq 80 \text{ nF}$$
$$= 100 \text{ nF preferred value.}$$

### Short-circuit Protection

If the output terminals of an a.f. power amplifier are inadvertently short-circuited a very large current could flow through the output transistors which would very likely destroy them. Often, a power amplifier is provided with *short-circuit protection* circuitry to protect it against such an eventuality.

A typical circuit is shown by Fig. 5.14. Transistor $T_1$ protects the amplifier on the positive half-cycles of the output waveform and $T_2$ provides the protection during the negative half-cycles. Consider the positive half-cycle. During normal operation the current that flows in $R_1$ is not large enough for the voltage across it to turn $T_1$ ON. If,

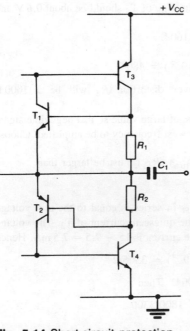

**Fig. 5.14** Short-circuit protection

however, the output terminals should be short-circuited the current in $R_1$ rises rapidly and the voltage across $R_1$ turns $T_1$ ON. A large part of the base current of $T_3$ is now diverted to become the collector current of $T_1$, and this reduces the collector current of $T_3$ below the values at which damage might occur.

### Integrated Circuit Audio-frequency Power Amplifiers

A wide variety of IC power amplifiers are offered by various manufacturers which are capable of providing output powers ranging from 1 W up to 21 W or more. The data sheet for each device gives details of the external components that are needed to operate the device. One commonly employed IC is the LM 380 whose pin connections are given in Fig. 5.15(a). (The data sheet is given in *Electronics IV*.) The IC has an internally set voltage gain of 50 and both inverting and non-inverting terminals. An earth referenced source can be connected to either input terminal with the unused input either (a) left disconnected, or (b) connected to earth directly, via a resistor, or via a capacitor. In most applications in which the non-inverting terminal is used the inverting terminal is left disconnected. When the inverting terminal is used the circuit may become unstable if the non-inverting terminal is left disconnected. The non-inverting terminal should be connected to earth directly if the source is of low resistance, via a resistor if the source resistance is of medium value, and via a capacitor if the source resistance is high. The $V_{CC}$ supply line should be decoupled with a 0.1 $\mu$F capacitor. To prevent oscillation at a frequency in the region of 5–10 MHz when the amplifier is supplying a high current to a load, a $CR$ series circuit should be connected between the output terminal and earth. Typically, $R_1 = 2.7 \, \Omega$ and $C_1 = 0.1 \, \mu$F. An output coupling capacitor of about 68 $\mu$F is necessary when the load resistance is 8 $\Omega$. The circuit is shown by Fig. 5.15(b).

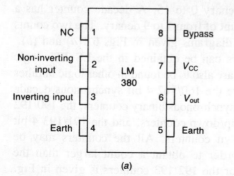

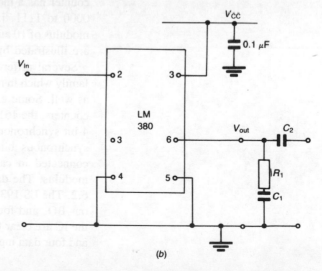

**Fig. 5.15** (a) LM 380 pin connections; (b) LM 380 as an a.f. power amplifier

# 6 MSI/LSI/VLSI Devices

A wide variety of digital circuits can be obtained in medium-scale integration (MSI), large-scale integration (LSI), and very large-scale integration (VSLI) packages. The vast majority of digital circuit applications can be carried out using one or more of these integrated circuits (ICs), but small-scale integration (SSI) devices are no longer employed for other than 'glue' logic, interfacing, and delay purposes; MSI circuits include counters, decoders, encoders, multiplexers, etc.; LSI devices include memories, display drivers and keyboard encoders, while VSLI circuits include large-capacity memories, microprocessors, floppy disc controllers and various peripheral devices such as the asynchronous communications interface adaptor (ACIA). Microprocessors and their peripheral circuits are beyond the scope of this book.

Many devices are available in more than one of the logic families. For, example, the 193 synchronous 4-bit binary up-down counter can be obtained in standard, LS, ALS and F TTL families and also in the HC, AC and ACT high-speed CMOS families.

**Counters**

The *modulus*, or *scale*, of a counter is the number of different states that can exist within the counter. A binary counter has a modulus of $2^n$, where $n$ is the number of stages. This means that a four-stage counter has a modulus of $2^4 = 16$ and hence its count will be from 0000 to 1111, i.e. from denary 0 to 15. A decade counter has a modulus of 10 and has a count of from 1 to 9 denary. The two counts are illustrated by the state diagrams given in Figs 6.1(a) and (b).

Several different counters can be obtained in the LS TTL logic family which in many cases are also to be found in other logic families as well. Some examples are the 160/162 4-bit synchronous decade counters, the 161/163 4-bit synchronous binary counters, the 190/192 4-bit synchronous decade up/down counters, and the 191/193 4-bit synchronous binary up/down counters. All the counters may be connected in cascade in order to obtain a count larger than the modulus. The data sheet for the 192/193 counters is given in Fig. 6.2. The LS 193 IC has six outputs; these are carry out $\overline{CO}$, borrow out $\overline{BO}$, and four stage outputs $Q_A$, $Q_B$, $Q_C$ and $Q_D$. The inputs to the IC are clear CLR, load $\overline{LOAD}$, count-up UP, count-down DOWN, and four data inputs A, B, C and D. The non-synchronous CLR input

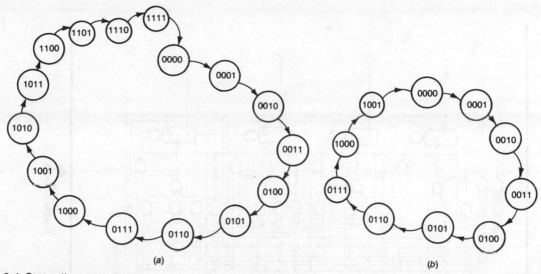

**Fig. 6.1** State diagrams for (a) a 4-bit binary counter and (b) a decade counter

resets the counter to 0000 whenever it is taken HIGH. The $\overline{\text{LOAD}}$ input will load the data present at the four data inputs when it is taken LOW. The direction of the count is determined by the voltages applied to the UP and DOWN inputs; if the UP pin is held HIGH and the clock is applied to the DOWN pin the circuit will count down, and vice versa. The carry out $\overline{\text{CO}}$ pin goes LOW only when the count is 1111 and the UP pin is LOW. Conversely, the borrow out $\overline{\text{BO}}$ pin goes LOW when the count is 0000 and the DOWN line is LOW. The $\overline{\text{CO}}$ and $\overline{\text{BO}}$ outputs are used when counters are connected in cascade.

The 193 counter has a modulus of 16 but this may be reduced to a lesser figure by the use of the non-synchronous control inputs. The timing diagram given in the data sheet shows how the counter may be operated in a count sequence in which it starts counting up from an initial value of 13, to follow the sequence 13, 14, 15, 0, 1, 2, and then it switches to count down through the sequence 1, 0, 15, 14 and 13. When the count passes through 15 to 0 a LOW pulse appears at the $\overline{\text{CO}}$ output; similarly, when the count passes from 0 to 15 a LOW pulse appears at the $\overline{\text{BO}}$ output.

### Example 6.1

Use an LS 193 to design a modulo 6 counter that will follow the count sequence 0, 1, 2, 7, 8, 9, 0, 1, etc.

### Solution

The required count sequence is shown in Table 6.1. The required circuit is shown by Fig. 6.3

Integrated circuit counters may be connected in cascade in order to increase their modulus and hence to increase the division ratio.

TYPES SN54192, SN54193, SN54LS192, SN54LS193
SN74192, SN74193, SN74LS192, SN74LS193
SYNCHRONOUS 4-BIT UP/DOWN COUNTERS (DUAL CLOCK WITH CLEAR)

DECEMBER 1972—REVISED DECEMBER 1983

- Cascading Circuitry Provided Internally
- Synchronous Operation
- Individual Preset to Each Flip-Flop
- Fully Independent Clear Input

| TYPES | TYPICAL MAXIMUM COUNT FREQUENCY | TYPICAL POWER DISSIPATION |
|---|---|---|
| '192, '193 | 32 MHz | 325 mW |
| 'LS192, 'LS193 | 32 MHz | 95 mW |

**description**

These monolithic circuits are synchronous reversible (up/down) counters having a complexity of 55 equivalent gates. The '192 and 'LS192 circuits are BCD counters and the '193 and 'LS193 are 4-bit binary counters. Synchronous operation is provided by having all flip-flops clocked simultaneously so that the outputs change coincidently with each other when so instructed by the steering logic. This mode of operation eliminates the output counting spikes which are normally associated with asynchronous (ripple-clock) counters.

The outputs of the four master-slave flip-flops are triggered by a low-to-high-level transition of either count (clock) input. The direction of counting is determined by which count input is pulsed while the other count input is high.

All four counters are fully programmable; that is, each output may be preset to either level by entering the desired data at the data inputs while the load input is low. The output will change to agree with the data inputs independently of the count pulses. This feature allows the counters to be used as modulo-N dividers by simply modifying the count length with the preset inputs.

A clear input has been provided which forces all outputs to the low level when a high level is applied. The clear function is independent of the count and load inputs. The clear, count, and load inputs are buffered to lower the drive requirements.

These counters were designed to be cascaded without the need for external circuitry. Both borrow and carry outputs are available to cascade both the up- and down-counting functions. The borrow output produces a pulse equal in width to the count-down input when the counter underflows. Similarly, the carry output produces a pulse equal in width to the count-up input when an overflow condition exists. The counters can then be easily cascaded by feeding the borrow and carry outputs to the count-down and count-up inputs respectively of the succeeding counter.

**SN54192, SN54193, SN54LS192, SN54LS193 ... J OR W PACKAGE**
**SN74192, SN74193 ... J OR N PACKAGE**
**SN74LS192, SN74LS193 ... D, J OR N PACKAGE**
(TOP VIEW)

```
        ┌──∪──┐
  B  [ 1    16 ] VCC
 QB  [ 2    15 ] A
 QA  [ 3    14 ] CLR
DOWN [ 4    13 ] BO
  UP [ 5    12 ] CO
 QC  [ 6    11 ] LOAD
 QD  [ 7    10 ] C
 GND [ 8     9 ] D
        └─────┘
```

**SN54LS192, SN54LS193**
**SN74LS192, SN74LS193 ... FK PACKAGE**
(TOP VIEW)

```
          QA  NC  CLR
          3   2   1  20 19
     QA [ 4              18 ] CLR
   DOWN [ 5              17 ] BO
     NC [ 6              16 ] NC
     UP [ 7              15 ] CO
     QC [ 8              14 ] LOAD
          9  10 11 12 13
          QC  GND NC  C  D
```

NC - No internal connection

**logic diagram**

Pin numbers shown on logic notation are for D, J or N packages.

**absolute maximum ratings over operating free-air temperature range (unless otherwise noted)**

| | SN54' | SN54LS' | SN74' | SN74LS' | UNIT |
|---|---|---|---|---|---|
| Supply voltage, $V_{CC}$ (see Note 1) | 7 | 7 | 7 | 7 | V |
| Input voltage | 5.5 | | 5.5 | | V |
| | | 7 | | 7 | V |
| Operating free-air temperature range | −55 to 125 | −55 to 125 | 0 to 70 | 0 to 70 | °C |
| Storage temperature range | −65 to 150 | −65 to 150 | −65 to 150 | −65 to 150 | °C |

NOTE 1: Voltage values are with respect to network ground terminal.

TEXAS INSTRUMENTS

**Fig. 6.2** Data sheet for the 192/3 counters (courtesy of Texas Instruments)

**logic symbols**

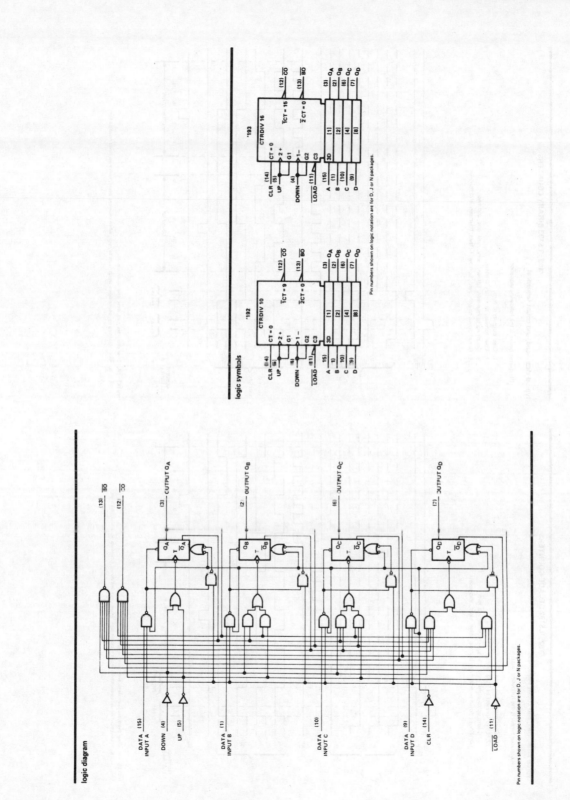

Pin numbers shown on logic notation are for D, J or N packages.

**logic diagram**

Pin numbers shown on logic notation are for D, J or N packages.

## '192, 'LS192 DECADE COUNTERS

**typical clear, load, and count sequences**

Illustrated below is the following sequence:

1. Clear outputs to zero.
2. Load (preset) to BCD seven.
3. Count up to eight, nine, carry, zero, one, and two.
4. Count down to one, zero, borrow, nine, eight, and seven.

## '193, 'LS193 BINARY COUNTERS

**typical clear, load, and count sequences**

Illustrated below is the following sequence:

1. Clear outputs to zero.
2. Load (preset) to binary thirteen.
3. Count up to fourteen, fifteen, carry, zero, one, and two.
4. Count down to one, zero, borrow, fifteen, fourteen, and thirteen.

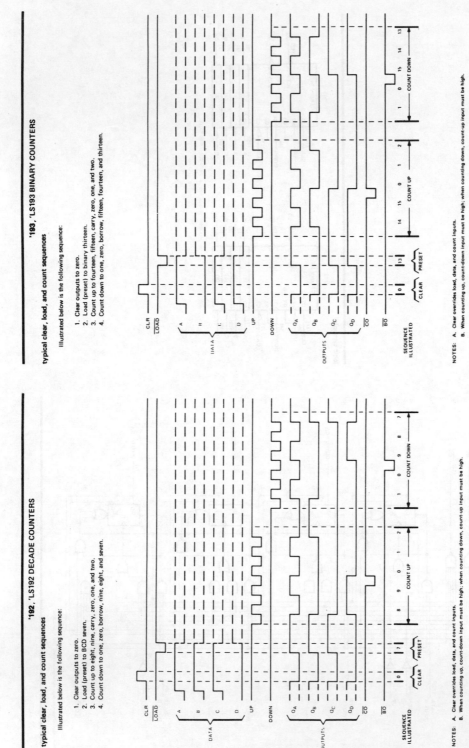

## recommended operating conditions

| | | SN54LS192 SN54LS193 | | | SN74LS192 SN74LS193 | | | UNIT |
|---|---|---|---|---|---|---|---|---|
| | | MIN | NOM | MAX | MIN | NOM | MAX | |
| $V_{CC}$ | Supply voltage | 4.5 | 5 | 5.5 | 4.75 | 5 | 5.25 | V |
| $I_{OH}$ | High-level output current | | | −400 | | | −400 | µA |
| $I_{OL}$ | Low-level output current | | | 4 | | | 8 | mA |
| $f_{clock}$ | Clock frequency | 0 | | 25 | 0 | | 25 | MHz |
| $t_w$ | Width of any input pulse | 20 | | | 20 | | | ns |
| $t_{su}$ | Clear inactive-state setup time | 15 | | | 15 | | | ns |
| | Load inactive-state setup time | 15 | | | 15 | | | ns |
| | Data setup time (see Figure 1) | 20 | | | 20 | | | ns |
| $t_h$ | Data hold time | 5 | | | 5 | | | ns |
| $T_A$ | Operating free-air temperature range | −55 | | 125 | 0 | | 70 | °C |

## electrical characteristics over recommended operating free-air temperature range (unless otherwise noted)

| PARAMETER | | TEST CONDITIONS | | SN54LS192 SN54LS193 | | | SN74LS192 SN74LS193 | | | UNIT |
|---|---|---|---|---|---|---|---|---|---|---|
| | | | | MIN | TYP‡ | MAX | MIN | TYP‡ | MAX | |
| $V_{IH}$ | High-level input voltage | | | 2 | | | 2 | | | V |
| $V_{IL}$ | Low-level input voltage | | | | | 0.7 | | | 0.8 | V |
| $V_{IK}$ | Input clamp voltage | $V_{CC}$ = MIN, | $I_I$ = −18 mA | | | −1.5 | | | −1.5 | V |
| $V_{OH}$ | High-level output voltage | $V_{CC}$ = MIN, $V_{IH}$ = 2 V, $V_{IL}$ = $V_{IL}$ max, $I_{OH}$ = −400 µA | | 2.5 | 3.4 | | 2.7 | 3.4 | | V |
| $V_{OL}$ | Low-level output voltage | $V_{CC}$ = MIN, $V_{IH}$ = 2 V, $V_{IL}$ = $V_{IL}$ max | $I_{OL}$ = 4 mA | | 0.25 | 0.4 | | 0.15 | 0.4 | V |
| | | | $I_{OL}$ = 8 mA | | 0.35 | 0.5 | | | | V |
| $I_I$ | Input current at maximum input voltage | $V_{CC}$ = MAX, | $V_I$ = 7 V | | | 0.1 | | | 0.1 | mA |
| $I_{IH}$ | High-level input current | $V_{CC}$ = MAX, | $V_I$ = 2.7 V | | | 20 | | | 20 | µA |
| $I_{IL}$ | Low-level input current | $V_{CC}$ = MAX, | $V_I$ = 0.4 V | | | −0.4 | | | −0.4 | mA |
| $I_{OS}$ | Short-circuit output current§ | $V_{CC}$ = MAX | | −20 | | −100 | −20 | | −100 | mA |
| $I_{CC}$ | Supply current | $V_{CC}$ = MAX, See Note 2 | | | 19 | 34 | | 19 | 34 | mA |

† For conditions shown as MIN or MAX, use the appropriate value specified under recommended operating conditions for the applicable type.
‡ All typical values are at $V_{CC}$ = 5 V, $T_A$ = 25°C.
§ Not more than one output should be shorted at a time, and duration of the short-circuit should not exceed one second.
NOTE 2: $I_{CC}$ is measured with all inputs open, and load inputs grounded, and all other inputs at 4.5 V.

## switching characteristics, $V_{CC}$ = 5 V, $T_A$ = 25°C

| PARAMETER | FROM INPUT | TO OUTPUT | TEST CONDITIONS | MIN | TYP | MAX | UNIT |
|---|---|---|---|---|---|---|---|
| $f_{max}$ | | | | 25 | 32 | | MHz |
| $t_{PLH}$ | UP | CO | | | 17 | 26 | ns |
| $t_{PHL}$ | | | | | 18 | 24 | ns |
| $t_{PLH}$ | DOWN | BO | $C_L$ = 15 pF, $R_L$ = 2 kΩ, See Figures 1 and 2 | | 16 | 24 | ns |
| $t_{PHL}$ | | | | | 15 | 24 | ns |
| $t_{PLH}$ | UP OR DOWN | Q | | | 27 | 38 | ns |
| $t_{PHL}$ | | | | | 30 | 47 | ns |
| $t_{PLH}$ | LOAD | Q | | | 24 | 40 | ns |
| $t_{PHL}$ | | | | | 25 | 40 | ns |
| $t_{PHL}$ | CLR | Q | | | 23 | 35 | ns |

---

## recommended operating conditions

| | | SN54192 SN54193 | | | SN74192 SN74193 | | | UNIT |
|---|---|---|---|---|---|---|---|---|
| | | MIN | NOM | MAX | MIN | NOM | MAX | |
| $V_{CC}$ | Supply voltage | 4.5 | 5 | 5.5 | 4.75 | 5 | 5.25 | V |
| $I_{OH}$ | High-level output current | | | −0.4 | | | −0.4 | mA |
| $I_{OL}$ | Low-level output current | | | 16 | | | 16 | mA |
| $f_{clock}$ | Clock frequency | 0 | | 25 | 0 | | 25 | MHz |
| $t_w$ | Width of any input pulse | 20 | | | 20 | | | ns |
| $t_{su}$ | Data setup time, (see Figure 1) | 20 | | | 20 | | | ns |
| $t_h$ | Hold time | 0 | | | 0 | | | ns |
| $T_A$ | Operating free-air temperature | −55 | | 125 | 0 | | 70 | °C |

## electrical characteristics over recommended operating free-air temperature range (unless otherwise noted)

| PARAMETER | | TEST CONDITIONS | | SN54192 SN54193 | | | SN74192 SN74193 | | | UNIT |
|---|---|---|---|---|---|---|---|---|---|---|
| | | | | MIN | TYP‡ | MAX | MIN | TYP‡ | MAX | |
| $V_{IH}$ | High-level input voltage | | | 2 | | | 2 | | | V |
| $V_{IL}$ | Low-level input voltage | | | | | 0.8 | | | 0.8 | V |
| $V_{IK}$ | Input clamp voltage | $V_{CC}$ = MIN, | $I_I$ = −12 mA | | | −1.5 | | | −1.5 | V |
| $V_{OH}$ | High-level output voltage | $V_{CC}$ = MIN, $V_{IH}$ = 2 V, $V_{IL}$ = 0.8 V, $I_{OH}$ = −0.4 mA | | 2.4 | 3.4 | | 2.4 | 3.4 | | V |
| $V_{OL}$ | Low-level output voltage | $V_{CC}$ = MIN, $V_{IH}$ = 2 V, $V_{IL}$ = 0.8 V, $I_{OL}$ = 16 mA | | | 0.2 | 0.4 | | 0.2 | 0.4 | V |
| $I_I$ | Input current at maximum input voltage | $V_{CC}$ = MAX, | $V_I$ = 5.5 V | | | 1 | | | 1 | mA |
| $I_{IH}$ | High-level input current | $V_{CC}$ = MAX, | $V_I$ = 2.4 V | | | 40 | | | 40 | µA |
| $I_{IL}$ | Low-level input current | $V_{CC}$ = MAX, | $V_I$ = 0.4 V | | | −1.6 | | | −1.6 | mA |
| $I_{OS}$ | Short-circuit output current§ | $V_{CC}$ = MAX | | −20 | | −65 | −18 | | −65 | mA |
| $I_{CC}$ | Supply current | $V_{CC}$ = MAX, See Note 2 | | | 65 | 89 | | 65 | 102 | mA |

† For conditions shown as MIN or MAX, use the appropriate value specified under recommended operating conditions for the applicable type.
‡ All typical values are at $V_{CC}$ = 5 V, $T_A$ = 25°C.
§ Not more than one output should be shorted at a time.
NOTE 2: $I_{CC}$ is measured with all outputs open, clear and load inputs grounded, and all other inputs at 4.5 V.

## switching characteristics, $V_{CC}$ = 5 V, $T_A$ = 25°C

| PARAMETER¶ | FROM INPUT | TO OUTPUT | TEST CONDITIONS | MIN | TYP | MAX | UNIT |
|---|---|---|---|---|---|---|---|
| $f_{max}$ | | | | 25 | 32 | | MHz |
| $t_{PLH}$ | UP | CO | | | 17 | 26 | ns |
| $t_{PHL}$ | | | | | 16 | 24 | ns |
| $t_{PLH}$ | DOWN | BO | $C_L$ = 15 pF, $R_L$ = 400 Ω, See Figures 1 and 2 | | 16 | 24 | ns |
| $t_{PHL}$ | | | | | 25 | 38 | ns |
| $t_{PLH}$ | UP OR DOWN | Q | | | 31 | 47 | ns |
| $t_{PHL}$ | | | | | 27 | 40 | ns |
| $t_{PLH}$ | LOAD | Q | | | 29 | 40 | ns |
| $t_{PHL}$ | CLR | Q | | | 22 | 35 | ns |

¶ $f_{max}$ maximum clock frequency
$t_{PLH}$ propagation delay time, low to high level output
$t_{PHL}$ propagation delay time, high to low level output

**Table 6.1**

| Data inputs | | | | UP clock pulse | Outputs | | | | Denary value |
|---|---|---|---|---|---|---|---|---|---|
| D | C | B | A | | $Q_D$ | $Q_C$ | $Q_B$ | $Q_A$ | |
| 0 | 0 | 0 | 0 | 0 | 0 | 0 | 0 | 0 | 0 |
| | | | | 1 | 0 | 0 | 0 | 1 | 1 |
| | | | | 2 | 0 | 0 | 1 | 0 | 2 |
| | | | | 3 | 0 | 0 | 1 | 1 | ⊓ Decode glitch to load 0111 (7) |
| 0 | 1 | 1 | 1 | 4 | 0 | 1 | 1 | 1 | 7 |
| | | | | 5 | 1 | 0 | 0 | 0 | 8 |
| | | | | 6 | 1 | 0 | 0 | 1 | 9 |
| | | | | 7 | 1 | 0 | 1 | 0 | ⊓ Decode to clear counter |

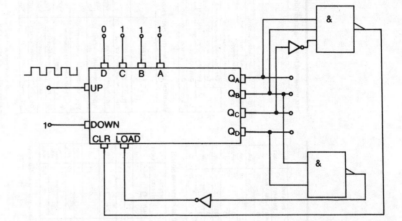

Fig. 6.3

For non-synchronous counters, such as the LS 90, the $Q_D$ output of one counter is connected to the clock input of the next counter in the cascade. When synchronous counters are to be cascaded the carry output of one stage is connected to the count input of the next counter. In the case of an up/down counter, like the LS 193, cascaded up-count is obtained by connecting the $\overline{CO}$ output of one counter to the UP input of the next counter; for cascaded down-count the $\overline{BO}$ output of one counter is connected to the DOWN input of the next counter.

Binary counters have a modulus of 16 so that two binary counters in cascade have a modulus of $16^2 = 256$, three counters in cascade have a modulus of $16^3 = 4096$, and so on. If the required count is not equal to one of these values then the overall modulus must be reduced.

There are two ways in which the count may be reduced:

(a) Decode the desired count and connect the decoder output to the CLR input. The count starts from 0000 and when it reaches the wanted value the decoded signal will clear the counter for the count to be repeated. Suppose that a count of 153 is wanted. This will require the connection of two 193 up-down counters in

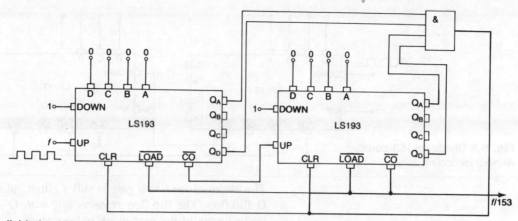

**Fig. 6.4** Basic divide-by-153 counter

cascade. A count of 153 means the counter is to count from 0 to $152_{10}$; $153_{10} = 99H = 1001\ 1001$ so that the $Q_A$ and $Q_D$ outputs of both counters are to be decoded to clear the counters. The circuit is shown by Fig. 6.4. The operation of the circuit may be somewhat unpredictable since the stages of the two counters may clear with different speeds. If any stage, say $Q_D$ in the second counter, is slower to clear than the other three stages then the output of the AND gate, and hence the CLR line, will go LOW before the slowest stage has cleared. It will then not clear at all and the counter will then be in the state 1000 0000 or 80H instead of 00H.

(b) To overcome this problem the CLR line must be held HIGH for a sufficiently long time to ensure that all stages in both counters are cleared and this means that the CLR pulse must be lengthened. To accomplish this an edge-triggered D flip-flop can be employed. The LS74 D flip-flop has its SET input active LOW and hence a NAND gate must be employed, instead of the AND gate, to decode the outputs. The improved circuit is shown in Fig. 6.5.

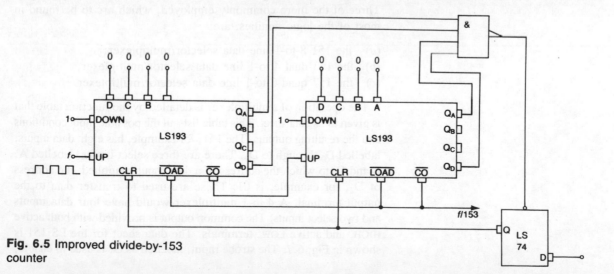

**Fig. 6.5** Improved divide-by-153 counter

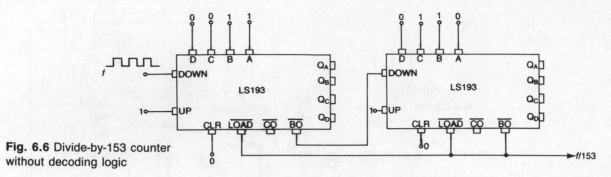

**Fig. 6.6** Divide-by-153 counter without decoding logic

The output of the NAND gate is still a glitch but now it sets the D flip-flop. The flip-flop remains set, with Q = 1, until the leading edge of the next clock pulse occurs.

The disadvantage of cascading up counters is the external logic circuitry that is required. A division ratio of 153 can be obtained without the use of any external logic if instead the 193 is operated as a down counter. When used in this way, the counter must be preset to an initial value of $16^2 - 153 = 103$; $103_{10} = 76H = 0110\ 0011$ and hence the most significant counter should be preset to 0110 and least significant counter should be preset to 0011. The circuit is shown by Fig. 6.6.

**Multiplexers/Data Selectors**

A *multiplexer*, or *data selector*, is a circuit that is able to accept data from any one of a number of sources and then output the data on to a single line. The circuit may be employed to multiplex data, to select data, or to generate logic functions. A multiplexer has both data and select input terminals and a single output terminal. The data that is to be routed to the single output terminal is applied to the data input terminals each of which has its own unique address. The address applied to the input select terminals addresses the multiplexer to determine which of the data inputs is switched to the common output.

A number of multiplexer circuits are available in MSI packages. Three of the more commonly employed, which are to be found in most of the logic families, are:

(a)   the 151 8-to-1 line data selector/multiplexer;
(b)   the 153 dual 4-to-1 line data selector/multiplexer;
(c)   the 157 quad 2-to-1 line data selector/multiplexer.

The operation of a multiplexer is detailed by the function table that is given in its data sheet. This table lists all the possible input conditions and the resulting outputs. The 151, for example, has eight data inputs, labelled $D_0$ through to $D_7$. There are three select inputs, labelled A, B and C, to which the address of a data input is applied; the address of $D_2$, for example, is 01. These are used to transfer data to the output terminal. A 4-to-1 multiplexer would have four data inputs and two select inputs. The common output is provided with both active HIGH, and active LOW, terminals. The data sheet for the LS 151 is shown in Fig. 6.7. The strobe input, labelled as $\overline{G}$, is used to enable,

## TYPES SN54150, SN54151A, SN54152A, SN54LS151, SN54LS152, SN54S151, SN74150, SN74151A, SN74LS151, SN74S151
### DATA SELECTORS/MULTIPLEXERS

SN54150 . . . J OR W PACKAGE
SN74150 . . . J OR N PACKAGE

DECEMBER 1972—REVISED DECEMBER 1983

- '150 Selects One-of-Sixteen Data Sources
- Others Select One-of-Eight Data Sources
- Performs Parallel-to-Serial Conversion
- Permits Multiplexing from N Lines to One Line
- Also For Use as Boolean Function Generator
- Input-Clamping Diodes Simplify System Design
- Fully Compatible with Most TTL Circuits

| TYPE | TYPICAL AVERAGE PROPAGATION DELAY TIME DATA INPUT TO W OUTPUT | TYPICAL POWER DISSIPATION |
|---|---|---|
| '150 | 13 ns | 200 mW |
| '151A | 8 ns | 145 mW |
| '152A | 8 ns | 130 mW |
| 'LS151 | 13 ns | 30 mW |
| 'LS152 | 13 ns | 28 mW |
| 'S151 | 4.5 ns | 225 mW |

### description

These monolithic data selectors/multiplexers contain full on-chip binary decoding to select the desired data source. The '150 selects one-of-sixteen data sources; the '151A, '152A, 'LS151, 'LS152, and 'S151 select one-of-eight data sources. The '150, '151A, 'LS151, and 'S151 have a strobe input which must be at a low logic level to enable these devices. A high level at the strobe forces the W output high, and the Y output (as applicable) low.

The '151A, 'LS151, and 'S151 feature complementary W and Y outputs whereas the '150, '152A, and 'LS152 have an inverted (W) output only.

The '151A and '152A incorporate address buffers which have symmetrical propagation delay times through the complementary paths. This reduces the possibility of transients occurring at the output(s) due to changes made at the select inputs, even when the '151A outputs are enabled (i.e., strobe low).

### logic

#### '150
**FUNCTION TABLE**

| INPUTS | | | | STROBE | OUTPUT |
|---|---|---|---|---|---|
| SELECT | | | | | |
| D | C | B | A | G | W |
| X | X | X | X | H | H |
| L | L | L | L | L | E0 |
| L | L | L | H | L | E1 |
| L | L | H | L | L | E2 |
| L | L | H | H | L | E3 |
| L | H | L | L | L | E4 |
| L | H | L | H | L | E5 |
| L | H | H | L | L | E6 |
| L | H | H | H | L | E7 |
| H | L | L | L | L | E8 |
| H | L | L | H | L | E9 |
| H | L | H | L | L | E10 |
| H | L | H | H | L | E11 |
| H | H | L | L | L | E12 |
| H | H | L | H | L | E13 |
| H | H | H | L | L | E14 |
| H | H | H | H | L | E15 |

#### '151A, 'LS151, 'S151
**FUNCTION TABLE**

| INPUTS | | | | OUTPUTS | |
|---|---|---|---|---|---|
| SELECT | | STROBE | | | |
| C | B | A | G | Y | W |
| X | X | X | H | L | H |
| L | L | L | L | D0 | D̄0 |
| L | L | H | L | D1 | D̄1 |
| L | H | L | L | D2 | D̄2 |
| L | H | H | L | D3 | D̄3 |
| H | L | L | L | D4 | D̄4 |
| H | L | H | L | D5 | D̄5 |
| H | H | L | L | D6 | D̄6 |
| H | H | H | L | D7 | D̄7 |

#### '152A, 'LS152
**FUNCTION TABLE**

| SELECT INPUTS | | | OUTPUT |
|---|---|---|---|
| C | B | A | W |
| L | L | L | D̄0 |
| L | L | H | D̄1 |
| L | H | L | D̄2 |
| L | H | H | D̄3 |
| H | L | L | D̄4 |
| H | L | H | D̄5 |
| H | H | L | D̄6 |
| H | H | H | D̄7 |

H = high level, L = low level, X = irrelevant
E0, E1 . . . E15 = the complement of the level of the respective E input
D0, D1 . . . D7 = the level of the D respective input

SN54151A, SN54LS151, SN54S151 . . . J OR W PACKAGE
SN74151A, SN74S151 . . . D, J OR N PACKAGE
(TOP VIEW)

SN54LS151, SN54S151 . . . FK PACKAGE
SN74LS151, SN74S151 . . .
(TOP VIEW)

NC - No internal connection

SN54152A, SN54152 . . . W PACKAGE
(TOP VIEW)

For SN54S132 Chip Carrier Information, Contact The Factory.

TEXAS
INSTRUMENTS

**Fig. 6.7** Data sheet for the 151 data selector/multiplexer (*courtesy of Texas Instruments*)

## recommended operating conditions

| | SN54' MIN | SN54' NOM | SN54' MAX | SN74' MIN | SN74' NOM | SN74' MAX | UNIT |
|---|---|---|---|---|---|---|---|
| Supply voltage, $V_{CC}$ | 4.5 | 5 | 5.5 | 4.75 | 5 | 5.25 | V |
| High-level output current, $I_{OH}$ | | | -800 | | | -800 | µA |
| Low-level output current, $I_{OL}$ | | | 16 | | | 16 | mA |
| Operating free-air temperature, $T_A$ | -55 | | 125 | 0 | | 70 | °C |

## electrical characteristics over recommended operating free-air temperature range (unless otherwise noted)

| PARAMETER | TEST CONDITIONS† | '150 MIN | '150 TYP‡ | '150 MAX | '151A, '152A MIN | '151A, '152A TYP‡ | '151A, '152A MAX | UNIT |
|---|---|---|---|---|---|---|---|---|
| $V_{IH}$ High-level input voltage | | 2 | | | 2 | | | V |
| $V_{IL}$ Low-level input voltage | | | | 0.8 | | | 0.8 | V |
| $V_{IK}$ Input clamp voltage | $V_{CC}$ = MIN, $I_I$ = -8 mA | | | -1.5 | | | | V |
| $V_{OH}$ High-level output voltage | $V_{CC}$ = MIN, $V_{IL}$ = 0.8 V, $I_{OH}$ = -800 µA | 2.4 | 3.4 | | 2.4 | 3.4 | | V |
| $V_{OL}$ Low-level output voltage | $V_{CC}$ = MIN, $V_{IH}$ = 2 V, $V_{IL}$ = 0.8 V, $I_{OL}$ = 16 mA | | 0.2 | 0.4 | | 0.2 | 0.4 | V |
| $I_I$ Input current at maximum input voltage | $V_{CC}$ = MAX, $V_I$ = 5.5 V | | | 1 | | | 1 | mA |
| $I_{IH}$ High-level input current | $V_{CC}$ = MAX, $V_I$ = 2.4 V | | | 40 | | | 40 | µA |
| $I_{IL}$ Low-level input current | $V_{CC}$ = MAX, $V_I$ = 0.4 V | | | -1.6 | | | -1.6 | mA |
| $I_{OS}$ Short-circuit output current§ | $V_{CC}$ = MAX  SN54' | -20 | | -55 | -20 | | -55 | mA |
| | SN74'  '151A / '152A | -18 | | -55 | -18 | | -55 | |
| $I_{CC}$ Supply current | See Note 3  '150 | | 40 | 68 | | | | mA |
| | '151A | | | | | 29 | 48 | |
| | '152A | | | | | 26 | 43 | |

† For conditions shown as MIN or MAX, use the appropriate value specified under recommended operating conditions for the applicable device type.
‡ All typical values are at $V_{CC}$ = 5 V, $T_A$ = 25°C.
§ Not more than one output should be shorted at a time.
NOTE 3: $I_{CC}$ is measured with the strobe and data select inputs at 4.5 V, all other inputs and outputs open.

## switching characteristics, $V_{CC}$ = 5 V, $T_A$ = 25°C

| PARAMETER¶ | FROM (INPUT) | TO (OUTPUT) | TEST CONDITIONS | '150 MIN | '150 TYP | '150 MAX | '151A, '152A MIN | '151A, '152A TYP | '151A, '152A MAX | UNIT |
|---|---|---|---|---|---|---|---|---|---|---|
| $t_{PLH}$ | A, B, or C (4 levels) | Y | $C_L$ = 15 pF, $R_L$ = 400 Ω, See Note 4 | | 23 | 35 | | 25 | 38 | ns |
| $t_{PHL}$ | | | | | 22 | 33 | | 25 | 38 | ns |
| $t_{PLH}$ | A, B, C, or D (3 levels) | W | | | | | | 17 | 26 | ns |
| $t_{PHL}$ | | | | | | | | 19 | 30 | ns |
| $t_{PLH}$ | Strobe $\overline{G}$ | Y | | | 15.5 | 24 | | 21 | 33 | ns |
| $t_{PHL}$ | | | | | 21 | 30 | | 22 | 33 | ns |
| $t_{PLH}$ | Strobe $\overline{G}$ | W | | | | | | 14 | 21 | ns |
| $t_{PHL}$ | | | | | | | | 15 | 23 | ns |
| $t_{PLH}$ | D0 thru D7 | Y | | | 8.5 | 14 | | 13 | 20 | ns |
| $t_{PHL}$ | | | | | 13 | 20 | | 18 | 27 | ns |
| $t_{PLH}$ | E0 thru E15, or D0 thru D7 | W | | | | | | 8 | 14 | ns |
| $t_{PHL}$ | | | | | | | | 8 | 14 | ns |

¶ $t_{PLH}$ ≡ propagation delay time, low-to-high-level output
$t_{PHL}$ ≡ propagation delay time, high-to-low-level output
NOTE 4: See General Information Section for load circuits and voltage waveforms.

## logic diagrams

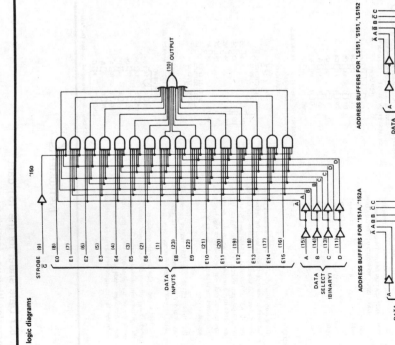

ADDRESS BUFFERS FOR '151A, '152A

ADDRESS BUFFERS FOR 'LS151, 'S151, 'LS152

Pin numbers shown on logic notation are for D, J or N packages.

## recommended operating conditions

| PARAMETER | SN54LS' MIN | NOM | MAX | SN74LS' MIN | NOM | MAX | UNIT |
|---|---|---|---|---|---|---|---|
| Supply voltage, $V_{CC}$ | 4.5 | 5 | 5.5 | 4.75 | 5 | 5.25 | V |
| High-level output current, $I_{OH}$ | | | -400 | | | -400 | µA |
| Low-level output current, $I_{OL}$ | | | 4 | | | 8 | mA |
| Operating free-air temperature, $T_A$ | -55 | | 125 | 0 | | 70 | °C |

## electrical characteristics over recommended operating free-air temperature range (unless otherwise noted)

| PARAMETER | TEST CONDITIONS† | SN54LS' MIN | TYP‡ | MAX | SN74LS' MIN | TYP‡ | MAX | UNIT |
|---|---|---|---|---|---|---|---|---|
| $V_{IH}$ High-level input voltage | | 2 | | | 2 | | | V |
| $V_{IL}$ Low-level input voltage | | | | 0.7 | | | 0.8 | V |
| $V_{IK}$ Input clamp voltage | $V_{CC}$ = MIN, $I_I$ = -18 mA | | | -1.5 | | | -1.5 | V |
| $V_{OH}$ High-level output voltage | $V_{CC}$ = MIN, $V_{IH}$ = 2 V, $V_{IL}$ = $V_{IL}$ max, $I_{OH}$ = -400 µA | 2.5 | 3.4 | | 2.7 | 3.4 | | V |
| $V_{OL}$ Low-level output voltage | $V_{CC}$ = MIN, $V_{IH}$ = 2 V, $V_{IL}$ = $V_{IL}$ max, $I_{OL}$ = 4 mA | | 0.25 | 0.4 | | 0.25 | 0.4 | V |
| | $I_{OL}$ = 8 mA | | | | | 0.35 | 0.5 | V |
| $I_I$ Input current at maximum input voltage | $V_{CC}$ = MAX, $V_I$ = 7 V | | | 0.1 | | | 0.1 | mA |
| $I_{IH}$ High-level input current | $V_{CC}$ = MAX, $V_I$ = 2.7 V | | | 20 | | | 20 | µA |
| $I_{IL}$ Low-level input current | $V_{CC}$ = MAX, $V_I$ = 0.4 V | | | -0.4 | | | -0.4 | mA |
| $I_{OS}$ Short-circuit output current§ | $V_{CC}$ = MAX | -20 | | -100 | | | -100 | mA |
| $I_{CC}$ Supply current | $V_{CC}$ = MAX, 'LS151 | | 6.0 | 10 | | 6.0 | 10 | mA |
| | All inputs at 4.5 V  'LS152 | | 5.6 | 9 | | | | mA |

†For conditions shown as MIN or MAX, use the appropriate value specified under recommended operating conditions for the applicable device type.
‡All typical values are at $V_{CC}$ = 5 V, $T_A$ = 25°C.
§Not more than one output should be shorted at a time and duration of short-circuit should not exceed one second.

## switching characteristics, $V_{CC}$ = 5 V, $T_A$ = 25°C

| PARAMETER† | FROM (INPUT) | TO (OUTPUT) | TEST CONDITIONS | SN54LS', SN74LS' MIN | TYP | MAX | UNIT |
|---|---|---|---|---|---|---|---|
| $t_{PLH}$ | A, B, or C (4 levels) | Y | | | 27 | 43 | ns |
| $t_{PHL}$ | A, B, or C (4 levels) | Y | | | 18 | 30 | ns |
| $t_{PLH}$ | A, B, or C (3 levels) | W | | | 14 | 23 | ns |
| $t_{PHL}$ | A, B, or C (3 levels) | W | $C_L$ = 15 pF, | | 20 | 32 | ns |
| $t_{PLH}$ | Strobe $\bar{G}$ | Y | $R_L$ = 2 kΩ, | | 26 | 42 | ns |
| $t_{PHL}$ | Strobe $\bar{G}$ | Y | See Note 4 | | 20 | 32 | ns |
| $t_{PLH}$ | Strobe $\bar{G}$ | W | | | 15 | 24 | ns |
| $t_{PHL}$ | Strobe $\bar{G}$ | W | | | 18 | 30 | ns |
| $t_{PLH}$ | Any D | Y | | | 20 | 32 | ns |
| $t_{PHL}$ | Any D | Y | | | 16 | 26 | ns |
| $t_{PLH}$ | Any D | W | | | 13 | 21 | ns |
| $t_{PHL}$ | Any D | W | | | 12 | 20 | ns |

†$t_{PLH}$ = propagation delay time, low-to-high-level output
$t_{PHL}$ = propagation delay time, high-to-low-level output
NOTE 4: See General Information Section for load circuits and voltage waveforms.

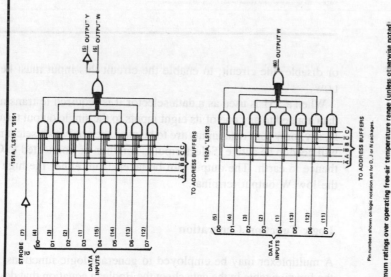

'151A, 'LS151, 'S151

STROBE $\bar{G}$ (7)

DATA INPUTS  D0 (4)  D1 (3)  D2 (2)  D3 (1)  D4 (15)  D5 (14)  D6 (13)  D7 (12)

OUTPUT Y (5)
OUTPUT W (6)

TO ADDRESS BUFFERS  A $\bar{A}$ B $\bar{B}$ C $\bar{C}$

'152A, 'LS152

DATA INPUTS  D0 (5)  D1 (4)  D2 (3)  D3 (2)  D4 (1)  D5 (13)  D6 (12)  D7 (11)

OUTPUT W (6)

TO ADDRESS BUFFERS  A $\bar{A}$ B $\bar{B}$ C $\bar{C}$

Pin numbers shown on logic notation are for D, J or N packages.

## absolute maximum ratings over operating free-air temperature range (unless otherwise noted)

Supply voltage, $V_{CC}$ (see Note 1) .......... 7 V
Input voltage (see Note 2): '150, '151A, '152A .......... 5.5 V
 'LS151, 'LS152 .......... 7 V
Operating free-air temperature range: SN54' .......... -55°C to 125°C
 SN74' .......... 0°C to 70°C
Storage temperature range .......... -65°C to 150°C

NOTES: 1. Voltage values are with respect to network ground terminal.
2. For the '150, input voltages must be zero or positive with respect to network ground terminal.

recommended operating conditions

| | SN54S151 | | | SN74S151 | | | UNIT |
|---|---|---|---|---|---|---|---|
| | MIN | NOM | MAX | MIN | NOM | MAX | |
| Supply voltage, $V_{CC}$ | 4.5 | 5 | 5.5 | 4.75 | 5 | 5.25 | V |
| High-level output current, $I_{OH}$ | | | −1 | | | −1 | mA |
| Low-level output current, $I_{OL}$ | | | 20 | | | 20 | mA |
| Operating free-air temperature, $T_A$ | −55 | | 125 | 0 | | 70 | °C |

electrical characteristics over recommended operating free-air temperature range (unless otherwise noted)

| | PARAMETER | TEST CONDITIONS[†] | | MIN | TYP[‡] | MAX | UNIT |
|---|---|---|---|---|---|---|---|
| $V_{IH}$ | High-level input voltage | | | 2 | | | V |
| $V_{IL}$ | Low-level input voltage | | | | | 0.8 | V |
| $V_{IK}$ | Input clamp voltage | $V_{CC}$ = MIN, $I_I$ = −18 mA | | | | −1.2 | V |
| $V_{OH}$ | High-level output voltage | $V_{CC}$ = MIN, $V_{IH}$ = 2 V, | SN54S' | 2.5 | 3.4 | | V |
| | | $V_{IL}$ = 0.8 V, $I_{OH}$ = −1 mA | SN74S' | 2.7 | 3.4 | | |
| $V_{OL}$ | Low-level output voltage | $V_{CC}$ = MIN, $V_{IH}$ = 2 V, $V_{IL}$ = 0.8 V, $I_{OL}$ = 20 mA | | | | 0.5 | V |
| $I_I$ | Input current at maximum input voltage | $V_{CC}$ = MAX, $V_I$ = 5.5 V | | | | 1 | mA |
| $I_{IH}$ | High-level input current | $V_{CC}$ = MAX, $V_I$ = 2.7 V | | | | 50 | μA |
| $I_{IL}$ | Low-level input current | $V_{CC}$ = MAX, $V_I$ = 0.5 V | | | | −2 | mA |
| $I_{OS}$ | Short-circuit output current[§] | $V_{CC}$ = MAX | | −40 | | −100 | mA |
| $I_{CC}$ | Supply current | $V_{CC}$ = MAX, All inputs at 4.5 V, All outputs open | | | 45 | 70 | mA |

[†]For conditions shown as MIN or MAX, use the appropriate value specified under recommended operating conditions for the applicable device type.
[‡]All typical values are at $V_{CC}$ = 5 V, $T_A$ = 25°C.
[§]Not more than one output should be shorted at a time, and duration of the short-circuit should not exceed one second.

switching characteristics, $V_{CC}$ = 5 V, $T_A$ = 25°C

| PARAMETER[¶] | FROM (INPUT) | TO (OUTPUT) | TEST CONDITIONS | SN54S151, SN74S151 | | | UNIT |
|---|---|---|---|---|---|---|---|
| | | | | MIN | TYP | MAX | |
| $t_{PLH}$ | A, B, or C | Y | | | 12 | 18 | ns |
| $t_{PHL}$ | (4 levels) | | | | 12 | 18 | |
| $t_{PLH}$ | A, B, or C | W | | | 10 | 15 | ns |
| $t_{PHL}$ | (3 levels) | | | | 9 | 13.5 | |
| $t_{PLH}$ | Any D | Y | $C_L$ = 15 pF, | | 8 | 12 | ns |
| $t_{PHL}$ | | | $R_L$ = 280 Ω, | | 8 | 12 | |
| $t_{PLH}$ | Any D | W | See Note 4 | | 4.5 | 7 | ns |
| $t_{PHL}$ | | | | | 4.5 | 7 | |
| $t_{PLH}$ | Strobe $\overline{G}$ | Y | | | 11 | 16.5 | ns |
| $t_{PHL}$ | | | | | 12 | 18 | |
| $t_{PLH}$ | Strobe $\overline{G}$ | W | | | 9 | 13 | ns |
| $t_{PHL}$ | | | | | 8.5 | 12 | |

[¶]$t_{PLH}$ ≡ Propagation delay time, low-to-high-level output
$t_{PHL}$ ≡ Propagation delay time, high-to-low-level output
NOTE 4: See General Information Section for load circuits and voltage waveforms.

or disable, the circuit; to enable the circuit this input must be held LOW.

When a 151 is used as a data selector it is required to transmit the data applied to any one of its eight inputs to the single output terminal. If, say, only four data inputs are to be used then only two select inputs will need to be active. Select input C can then be connected to G and thence to earth. The output can be taken from either the high Y or the low W output terminal.

## Logic Function Generation

A multiplexer may be employed to generate logic functions. From the function table in the data sheet the Boolean equation that describes the operation of the circuit is

$$F = \overline{A}\overline{B}\overline{C}D_0 + A\overline{B}\overline{C}D_1 + \overline{A}B\overline{C}D_2 + AB\overline{C}D_3 + \overline{A}\overline{B}CD_4$$
$$+ A\overline{B}CD_5 + \overline{A}BCD_6 + ABCD_7 \tag{6.1}$$

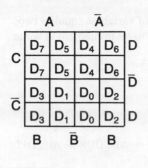

The Karnaugh mapping of this equation is shown alongside. The output state is made equal to the data input. The chosen multiplexer must have $n - 1$ select lines where $n$ is the number of variables. The input data lines correspond with the $n - 1$ MSBs of the function to be implemented. A given logic equation can be implemented by comparing the equations term by term with equation (6.1), or by comparing the Karnaugh mappings.

### Example 6.2

Implement the function $F = \overline{A}\overline{B}\overline{C} + \overline{A}B\overline{C} + A\overline{B}\overline{C} + \overline{A}BC + A\overline{B}C$ using the LS 151 multiplexer.

### Solution

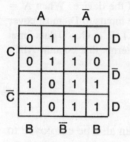

The Karnaugh mapping of the function is shown alongside. Comparing it with the mapping of the multiplexer gives

$$D_0 = D_2 = D_3 = D_4 = D_5 = 1$$
$$D_1 = D_6 = D_7 = 0 \qquad (Ans.)$$

When the number of variables is one more than the number of select inputs one of the variables will have to be connected to some of the data inputs.

### Example 6.3

Implement the function $F = \overline{A}\overline{B}\overline{C}\overline{D} + A\overline{B}\overline{C}D + AB\overline{C}\overline{D} + \overline{A}\overline{B}CD + ABCD$ using the LS 151 multiplexer.

### Solution

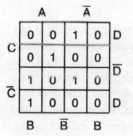

The Karnaugh mapping of this equation is shown alongside. Comparing it with the Karnaugh map of the multiplexer gives

$$D_0 = D_5 = \overline{D} \qquad D_1 = D_2 = D_6 = D_7 = 0$$
$$D_3 = 1 \quad \text{and} \quad D_4 = D$$

Alternatively, draw up Table 6.2.

**Table 6.2**

| Select input | | | Data input | Terms covered | Wanted terms | $D_n$ |
|---|---|---|---|---|---|---|
| C | B | A | | | | |
| 0 | 0 | 0 | 0 | $\overline{A}\overline{B}\overline{C}$ $(D + \overline{D})$ | $\overline{A}\overline{B}\overline{C}\overline{D}$ | $\overline{D}$ |
| 0 | 0 | 1 | 1 | $A\overline{B}\overline{C}$ $(D + \overline{D})$ | — | 0 |
| 0 | 1 | 0 | 2 | $\overline{A}B\overline{C}$ $(D + \overline{D})$ | — | 0 |
| 0 | 1 | 1 | 3 | $AB\overline{C}$ $(D + \overline{D})$ | $AB\overline{C}D$, $AB\overline{C}\overline{D}$ | $D + \overline{D} = 1$ |
| 1 | 0 | 0 | 4 | $\overline{A}\overline{B}C$ $(D + \overline{D})$ | $\overline{A}\overline{B}CD$ | $D$ |
| 1 | 0 | 1 | 5 | $A\overline{B}C$ $(D + \overline{D})$ | $A\overline{B}C\overline{D}$ | $\overline{D}$ |
| 1 | 1 | 0 | 6 | $\overline{A}BC$ $(D + \overline{D})$ | — | 0 |
| 1 | 1 | 1 | 7 | $ABC$ $(D + \overline{D})$ | — | 0 |

When a logic function containing a number of variables equal to two, or more, than the number of select inputs is to be implemented some random logic will be necessary.

*Example 6.4*

Implement the function

$$F = \overline{A}\overline{B}\overline{C}\overline{D}\overline{E}F + A\overline{B}\overline{C}\overline{D}\overline{E}F + A\overline{B}\overline{C}D\overline{E}F + \overline{A}\overline{B}CD\overline{E}F + ABC\overline{D}\overline{E}F + ABC\overline{D}\overline{E}\overline{F}$$

using a LS 151 multiplexer.

*Solution*

Connect inputs A, B and C to the select inputs of the device. When A = B = C = 0 data input $D_0$ is selected and hence the inputs to $D_0$ must cover any terms that contain $\overline{A}\overline{B}\overline{C}$. Similarly, when A = 1, B = C = 0 data input $D_1$ is selected and the inputs to $D_1$ must include terms that contain $A\overline{B}\overline{C}$, and so on. Hence

$$D_0 = \overline{D}\overline{E}F \qquad D_1 = \overline{D}\overline{E}F \qquad D_3 = D\overline{E}F$$
$$D_4 = D\overline{E}F \qquad D_7 = \overline{D}\overline{E}F + \overline{D}\overline{E}\overline{F} = \overline{D}\overline{E}$$

A 4-to-1 line multiplexer, i.e. the 74153, can also be employed to implement logic functions. The Boolean equation describing the operation of the circuit is given by equation (6.2).

$$F = \overline{A}\overline{B}C_0 + A\overline{B}C_1 + \overline{A}BC_2 + ABC_3 \tag{6.2}$$

**Decoders/De-multiplexers**

The function of a decoder/de-multiplexer is to route input data to any one of the several output lines. Decoders/de-multiplexers in common use include the 137/8 3-to-8 line decoder, the 139 dual 2-to-4 line decoder and the 154 4-to-16 line decoder. Figure 6.8 shows the data sheet for the LS 138 3-to-8 line decoder/de-multiplexer. The operation of the circuit is summarized by its function table. When the IC is used as a decoder its inputs, $G_1$, $\overline{G_{2A}}$ and $\overline{G_{2B}}$, function as enable inputs, but when the device is used as a de-multiplexer the input data is applied to any one of these three inputs. $G_1$ is active HIGH but the other two is are active LOW. The address of the output to which the data is to be directed is applied to the select inputs A, B and C (A = LSB). The Boolean equation that describes the operation of the circuit can be derived from the function table, and is

$$F = \overline{A}\overline{B}\overline{C}Y_0 + A\overline{B}\overline{C}Y_1 + \overline{A}B\overline{C}Y_2 + AB\overline{C}Y_3 + \overline{A}\overline{B}CY_4 + A\overline{B}CY_5 + \overline{A}BCY_6 + ABCY_7 \tag{6.3}$$

When the device is used as a de-multiplexer its action is to route a single input data line to any one of the eight output lines $Y_0$ through to $Y_7$. Figure 6.9 shows how the LS 138 is connected for this purpose. The input data is shown to be connected to the $\overline{G_2}$ inputs so $G_1$ must be held HIGH. The outputs are then active LOW. If the

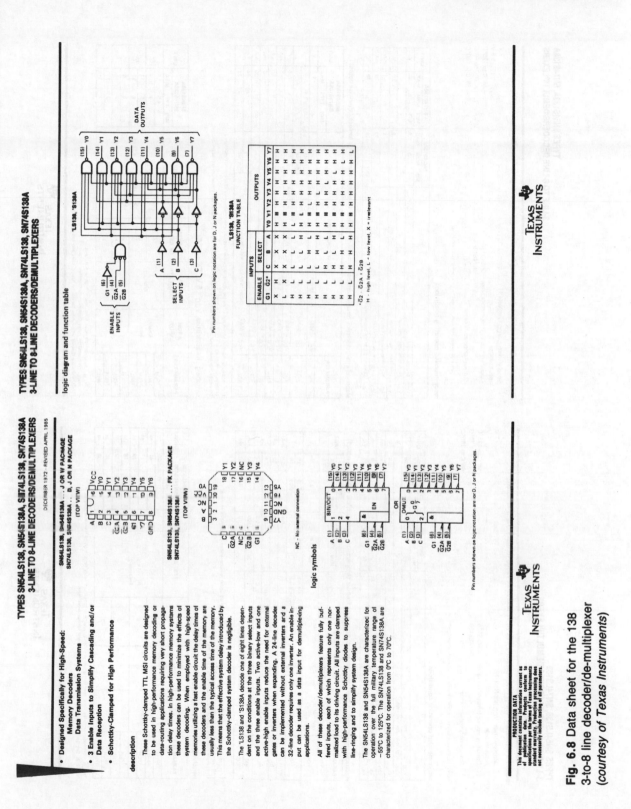

**Fig. 6.8** Data sheet for the 138 3-to-8 line decoder/de-multiplexer (courtesy of Texas Instruments)

## TYPES SN54LS138, SN74LS138
## 3-LINE TO 8-LINE DECODERS/DEMULTIPLEXERS

**absolute maximum ratings over operating free-air temperature range (unless otherwise noted)**

Supply voltage, $V_{CC}$ (see Note 1) ............ 7 V
Input voltage ............ 7 V
Operating free-air temperature range: SN54LS138 ............ −55°C to 125°C
   SN74LS138 ............ 0°C to 70°C
Storage temperature range ............ −65°C to 150°C

NOTE 1: Voltage values are with respect to network ground terminal.

**recommended operating conditions**

| | | SN54LS138 | | | SN74LS138 | | | UNIT |
|---|---|---|---|---|---|---|---|---|
| | | MIN | NOM | MAX | MIN | NOM | MAX | |
| $V_{CC}$ | Supply voltage | 4.5 | 5 | 5.5 | 4.75 | 5 | 5.25 | V |
| $V_{IH}$ | High-level input voltage | 2 | | | 2 | | | V |
| $V_{IL}$ | Low-level input voltage | | | 0.7 | | | 0.8 | V |
| $I_{OH}$ | High-level output current | | | −0.4 | | | −0.4 | mA |
| $I_{OL}$ | Low-level output current | | | 4 | | | 8 | mA |
| $T_A$ | Operating free-air temperature | −55 | | 125 | 0 | | 70 | °C |

**electrical characteristics over recommended operating free-air temperature range (unless otherwise noted)**

| PARAMETER | TEST CONDITIONS† | | SN54LS138 | | | SN74LS138 | | | UNIT |
|---|---|---|---|---|---|---|---|---|---|
| | | | MIN | TYP‡ | MAX | MIN | TYP‡ | MAX | |
| $V_{IK}$ | $V_{CC}$ = MIN, $I_I$ = −18 mA | | | | −1.5 | | | −1.5 | V |
| $V_{OH}$ | $V_{CC}$ = MIN, $V_{IH}$ = 2 V, $V_{IL}$ = MAX, $I_{OH}$ = −0.4 mA | | 2.5 | 3.4 | | 2.7 | 3.4 | | V |
| $V_{OL}$ | $V_{CC}$ = MIN, $V_{IH}$ = 2 V, $V_{IL}$ = MAX | $I_{OL}$ = 4 mA | | 0.25 | 0.4 | | 0.25 | 0.4 | V |
| | | $I_{OL}$ = 8 mA | | | | | 0.35 | 0.5 | |
| $I_I$ | $V_{CC}$ = MAX, $V_I$ = 7 V | | | | 0.1 | | | 0.1 | mA |
| $I_{IH}$ | $V_{CC}$ = MAX, $V_I$ = 2.7 V | | | | 20 | | | 20 | µA |
| $I_{IL}$ | $V_{CC}$ = MAX, $V_I$ = 0.4 V | Enable | | | −0.4 | | | −0.4 | mA |
| | | A, B, C | | | −0.2 | | | −0.2 | |
| $I_{OS}$§ | $V_{CC}$ = MAX | | −20 | | −100 | −20 | | −100 | mA |
| $I_{CC}$ | $V_{CC}$ = MAX, Outputs enabled and open | | | 6.3 | 10 | | 6.3 | 10 | mA |

†For conditions shown as MIN or MAX, use the appropriate value specified under recommended operating conditions.
‡All typical values are at $V_{CC}$ = 5 V, $T_A$ = 25°C.
§Not more than one output should be shorted at a time.

**switching characteristics, $V_{CC}$ = 5 V, $T_A$ = 25°C**

| PARAMETER¶ | FROM (INPUT) | TO (OUTPUT) | LEVELS OF DELAY | TEST CONDITIONS | SN54LS138 SN74LS138 | | | UNIT |
|---|---|---|---|---|---|---|---|---|
| | | | | | MIN | TYP | MAX | |
| $t_{PLH}$ | Binary | Any | 2 | | | 11 | 20 | ns |
| $t_{PHL}$ | | | | | | 18 | 41 | ns |
| $t_{PLH}$ | Select | | 3 | | | 21 | 27 | ns |
| $t_{PHL}$ | | | | $R_L$ = 2 kΩ, $C_L$ = 15 pF, See Note 2 | | 20 | 39 | ns |
| $t_{PLH}$ | Enable | Any | 2 | | | 12 | 18 | ns |
| $t_{PHL}$ | | | | | | 20 | 32 | ns |
| $t_{PLH}$ | | | 3 | | | 14 | 26 | ns |
| $t_{PHL}$ | | | | | | 13 | 38 | ns |

¶$t_{PLH}$ = propagation delay time, low-to-high-level output; $t_{PHL}$ = propagation delay time, high-to-low-level output.
NOTE 2: See General Information Section for load circuits and voltage waveforms.

TEXAS INSTRUMENTS

---

## TYPES SN54S138A, SN74S138A
## 3-LINE TO 8-LINE DECODERS/DEMULTIPLEXERS

**absolute maximum ratings over operating free-air temperature range (unless otherwise noted)**

Supply voltage, $V_{CC}$ (see Note 1) ............ 7 V
Input voltage ............ 5.5 V
Operating free-air temperature range: SN54S138A ............ −55°C to 125°C
   SN74S138A ............ 0°C to 70°C
Storage temperature range ............ −65°C to 150°C

NOTE 1: Voltage values are with respect to network ground terminal.

**recommended operating conditions**

| | | SN54S138A | | | SN74S138A | | | UNIT |
|---|---|---|---|---|---|---|---|---|
| | | MIN | NOM | MAX | MIN | NOM | MAX | |
| $V_{CC}$ | Supply voltage | 4.5 | 5 | 5.5 | 4.75 | 5 | 5.25 | V |
| $V_{IH}$ | High-level input voltage | 2 | | | 2 | | | V |
| $V_{IL}$ | Low-level input voltage | | | 0.8 | | | 0.8 | V |
| $I_{OH}$ | High-level output current | | | −1 | | | −1 | mA |
| $I_{OL}$ | Low-level output current | | | 20 | | | 20 | mA |
| $T_A$ | Operating free-air temperature | −55 | | 125 | 0 | | 70 | °C |

**electrical characteristics over recommended operating free-air temperature range (unless otherwise noted)**

| PARAMETER | TEST CONDITIONS† | | SN54S138A SN74S138A | | | UNIT |
|---|---|---|---|---|---|---|
| | | | MIN | TYP‡ | MAX | |
| $V_{IK}$ | $V_{CC}$ = MIN, $I_I$ = −18 mA | | | | −1.2 | V |
| $V_{OH}$ | $V_{CC}$ = MIN, $V_{IH}$ = 2 V, $V_{IL}$ = 0.8 V, $I_{OH}$ = −1 mA | SN54S' | 2.5 | 3.4 | | V |
| | | SN74S' | 2.7 | 3.4 | | |
| $V_{OL}$ | $V_{CC}$ = MIN, $V_{IH}$ = 2 V, $V_{IL}$ = 0.8 V, $I_{OL}$ = 20 mA | | | | 0.5 | V |
| $I_I$ | $V_{CC}$ = MAX, $V_I$ = 5.5 V | | | | 1 | mA |
| $I_{IH}$ | $V_{CC}$ = MAX, $V_I$ = 2.7 V | | | | 50 | µA |
| $I_{IL}$ | $V_{CC}$ = MAX, $V_I$ = 0.5 V | | | | −2 | mA |
| $I_{OS}$§ | $V_{CC}$ = MAX | | −40 | | −100 | mA |
| | | | | | −2 | |
| $I_{CC}$ | $V_{CC}$ = MAX, Outputs enabled and open | SN54S' | | 49 | 74 | mA |
| | | SN74S' | | 49 | 74 | |

†For conditions shown as MIN or MAX, use the appropriate value specified under recommended operating conditions.
‡All typical values are at $V_{CC}$ = 5 V, $T_A$ = 25°C.
§Not more than one output should be shorted at a time, and duration of the short circuit test should not exceed one second.

**switching characteristics, $V_{CC}$ = 5 V, $T_A$ = 25°C**

| PARAMETER¶ | FROM (INPUT) | TO (OUTPUT) | LEVELS OF DELAY | TEST CONDITIONS | SN54S138A SN74S138A | | | UNIT |
|---|---|---|---|---|---|---|---|---|
| | | | | | MIN | TYP | MAX | |
| $t_{PLH}$ | Binary | Any | 2 | | | 4.5 | 7 | ns |
| $t_{PHL}$ | | | | | | 7 | 10.5 | ns |
| $t_{PLH}$ | Select | | 3 | | | 7.5 | 12 | ns |
| $t_{PHL}$ | | | | $R_L$ = 280 Ω, $C_L$ = 15 pF, See Note 2 | | 8 | 12 | ns |
| $t_{PLH}$ | Enable | Any | 2 | | | 5 | 8 | ns |
| $t_{PHL}$ | | | | | | 7 | 11 | ns |
| $t_{PLH}$ | | | 3 | | | 7 | 11 | ns |
| $t_{PHL}$ | | | | | | 7 | 11 | ns |

¶$t_{PLH}$ = propagation delay time, low-to-high-level output
$t_{PHL}$ = propagation delay time, high-to-low-level output
NOTE 2: See General Information Section for load circuits and voltage waveforms.

TEXAS INSTRUMENTS

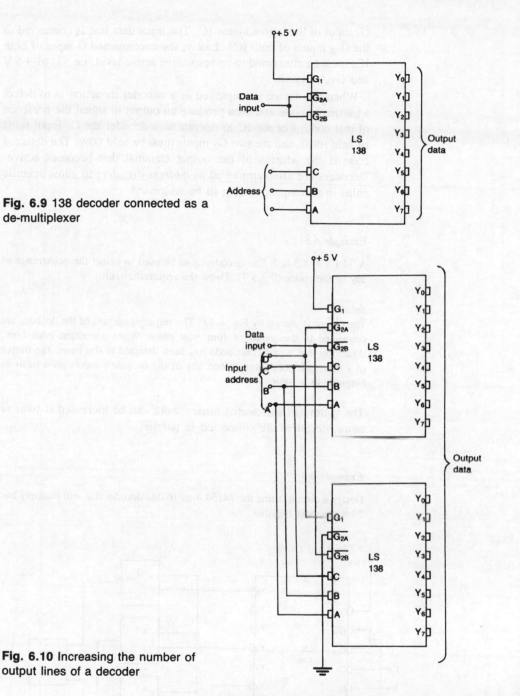

**Fig. 6.9** 138 decoder connected as a de-multiplexer

**Fig. 6.10** Increasing the number of output lines of a decoder

input data was to be applied to the $G_1$ input instead then the $\overline{G}_2$ inputs should be connected to earth and the outputs will be active HIGH.

The number of output lines can be increased by connecting two 138s together as shown by Fig. 6.10. The address lines A, B and C are connected to the A, B and C inputs of both ICs. The address line D is connected to the $\overline{G_{2A}}$ input of the low-address IC and to the

$G_1$ input of the high-address IC. The input data line is connected to the $\overline{G_{2B}}$ inputs of both ICs. Lastly, the unconnected G input of both ICs must be connected to its respective active level, i.e. $G_1$ to $+5$ V and $\overline{G_{2A}}$ to earth.

When the device is employed as a decoder its action is to detect a particular code and then produce an output to signal the presence of that code. For the IC to operate as a decoder the $G_1$ input must be held HIGH, and the two $\overline{G_2}$ inputs must be held LOW. The detected code is the address of the output terminal that becomes active. Decoders are often employed as *address decoders* to allow specific chips in a complex system to be addressed.

### Example 6.5

A 74 LS 138 3-to-8 line decoder is to be used to detect the occurrence of any of the codes (0,3,5,7). Draw the required circuit.

### Solution

The circuit is shown by Fig. 6.11. The required outputs of the decoder are connected to the inputs of four NOR gates. When an output goes LOW, signalling that a particular code has been detected at the input, the output of a NOR gate goes HIGH. When any of the OR gate's inputs goes HIGH its output also is HIGH.

The width of the detected binary word can be increased if two, or more, decoders are connected in parallel.

### Example 6.6

Design a circuit, using the 74154 4-to-16 line decoder that will multiply two 2-bit numbers together.

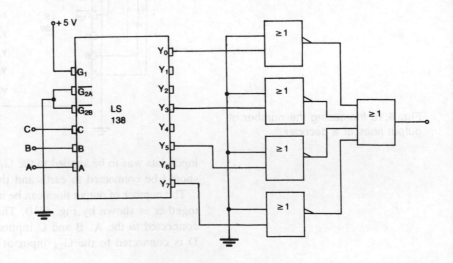

**Fig. 6.11**

**Table 6.3**

| Inputs | | | | | | | Multiplication | | | | |
| Denary | | Binary | | | | | Denary | Binary | | | |
| A | B | $A_1$ | $A_0$ | $B_1$ | $B_0$ | Output | | $P_4$ | $P_3$ | $P_2$ | $P_1$ |
|---|---|---|---|---|---|---|---|---|---|---|---|
| 0 | 0 | 0 | 0 | 0 | 0 | 0 | 0 | 0 | 0 | 0 | 0 |
| 1 | 0 | 0 | 1 | 0 | 0 | 1 | 0 | 0 | 0 | 0 | 0 |
| 2 | 0 | 1 | 0 | 0 | 0 | 2 | 0 | 0 | 0 | 0 | 0 |
| 3 | 0 | 1 | 1 | 0 | 0 | 3 | 0 | 0 | 0 | 0 | 0 |
| 0 | 1 | 0 | 0 | 0 | 1 | 4 | 0 | 0 | 0 | 0 | 0 |
| 1 | 1 | 0 | 1 | 0 | 1 | 5 | 1 | 0 | 0 | 0 | 1 |
| 2 | 1 | 1 | 0 | 0 | 1 | 6 | 2 | 0 | 0 | 1 | 0 |
| 3 | 1 | 1 | 1 | 0 | 1 | 7 | 3 | 0 | 0 | 1 | 1 |
| 0 | 2 | 0 | 0 | 1 | 0 | 8 | 0 | 0 | 0 | 0 | 0 |
| 1 | 2 | 0 | 1 | 1 | 0 | 9 | 2 | 0 | 0 | 1 | 0 |
| 2 | 2 | 1 | 0 | 1 | 0 | 10 | 4 | 0 | 1 | 0 | 0 |
| 3 | 2 | 1 | 1 | 1 | 0 | 11 | 6 | 0 | 1 | 1 | 0 |
| 0 | 3 | 0 | 0 | 1 | 1 | 12 | 0 | 0 | 0 | 0 | 0 |
| 1 | 3 | 0 | 1 | 1 | 1 | 13 | 3 | 0 | 0 | 1 | 1 |
| 2 | 3 | 1 | 0 | 1 | 1 | 14 | 6 | 0 | 1 | 1 | 0 |
| 3 | 3 | 1 | 1 | 1 | 1 | 15 | 9 | 1 | 0 | 0 | 1 |

*Solution*

The 74154 has 4 data input terminals, 16 active LOW outputs, labelled 0 through to 15, and 2 strobe inputs $\overline{G}_1$ and $\overline{G}_2$. Whenever either strobe input is HIGH all the outputs are also HIGH. The truth table of the required operation is given by Table 6.3. From Table 6.3,

$$P_1 = \overline{5} + \overline{7} + \overline{13} + \overline{15} = \overline{5.7.13.15}$$
$$P_2 = \overline{6} + \overline{7} + \overline{9} + \overline{11} + \overline{13} + \overline{14} = \overline{6.7.9.11.13.14}$$
$$P_3 = \overline{10} + \overline{11} + \overline{14} = \overline{10.11.14}$$
$$P_4 = \overline{15}$$

The required circuit is shown by Fig. 6.12.

The 74154 uses a 24-pin package and it is not available in an LS version (although it is in ALS, HC, AC and ACT). This may be a disadvantage that can be overcome by combining two 138 3-to-8 line decoders to act as a single 4-to-16 line decoder (see Fig. 6.10).

**Encoders**

An encoder performs the opposite function to a decoder. Only one of its inputs is active at a time and each one of them produces a specific code at the outputs. The main application for encoders is in connection with keyboards where the pressing of a single key must produce the code combination that represents that character or number.

The 74147 is a 10-to-4 line priority encoder that converts from decimal to BCD. The device has active LOW inputs and outputs. The

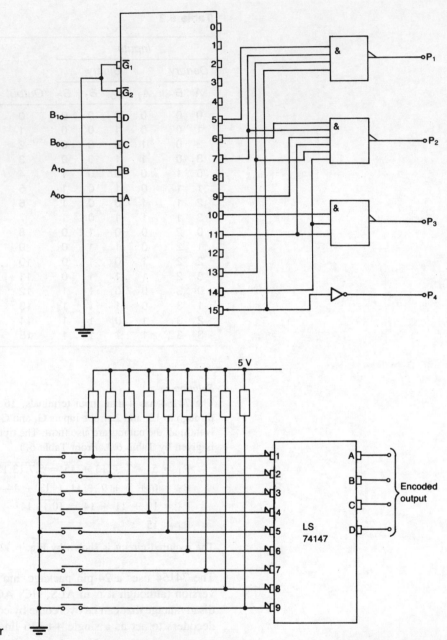

**Fig. 6.12**

**Fig. 6.13** Keyboard encoder

'priority' term means that a number of greater magnitude is given priority over numbers of lesser magnitude. There are only nine inputs since 0 is assumed if there is no other input. The BCD value of the highest input number present at the input is produced at the four outputs.

The 74147 can be used as a keypad encoder. A keypad, or keyboard, is a matrix of push-to-make switches that is used as an input device to send characters to a computer, or other digital system. Figure 6.13

shows the basic arrangement; as each key is pressed the decimal number it represents must be encoded into the corrresponding binary number. Since the IC has active LOW inputs each keyline is connected to + 5 V via a pull-up resistor. The encoder for a computer keyboard is more complex because of the greater number of characters to be encoded using the ASCII code. A VSLI device, such as the AY 3-8910 PSG, would be used in conjunction with a more complex matrix than that shown in Fig. 6.13. When a key is pressed a column line is connected to a row line and this connection signals to the keyboard encoder which character is to be encoded and passed on to the computer.

## Binary Full Adders

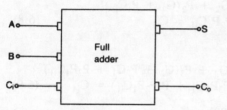

**Fig. 6.14** Full adder

A *binary full adder* is a circuit that adds two input bits A and B together with any carry bit from a previous stage. The block diagram of a full adder is given by Fig. 6.14. The equations for the sum S and the carry-out $C_o$ are:

$$S = (A\bar{B} + \bar{A}B)\bar{C}_i + (\overline{\bar{A}B + AB})C_i$$
$$= (A \oplus B)\bar{C}_i + \overline{(A \oplus B)}C_i \tag{6.4}$$
$$C_o = AB + C_i(A\bar{B} + \bar{A}B) = AB + C_i(A \oplus B) \tag{6.5}$$

When two 4-bit numbers are to be added four full-adders must be connected as shown by Fig. 6.15 to give a *parallel adder*. Two 4-bit numbers A and B are applied to the A and B inputs of the four full adders. Any carry from a previous stage is applied to the $C_i$ terminal of the least significant adder and the carry-out $C_o$ terminal of each adder is directly connected to the $C_i$ terminal of the next more significant adder. The $C_o$ terminal of the most significant adder gives the carry-out for the complete circuit. The carry-out bits ripple through the system stage by stage and an individual sum will only be correct when the carry from the previous stage has arrived. Hence the output sum and carry of the circuit must not be read until enough time has elapsed for all the carry bits to have passed through the circuit. This means that ripple addition is slow if large numbers are to be added and a technique known as *look-ahead carry* is often employed. This technique involves obtaining the final carry-out of the addition using separate circuitry.

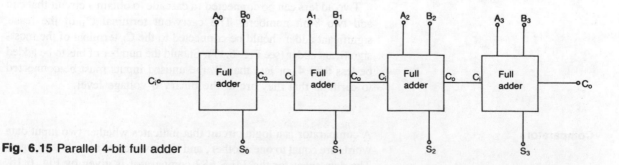

**Fig. 6.15** Parallel 4-bit full adder

At each stage a carry-out will occur if both the bits that are being added are 1, or if either bit is a 1 and there is a carry-in from the previous stage. Therefore

$$C_{on} = A_n B_n + (A_n \oplus B_n)C_{in}$$

The term $A_n B_n$ is known as the *carry-generate* $G_n$ and the term $(A_n \oplus B_n)$ is known as the *carry-propagate* $P_n$.

For each stage in the circuit,

$$C_{on} = G_n + P_n C_{in} = G_n + P_n C_{o(n-1)} \tag{6.6}$$

For the first (least significant) stage,

$$C_{o1} = G_1 + P_1 C_{i1} = C_{i2} \tag{6.7}$$

For the second stage,

$$C_{o2} = G_2 + P_2 C_{i2} = G_2 + P_2(G_1 + P_1 C_{i1})$$
$$= G_2 + P_2 G_1 + P_2 P_1 C_{i1} = C_{i3} \tag{6.8}$$

For the third stage,

$$C_{o3} = G_3 + P_3 C_{i3} = G_3 + P_3(G_2 + P_2 G_1 + P_2 P_1 C_{i1})$$
$$= G_3 + P_3 G_2 + P_3 P_2 G_1 + P_3 P_2 P_1 C_{i1}) = C_{i4} \tag{6.9}$$

For the most-significant stage,

$$C_{o4} = G_4 + P_4 C_{i4}$$
$$= G_4 + P_4(G_3 + P_3 G_2 + P_3 P_2 G_1 + P_3 P_2 P_1 C_{i1})$$
$$= G_4 + P_4 G_3 + P_4 P_3 G_2 + P_4 P_3 P_2 G_1 + P_4 P_3 P_2 P_1 C_{i1} \tag{6.10}$$

It can be seen that although the expressions get longer as the bit significance increases, the overall delay in producing the carry-out remains constant. This means that there are no cumulative effects and so the ripple effect is eliminated. The look-ahead circuitry is required to implement equations (6.6)–(6.10).

The 283 4-bit full adder includes the look-ahead circuitry and its data sheet is given in Fig. 6.16. To employ the device to add two 4-bit numbers A and B together, the numbers are applied to the appropriate inputs, $A_1$ to $A_4$ and $B_1$ to $B_4$, and the $C_o$ terminal is connected to earth. The result of the addition then appears at terminals $\Sigma_1$ through to $\Sigma_4$ and any carry-out appears at terminal $C_4$.

Two adders can be connected in cascade to obtain a circuit that can add two 8-bit numbers. The carry-out terminal $C_4$ of the least-significant adder should be connected to the $C_o$ terminal of the most-significant adder (see Fig. 6.17). Should the number of bits to be added be less than 4, or less than 8, the unused inputs must be connected to earth so that they are at the binary 0 voltage level.

**Comparators**

A comparator is a logic circuit that indicates whether two input data words are equal to one another, and if not which of them is the larger. The data sheet for the 74LS 682 comparator is given by Fig. 6.18.

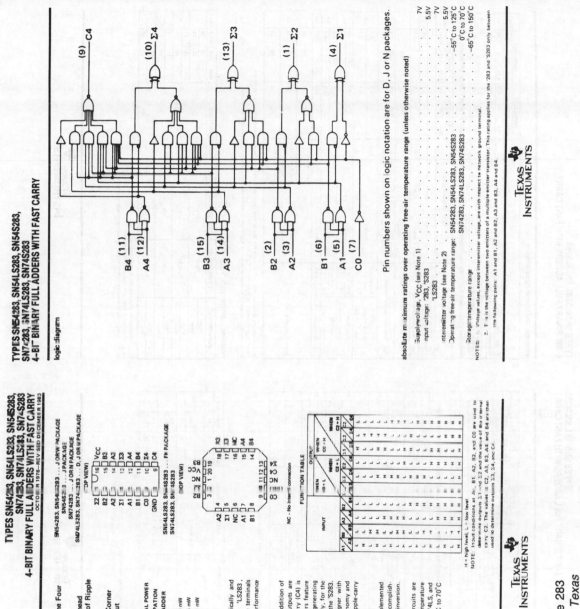

**Fig. 6.16** Data sheet for the 283 4-bit full adder (*courtesy of Texas Instruments*)

## TYPES SN54283, SN74283
## 4-BIT BINARY FULL ADDERS WITH FAST CARRY

### recommended operating conditions

| | | SN54283 MIN | NOM | MAX | SN74283 MIN | NOM | MAX | UNIT |
|---|---|---|---|---|---|---|---|---|
| Supply Voltage, $V_{CC}$ | | 4.5 | 5 | 5.5 | 4.75 | 5 | 5.25 | V |
| High-level output current, $I_{OH}$ | Any output except C4 | | | -800 | | | -800 | μA |
| | Output C4 | | | -400 | | | -400 | |
| Low-level output current, $I_{OL}$ | | | | 16 | | | 16 | mA |
| | Any output except C4 | | | 8 | | | 8 | |
| Operating free-air temperature, $T_A$ | | -55 | | 125 | 0 | | 70 | °C |

### electrical characteristics over recommended operating free-air temperature range (unless otherwise noted)

| PARAMETER | | TEST CONDITIONS | SN54283 MIN | TYP‡ | MAX | SN74283 MIN | TYP‡ | MAX | UNIT |
|---|---|---|---|---|---|---|---|---|---|
| $V_{IH}$ High-level input voltage | | | 2 | | | 2 | | | V |
| $V_{IL}$ Low-level input voltage | | | | | 0.8 | | | 0.8 | V |
| $V_{IK}$ Input clamp voltage | | $V_{CC}$ = MIN, $I_I$ = -12 mA | | | -1.5 | | | -1.5 | V |
| $V_{OH}$ High-level output voltage | | $V_{CC}$ = MIN, $V_{IH}$ = 2 V, $V_{IL}$ = 0.8 V, $I_{OH}$ = MAX | 2.4 | 3.6 | | 2.4 | 3.6 | | V |
| $V_{OL}$ Low-level output voltage | | $V_{CC}$ = MIN, $V_{IH}$ = 2 V, $V_{IL}$ = 0.8 V, $I_{OL}$ = MAX | | 0.2 | 0.4 | | 0.2 | 0.4 | V |
| $I_I$ Input current at maximum input voltage | | $V_{CC}$ = MAX, $V_I$ = 5.5 V | | | 1 | | | 1 | mA |
| $I_{IH}$ High-level input current | | $V_{CC}$ = MAX, $V_I$ = 2.4 V | | | 40 | | | 40 | μA |
| $I_{IL}$ Low-level input current | | $V_{CC}$ = MAX, $V_I$ = 0.4 V | | | -1.6 | | | -1.6 | mA |
| $I_{OS}$ Short-circuit output current§ | Any output except C4 | $V_{CC}$ = MAX | -20 | | -55 | -18 | | -55 | mA |
| | Output C4 | | -20 | | -70 | -18 | | -70 | |
| $I_{CC}$ Supply current | | $V_{CC}$ = MAX, Outputs open, All B low, other inputs at 4.5 V | | 56 | | | 56 | | mA |
| | | All inputs at 4.5 V | 66 | | 99 | 66 | | 110 | |

† For conditions shown as MIN or MAX, use the appropriate value specified under recommended operating conditions.
‡ All typical values are at $V_{CC}$ = 5 V, $T_A$ = 25 C.
§ Only one output should be shorted at a time.

### switching characteristics, $V_{CC}$ = 5 V, $T_A$ = 25°C

| PARAMETER¶ | FROM (INPUT) | TO (OUTPUT) | TEST CONDITIONS | MIN | TYP | MAX | UNIT |
|---|---|---|---|---|---|---|---|
| $t_{PLH}$ | CO | Any Σ | $C_L$ = 15 pF, $R_L$ = 400 Ω, See Note 3 | | 14 | 21 | ns |
| $t_{PHL}$ | CO | Any Σ | | | 12 | 21 | ns |
| $t_{PLH}$ | $A_i$ or $B_i$ | $Σ_i$ | | | 16 | 24 | ns |
| $t_{PHL}$ | $A_i$ or $B_i$ | $Σ_i$ | | | 16 | 24 | ns |
| $t_{PLH}$ | CO | C4 | $C_L$ = 15 pF, $R_L$ = 780 Ω, See Note 3 | | 9 | 14 | ns |
| $t_{PHL}$ | CO | C4 | | | 11 | 16 | ns |
| $t_{PLH}$ | $A_i$ or $B_i$ | C4 | | | 9 | 14 | ns |
| $t_{PHL}$ | $A_i$ or $B_i$ | C4 | | | 11 | 16 | ns |

¶ $t_{PLH}$ = Propagation delay time, low-to-high-level output
$t_{PHL}$ = Propagation delay time, high-to-low-level output
NOTE 3: See General Information Section for load circuits and voltage waveforms.

---

## TYPES SN54LS283, SN74LS283
## 4-BIT BINARY FULL ADDERS WITH FAST CARRY

### recommended operating conditions

| | SN54LS283 MIN | NOM | MAX | SN74LS283 MIN | NOM | MAX | UNIT |
|---|---|---|---|---|---|---|---|
| Supply voltage, $V_{CC}$ | 4.5 | 5 | 5.5 | 4.75 | 5 | 5.25 | V |
| High-level output current, $I_{OH}$ | | | -400 | | | -400 | μA |
| Low-level output current, $I_{OL}$ | | | 4 | | | 8 | mA |
| Operating free-air temperature, $T_A$ | -55 | | 125 | 0 | | 70 | C |

### electrical characteristics over recommended operating free-air temperature range (unless otherwise noted)

| PARAMETER | | TEST CONDITIONS | SN54LS283 MIN | TYP‡ | MAX | SN74LS283 MIN | TYP‡ | MAX | UNIT |
|---|---|---|---|---|---|---|---|---|---|
| $V_{IH}$ High-level input voltage | | | 2 | | | 2 | | | V |
| $V_{IL}$ Low-level input voltage | | | | | 0.7 | | | 0.8 | V |
| $V_{IK}$ Input clamp voltage | | $V_{CC}$ = MIN, $I_I$ = -18 mA | | | -1.5 | | | -1.5 | V |
| $V_{OH}$ High-level output voltage | | $V_{CC}$ = MIN, $V_{IH}$ = 2 V, $V_{IL}$ = $V_{IL}$ max, $I_{OH}$ = -400 μA | 2.5 | 3.4 | | 2.7 | 3.4 | | V |
| $V_{OL}$ Low-level output voltage | | $V_{CC}$ = MIN, $V_{IH}$ = 2 V, $V_{IL}$ = $V_{IL}$ max | $I_{OL}$ = 4 mA | 0.25 | 0.4 | | 0.25 | 0.4 | V |
| | | $I_{OL}$ = 8 mA | | | | 0.35 | 0.5 | | |
| $I_I$ Input current at maximum input voltage | Any A or B | $V_{CC}$ = MAX, $V_I$ = 7 V | | | 0.2 | | | 0.2 | mA |
| | CO | | | | 0.1 | | | 0.1 | |
| $I_{IH}$ High-level input current | Any A or B | $V_{CC}$ = MAX, $V_I$ = 2.7 V | | | 20 | | | 20 | μA |
| | CO | | | | 40 | | | 40 | |
| $I_{IL}$ Low-level input current | Any A or B | $V_{CC}$ = MAX, $V_I$ = 0.4 V | | | -0.4 | | | -0.4 | mA |
| | CO | | | | -0.8 | | | -0.8 | |
| $I_{OS}$ Short-circuit output current§ | | $V_{CC}$ = MAX | -20 | | -100 | -20 | | -100 | mA |
| $I_{CC}$ Supply current | | $V_{CC}$ = MAX, Outputs open, All inputs grounded | | 22 | 39 | | 22 | 39 | mA |
| | | All B low, other inputs at 4.5 V | | 19 | 34 | | 19 | 34 | |
| | | All inputs at 4.5 V | | 19 | 34 | | 19 | 34 | |

† For conditions shown as MIN or MAX, use the appropriate value specified under recommended operating conditions.
‡ All typical values are at $V_{CC}$ = 5 V, $T_A$ = 25°C.
§ Only one output should be shorted at a time and duration of the short-circuit should not exceed one second.

### switching characteristics, $V_{CC}$ = 5 V, $T_A$ = 25°C

| PARAMETER¶ | FROM (INPUT) | TO (OUTPUT) | TEST CONDITIONS | MIN | TYP | MAX | UNIT |
|---|---|---|---|---|---|---|---|
| $t_{PLH}$ | CO | Any Σ | $R_L$ = 2 kΩ, $C_L$ = 15 pF, See Note 3 | | 16 | 24 | ns |
| $t_{PHL}$ | CO | Any Σ | | | 15 | 24 | ns |
| $t_{PLH}$ | $A_i$ or $B_i$ | $Σ_i$ | | | 15 | 24 | ns |
| $t_{PHL}$ | $A_i$ or $B_i$ | $Σ_i$ | | | 11 | 17 | ns |
| $t_{PLH}$ | CO | C4 | | | 11 | 22 | ns |
| $t_{PHL}$ | CO | C4 | | | 11 | 17 | ns |
| $t_{PLH}$ | $A_i$ or $B_i$ | C4 | | | 12 | 17 | ns |
| $t_{PHL}$ | $A_i$ or $B_i$ | C4 | | | 12 | 17 | ns |

¶ $t_{PLH}$ = Propagation delay time, low-to-high-level output
$t_{PHL}$ = Propagation delay time, high-to-low-level output
NOTE 3: See General Information Section for load circuits and voltage waveforms.

TEXAS INSTRUMENTS

TYPES SN54S283, SN74S283
4-BIT BINARY FULL ADDERS WITH FAST CARRY

**recommended operating conditions**

| | | SN54S283 | | | SN74S283 | | | UNIT |
|---|---|---|---|---|---|---|---|---|
| | | MIN | NOM | MAX | MIN | NOM | MAX | |
| Supply voltage, $V_{CC}$ | | 4.5 | 5 | 5.5 | 4.75 | 5 | 5.25 | V |
| High-level output current, $I_{OH}$ | Any output except C4 | | | −1 | | | −1 | mA |
| | Output C4 | | | −500 | | | −500 | μA |
| Low-level output current, $I_{OL}$ | Any output except C4 | | | 20 | | | 20 | mA |
| | Output C4 | | | 10 | | | 10 | |
| Operating free-air temperature, $T_A$ | | −55 | | 125 | 0 | | 70 | °C |

**electrical characteristics over recommended operating free-air temperature range (unless otherwise noted)**

| | PARAMETER | | TEST CONDITIONS[†] | | MIN | TYP[‡] | MAX | UNIT |
|---|---|---|---|---|---|---|---|---|
| $V_{IH}$ | High-level input voltage | | | | 2 | | | V |
| $V_{IL}$ | Low-level input voltage | | | | | | 0.8 | V |
| $V_{IK}$ | Input clamp voltage | | $V_{CC}$ = MIN, | $I_I$ = −18 mA | | | −1.2 | V |
| $V_{OH}$ | High-level output voltage | SN54S283 | $V_{CC}$ = MIN, | $V_{IH}$ = 2 V | 2.5 | 3.4 | | V |
| | | SN74S283 | $V_{IL}$ = 0.8 V, | $I_{OH}$ = MAX | 2.7 | 3.4 | | |
| $V_{OL}$ | Low-level output voltage | | $V_{CC}$ = MIN, $V_{IH}$ = 2 V, $V_{IL}$ = 0.8 V, $I_{OL}$ = MAX | | | | 0.5 | V |
| $I_I$ | Input current at maximum input voltage | | $V_{CC}$ = MAX, $V_I$ = 5.5 V | | | | 1 | mA |
| $I_{IH}$ | High-level input current | | $V_{CC}$ = MAX, $V_I$ = 2.7 V | | | | 50 | μA |
| $I_{IL}$ | Low-level input current | | $V_{CC}$ = MAX, $V_I$ = 0.5 V | | | | −2 | mA |
| $I_{OS}$ | Short-circuit output current[§] | Any output except C4 | $V_{CC}$ = MAX | | −40 | | −100 | mA |
| | | Output C4 | | | −20 | | −100 | |
| $I_{CC}$ | Supply current | | $V_{CC}$ = MAX, Outputs open | All B low, other inputs at 4.5 V | | 80 | | mA |
| | | | | All inputs at 4.5 V | | 95 | 160 | |

[†]For conditions shown as MIN or MAX, use the appropriate value specified under recommended operating conditions for the applicable device type.
[‡]All typical values are at $V_{CC}$ = 5 V, $T_A$ = 25°C.
[§] Only one output should be shorted at a time, and duration of the short-circuit should not exceed one second.

**switching characteristics, $V_{CC}$ = 5 V, $T_A$ = 25°C**

| PARAMETER[¶] | FROM (INPUT) | TO (OUTPUT) | TEST CONDITIONS | MIN | TYP | MAX | UNIT |
|---|---|---|---|---|---|---|---|
| $t_{PLH}$ | C0 | Any Σ | $C_L$ = 15 pF, $R_L$ = 280 Ω, See Note 3 | | 11 | 18 | ns |
| $t_{PHL}$ | | | | | 12 | 18 | |
| $t_{PLH}$ | $A_i$ or $B_i$ | $\Sigma_i$ | | | 12 | 18 | ns |
| $t_{PHL}$ | | | | | 11.5 | 18 | |
| $t_{PLH}$ | C0 | C4 | $C_L$ = 15 pF, $R_L$ = 560 Ω, See Note 3 | | 6 | 11 | ns |
| $t_{PHL}$ | | | | | 7.5 | 11 | |
| $t_{PLH}$ | $A_i$ or $B_i$ | C4 | | | 7.5 | 12 | ns |
| $t_{PHL}$ | | | | | 8.5 | 12 | |

[¶]$t_{PLH}$ = Propagation delay time, low-to-high-level output
$t_{PHL}$ = Propagation delay time, high-to-low-level output
NOTE 3: See General Information Section for load circuits and voltage waveforms.

TEXAS
INSTRUMENTS

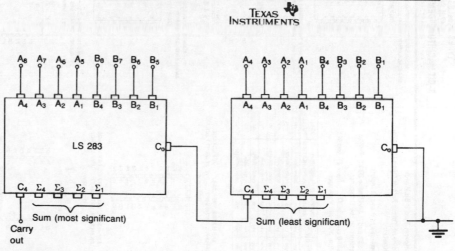

Fig. 6.17 Cascading 283 full adders
to allow two 8-bit numbers to be
added

## TYPES SN54LS682 THRU SN54LS689, SN74LS682 THRU SN74LS689
## 8-BIT MAGNITUDE/IDENTITY COMPARATORS

D2617, JANUARY 1981—REVISED DECEMBER 1983

- Compares Two 8-Bit Words
- Choice of Totem-Pole or Open-Collector Outputs
- Hysteresis at P and Q Inputs
- 'LS682 and 'LS683 have 20-kΩ Pullup Resistors on the Q Inputs
- 'LS686 and 'LS687 . . . New JT and NT Packages
- 'LS686 and 'LS687 . . . New JT and NT 24-Pin, 3000-Mil Packages

| TYPE | P = Q̄ | P > Q̄ | OUTPUT ENABLE | OUTPUT CONFIGURATION | 20-kΩ PULLUP |
|---|---|---|---|---|---|
| 'LS682 | yes | yes | no | totem-pole | yes |
| 'LS683 | yes | yes | no | open-collector | yes |
| 'LS684 | yes | yes | no | totem-pole | no |
| 'LS685 | yes | yes | no | open-collector | no |
| 'LS686 | yes | yes | yes | totem-pole | no |
| 'LS687 | yes | yes | yes | open-collector | no |
| 'LS688 | yes | yes | yes | totem-pole | no |
| 'LS689 | yes | no | yes | open-collector | no |

**description**

These magnitude comparators perform comparisons of two eight-bit binary or BCD words. All types provide P = Q̄ outputs and the 'LS682 thru 'LS687 provide P̄ > Q̄ outputs as well. The 'LS682, 'LS684, 'LS686, and 'LS688 have totem-pole outputs, while the 'LS683, 'LS685, 'LS687, and 'LS689 have open-collector outputs. The 'LS682 and 'LS683 feature 20-kΩ pullup termination resistors on the Q inputs for analog or switch data.

**FUNCTION TABLE**

| INPUTS | | | | OUTPUTS | |
|---|---|---|---|---|---|
| DATA | ENABLES | | | | |
| P, Q | Ḡ, Ḡ1 | Ḡ2 | | P = Q̄ | P̄ > Q̄ |
| P = Q | L | X | | L | H |
| P > Q | L | X | | H | L |
| P < Q | L | X | | H | H |
| P > Q | X | H | | H | H |
| P > Q | H | X | | H | H |
| X | H | H | | H | H |

**NOTES:**
1. The last three lines of the function table applies only to the devices having enable inputs, i.e., 'LS686 thru 'LS689.
2. The P = Q̄ or P > Q̄ function can be generated by applying the P = Q̄ and P̄ > Q̄ outputs to a 2-input NAND gate.
3. For 'LS686, 'LS687 G1 enables P = Q̄, and G2 enables P̄ > Q̄.

**logic symbols**

Pin numbers shown on logic notation are for DW, J, JT, N or NT packages.

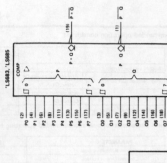
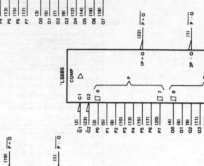

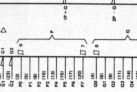
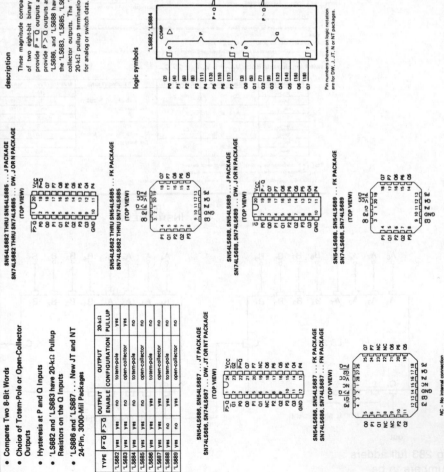

SN54LS682 THRU SN54LS685 . . . J PACKAGE
SN74LS682 THRU SN74LS685 . . . DW, J OR N PACKAGE
(TOP VIEW)

SN54LS682 THRU SN54LS685
SN74LS682 THRU SN74LS685 . . . FK PACKAGE
(TOP VIEW)

SN54LS686, SN54LS689 . . . J PACKAGE
SN74LS686, SN74LS689 . . . DW, J OR N PACKAGE
(TOP VIEW)

SN54LS688, SN54LS689
SN74LS688, SN74LS689 . . . FK PACKAGE
(TOP VIEW)

SN54LS686, SN54LS687 . . . JT PACKAGE
SN74LS686, SN74LS687 . . . DW, JT OR NT PACKAGE
(TOP VIEW)

SN54LS688, SN54LS687 . . . FK PACKAGE
SN74LS688, SN74LS687 . . . FN PACKAGE
(TOP VIEW)

NC — No internal connection

TEXAS INSTRUMENTS

**Fig. 6.18** Data sheet for the 682 comparator (*courtesy of Texas Instruments*)

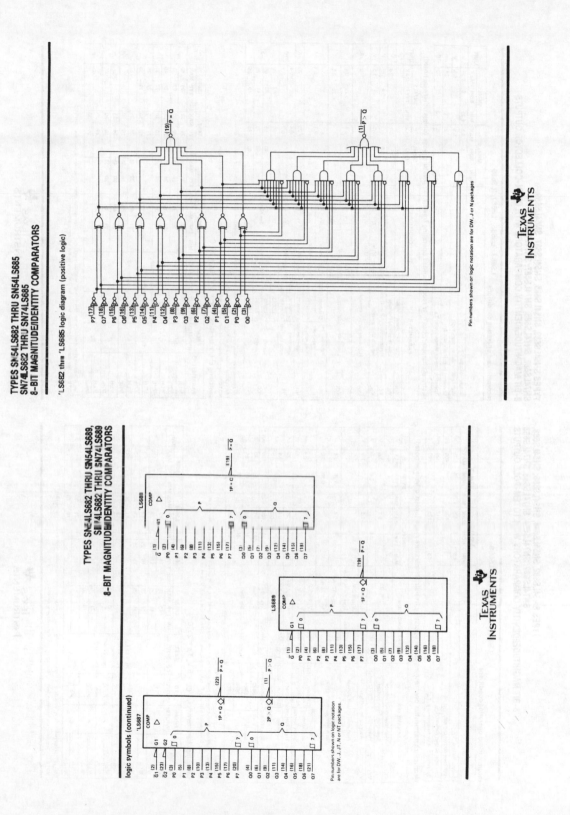

TYPES SN54LS682 THRU SN54LS685
SN74LS682 THRU SN74LS685
8–BIT MAGNITUDE/IDENTITY COMPARATORS

'LS682 thru 'LS685 logic diagram (positive logic)

Pin numbers shown on logic notation are for DW, J or N packages

TEXAS
INSTRUMENTS

TYPES SN54LS682 THRU SN54LS689,
SN74LS682 THRU SN74LS689
8–BIT MAGNITUDE/IDENTITY COMPARATORS

logic symbols (continued)

Pin numbers shown on logic notation
are for DW, J, JT, N or NT packages.

TEXAS
INSTRUMENTS

## TYPES SN54LS683, SN54LS685, SN54LS687, SN54LS689, SN74LS683, SN74LS685, SN74LS687, SN74LS689
## 8-BIT MAGNITUDE/IDENTITY COMPARATORS WITH OPEN-COLLECTOR OUTPUTS

**recommended operating conditions**   'LS683, 'LS685, 'LS687, 'LS689

| | SN54LS' MIN | NOM | MAX | SN74LS' MIN | NOM | MAX | UNIT |
|---|---|---|---|---|---|---|---|
| Supply voltage, V_CC | 4.5 | 5 | 5.5 | 4.75 | 5 | 5.25 | V |
| High-level output voltage, V_OH | | | 5.5 | | | 5.5 | V |
| Low-level output current, I_OL | | | 12 | | | 24 | mA |
| Operating free-air temperature, T_A | -55 | | 125 | 0 | | 70 | °C |

**electrical characteristics over recommended operating free-air temperature range (unless otherwise noted)**

| PARAMETER | | TEST CONDITIONS† | SN54LS' MIN | TYP‡ | MAX | SN74LS' MIN | TYP‡ | MAX | UNIT |
|---|---|---|---|---|---|---|---|---|---|
| V_IH | High-level input voltage | | 2 | | | 2 | | | V |
| V_IL | Low-level input voltage | | | | 0.7 | | | 0.8 | V |
| V_T+ − V_T− | Hysteresis  P or Q inputs | V_CC = MIN | | 0.4 | | | 0.4 | | V |
| V_IK | Input clamp voltage | V_CC = MIN, I_I = −18 mA | | | −1.5 | | | −1.5 | V |
| I_OH | | V_CC = MIN, V_IH = 2 V, V_OH = 5.5 V | | 250 | | | 100 | | µA |
| V_OL | High-level output voltage | V_CC = MIN, V_IH = 2 V, V_IL = V_IL max, I_OL = 12 mA | 0.25 | 0.4 | | 0.25 | 0.4 | | V |
| | Low-level output voltage | V_IH = 2 V, V_IL = V_IL max, I_OL = 24 mA | 0.35 | 0.5 | | 0.35 | 0.5 | | V |
| I_I | Input current at maximum input voltage  Q inputs, 'LS683 | V_CC = MAX, V_I = 5.5 V | | | 0.1 | | | 0.1 | mA |
| | All other inputs | V_CC = MAX, V_I = 7 V | | | | | | | |
| I_IH | High-level input current  Q inputs, 'LS683 | V_CC = MAX, V_I = 2.7 V | | | 20 | | | 20 | µA |
| | All other inputs | | | | −0.4 | | | −0.4 | mA |
| I_IL | Low-level input current  All other inputs | V_CC = MAX, V_I = 0.4 V | | | −0.2 | | | −0.2 | mA |
| I_CC | Supply current  'LS683 | V_CC = MAX, See Note 2 | | 42 | 70 | | 42 | 70 | mA |
| | 'LS687 | | | 40 | 65 | | 40 | 65 | |
| | 'LS689 | | | 44 | 75 | | 44 | 75 | |
| | | | | 40 | 65 | | 40 | 65 | |

† For conditions shown as MIN or MAX, use the appropriate value specified under recommended operating conditions.
‡ All typical values are at V_CC = 5 V, T_A = 25°C.
NOTE 2: I_CC is measured with any G inputs grounded, all other inputs at 4.5 V, and all outputs open.

**switching characteristics, V_CC = 5 V, T_A = 25°C**

| PARAMETER# | FROM (INPUTS) | TO (OUTPUT) | TEST CONDITIONS | 'LS683 MIN | TYP | MAX | 'LS685 MIN | TYP | MAX | 'LS687 MIN | TYP | MAX | 'LS689 MIN | TYP | MAX | UNIT |
|---|---|---|---|---|---|---|---|---|---|---|---|---|---|---|---|---|
| t_PLH | P | P=Q̄ | R_L = 667 Ω, C_L = 45 pF, All other inputs low, See Note 3 | | 20 | 30 | | 30 | 45 | | 20 | 30 | | 24 | 35 | ns |
| t_PHL | | | | | 24 | 35 | | 19 | 35 | | 24 | 35 | | 22 | 35 | |
| t_PLH | Q | P=Q̄ | | | 23 | 35 | | 23 | 35 | | 24 | 35 | | 22 | 35 | ns |
| t_PHL | | | | | 21 | 35 | | 18 | 35 | | 18 | 30 | | 19 | 30 | |
| t_PLH | G, G1 | P>Q | | | 31 | 45 | | 32 | 45 | | 24 | 35 | | | | ns |
| t_PHL | | | | | 17 | 30 | | 16 | 35 | | 16 | 30 | | | | |
| t_PLH | Q | P>Q | | | 30 | 45 | | 30 | 45 | | 24 | 35 | | | | ns |
| t_PHL | | | | | 21 | 30 | | 20 | 35 | | 16 | 30 | | | | |
| t_PLH | G2 | P>Q | | | | | | 24 | 35 | | | | | | | ns |
| t_PHL | | | | | | | | 15 | 30 | | | | | | | |

# t_PLH = propagation delay time, low-to-high-level output; t_PHL = propagation delay time, high-to-low-level output.
NOTE 3: See General Information Section for load circuits and voltage waveforms.

**TEXAS INSTRUMENTS**

---

## TYPES SN54LS682, SN54LS684, SN54LS686, SN54LS688, SN74LS682, SN74LS684, SN74LS686, SN74LS688
## 8-BIT MAGNITUDE/IDENTITY COMPARATORS WITH TOTEM-POLE OUTPUTS

**recommended operating conditions**   'LS682, 'LS684, 'LS686, 'LS688

| | SN54LS' MIN | NOM | MAX | SN74LS' MIN | NOM | MAX | UNIT |
|---|---|---|---|---|---|---|---|
| Supply voltage, V_CC | 4.5 | 5 | 5.5 | 4.75 | 5 | 5.25 | V |
| High-level output current, I_OH | | | −400 | | | −400 | µA |
| Low-level output current, I_OL | | | 12 | | | 24 | mA |
| Operating free-air temperature, T_A | -55 | | 125 | 0 | | 70 | °C |

**electrical characteristics over recommended operating free-air temperature range (unless otherwise noted)**

| PARAMETER | | TEST CONDITIONS† | SN54LS' MIN | TYP‡ | MAX | SN74LS' MIN | TYP‡ | MAX | UNIT |
|---|---|---|---|---|---|---|---|---|---|
| V_IH | High-level input voltage | | 2 | | | 2 | | | V |
| V_IL | Low-level input voltage | | | | 0.7 | | | 0.8 | V |
| V_T+ − V_T− | Hysteresis  P or Q inputs | V_CC = MIN | | 0.4 | | | 0.4 | | V |
| V_IK | Input clamp voltage | V_CC = MIN, I_I = −18 mA | | | −1.5 | | | −1.5 | V |
| V_OH | High-level output voltage | V_CC = MIN, V_IL = V_IL max, V_IH = 2 V, I_OH = −400 µA | 2.5 | | | 2.7 | | | V |
| V_OL | Low-level output voltage | V_CC = MIN, V_IH = 2 V, V_IL = V_IL max, I_OL = 12 mA | 0.25 | 0.4 | | 0.25 | 0.4 | | V |
| | | I_OL = 24 mA | 0.35 | 0.5 | | 0.35 | 0.5 | | V |
| I_I | Input current at maximum input voltage  Q inputs, 'LS682 | V_CC = MAX, V_I = 5.5 V | | | 0.1 | | | 0.1 | mA |
| | All other inputs | V_CC = MAX, V_I = 7 V | | | | | | | |
| I_IH | High-level input current  Q inputs, 'LS682 | V_CC = MAX, V_I = 2.7 V | | | 20 | | | 20 | µA |
| | All other inputs | | | | −0.4 | | | −0.4 | mA |
| I_IL | Low-level input current  All other inputs | V_CC = MAX, V_I = 0.4 V | | | −0.2 | | | −0.2 | mA |
| I_OS § | Short-circuit output current | V_CC = MAX, V_O = 0 | −20 | | −100 | −20 | | −100 | mA |
| I_CC | Supply current  'LS682 | V_CC = MAX, See Note 2 | | 42 | 70 | | 42 | 70 | mA |
| | 'LS684 | | | 40 | 65 | | 40 | 65 | |
| | 'LS686 | | | 44 | 75 | | 44 | 75 | |
| | 'LS688 | | | 40 | 65 | | 40 | 65 | |

‡ All typical values are at V_CC = 5 V, T_A = 25°C.
§ Not more than one output should be shorted at a time, and duration of the short-circuit should not exceed one second.
NOTE 2: I_CC is measured with any G inputs grounded, all other inputs at 4.5 V, and all outputs open.

**switching characteristics, V_CC = 5 V, T_A = 25°C**

| PARAMETER# | FROM (INPUTS) | TO (OUTPUT) | TEST CONDITIONS | 'LS682 MIN | TYP | MAX | 'LS684 MIN | TYP | MAX | 'LS686 MIN | TYP | MAX | 'LS688 MIN | TYP | MAX | UNIT |
|---|---|---|---|---|---|---|---|---|---|---|---|---|---|---|---|---|
| t_PLH | P | P=Q̄ | R_L = 667 Ω, C_L = 45 pF, All other inputs low, See Note 3 | | 13 | 25 | | 15 | 25 | | 13 | 25 | | 18 | 27 | ns |
| t_PHL | | | | | 15 | 25 | | 17 | 25 | | 20 | 30 | | 20 | 30 | |
| t_PLH | Q | P=Q̄ | | | 14 | 25 | | 16 | 25 | | 13 | 25 | | 18 | 27 | ns |
| t_PHL | | | | | 15 | 25 | | 15 | 25 | | 11 | 20 | | 20 | 30 | |
| t_PLH | G, G1 | P>Q | | | | | | 22 | 30 | | 19 | 30 | | 12 | 18 | ns |
| t_PHL | | | | | 20 | 30 | | 17 | 30 | | 15 | 30 | | 13 | 20 | |
| t_PLH | P | P>Q | | | 15 | 30 | | 21 | 30 | | 18 | 30 | | | | ns |
| t_PHL | | | | | 21 | 30 | | 24 | 30 | | 19 | 30 | | | | |
| t_PLH | Q | P>Q | | | 19 | | | 20 | 30 | | 21 | 30 | | | | ns |
| t_PHL | G2 | P>Q | | | | | | | | | 16 | 25 | | | | |

# t_PLH = propagation delay time, low-to-high-level output; t_PHL = propagation delay time, high-to-low-level output.
NOTE 3: See General Information Section for load circuits and voltage waveforms.

**TEXAS INSTRUMENTS**

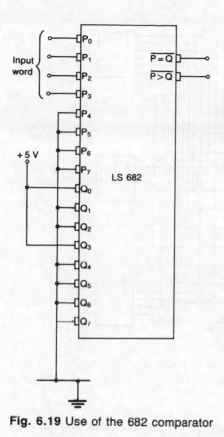

**Fig. 6.19** Use of the 682 comparator

*Example 6.7*

Design a circuit that will indicate when an input digital word is not a valid BCD code.

*Solution*

Valid BCD numbers are 0 through to 9. Any number greater than 9 is not valid. Denary 9 = 1001 in binary. If the data word to be checked is connected to the inputs $P_0$ through to $P_3$, then the inputs $P_4$ through to $P_7$ must be connected to earth. The binary number 1001 must then be set up on the Q input. This means that $Q_0$ and $Q_3$ are connected to +5 V and all the other Q inputs are connected to earth (see Fig. 6.19). If the input word is equal to denary 10 or more the P > Q output will go LOW to indicate that an invalid BCD word is present.

A decoder/demultiplexer can also be used to act as a comparator for 2-bit numbers. The 74LS 154 4-to-16 line decoder has four input bits which can be grouped into 2-bit numbers X and Y, where X = A and B and Y = C and D. From the truth table of the device Table 6.4 has been obtained. The inputs A,B,C and D are exactly the same as in the data sheet except that high/low has been replaced by 1/0.

When either X < Y, or Y < X or X = Y, a 1 has been entered in the appropriate right-hand column. From this right-hand column:

$$X < Y = \overline{4} + \overline{8} + \overline{9} + \overline{12} + \overline{13} + \overline{14} = \overline{4.8.9.12.13.14}$$
$$Y < X = \overline{1} + \overline{2} + \overline{3} + \overline{6} + \overline{7} + \overline{11} = \overline{1.2.3.6.7.11}$$
$$X = Y = \overline{0} + \overline{5} + \overline{10} + \overline{15} = \overline{0.5.10.15}$$

The circuit of the decoder-implemented comparator is shown by Fig. 6.20.

**Table 6.4**

| Y | | X | | Output | Y < X | X < Y | X = Y |
|---|---|---|---|--------|-------|-------|-------|
| 0 | 0 | 0 | 0 | 0 | 0 | 0 | 1 |
| 0 | 0 | 0 | 1 | 1 | 1 | 0 | 0 |
| 0 | 0 | 1 | 0 | 2 | 1 | 0 | 0 |
| 0 | 0 | 1 | 1 | 3 | 1 | 0 | 0 |
| 0 | 1 | 0 | 0 | 4 | 0 | 1 | 0 |
| 0 | 1 | 0 | 1 | 5 | 0 | 0 | 1 |
| 0 | 1 | 1 | 0 | 6 | 1 | 0 | 0 |
| 0 | 1 | 1 | 1 | 7 | 1 | 0 | 0 |
| 1 | 0 | 0 | 0 | 8 | 0 | 1 | 0 |
| 1 | 0 | 0 | 1 | 9 | 0 | 1 | 0 |
| 1 | 0 | 1 | 0 | 10 | 0 | 0 | 1 |
| 1 | 0 | 1 | 1 | 11 | 1 | 0 | 0 |
| 1 | 1 | 0 | 0 | 12 | 0 | 1 | 0 |
| 1 | 1 | 0 | 1 | 13 | 0 | 1 | 0 |
| 1 | 1 | 1 | 0 | 14 | 0 | 1 | 0 |
| 1 | 1 | 1 | 1 | 15 | 0 | 0 | 1 |

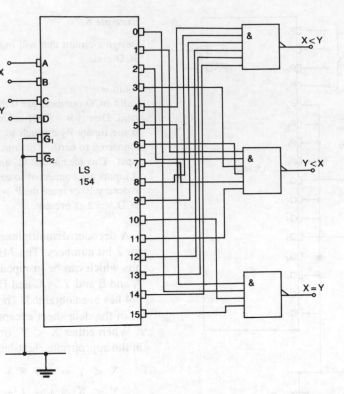

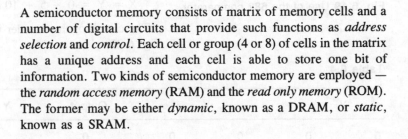

**Fig. 6.20** Comparator implemented using the 682 decoder

## Memories

A semiconductor memory consists of matrix of memory cells and a number of digital circuits that provide such functions as *address selection* and *control*. Each cell or group (4 or 8) of cells in the matrix has a unique address and each cell is able to store one bit of information. Two kinds of semiconductor memory are employed — the *random access memory* (RAM) and the *read only memory* (ROM). The former may be either *dynamic*, known as a DRAM, or *static*, known as a SRAM.

### Random Access Memory

The memory cells in a DRAM each consist of a single MOSFET and capacitance as shown by Fig. 6.21. Binary 1 is represented by a certain charge being stored in the capacitance; binary 0 is represented by a smaller charge which is in some DRAMs very nearly zero. Row and column decoders select one cell, or group of cells, in the matrix according to the input address (see Fig. 6.22). The row decoder selects one of the word lines by taking it HIGH and data from all the cells on this line are connected to the sense amplifiers. A particular sense amplifier is selected by using the column decoder to address a data line. The selected sense amplifier is connected to the output circuit. When a DRAM is read the data in the addressed cell are connected

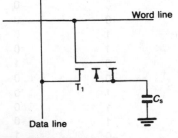

**Fig. 6.21** DRAM cell

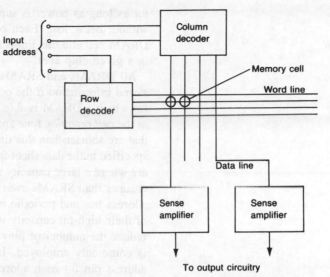

**Fig. 6.22** Selection of a cell in a DRAM

to the selected word line; this is HIGH and so the transistor $T_1$ turns ON, allowing the charge stored in capacitance $C_s$ to appear on the data line. The selected sense amplifier then detects the change in the data line's voltage that indicates whether the stored information was binary 1 or 0 and outputs the correct voltage level.

The charge stored in the capacitance of a cell will leak away in a few milliseconds and the stored data will be lost. It is therefore necessary for each cell to have its charge restored, or *refreshed*, every few milliseconds. The internal circuitry of a DRAM includes the necessary refreshing circuitry.

The memory cells in a SRAM each consist of a flip-flop that is set to store binary 1 and reset to store binary 0. The basic circuit of a CMOS cell is shown by Fig. 6.23. The data stored in a cell is retained

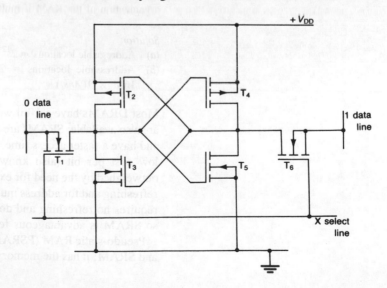

**Fig. 6.23** CMOS SRAM cell

for as long as power is supplied to the chip and it can be read out without being lost. Each cell occupies a greater area than does a DRAM cell and this limits the number of cells that can be provided in a given chip area.

All DRAMs and SRAMs are volatile devices, i.e. they lose their stored information if the power supplies are switched off. The time for which a DRAM is able to retain its stored information is known as the *cell retention time* and it must be refreshed at regular intervals that are shorter than this time. The minimum *refresh rate* is usually specified in the data sheet for a DRAM. The applications of DRAMs are where a large capacity memory is required since they tend to be cheaper than SRAMs even though there is a need for a multiplexed address bus and periodic refreshing. The former are used because of their high bit capacity which is due to their small cell area. To reduce the number of pins on the IC package address multiplexing is commonly employed. In a small-capacity memory there is an address pin for each address line. A memory with 1k addressable locations requires 10 address pins, for example. As the size of the memory is increased it becomes increasingly difficult to provide enough address pins and so multiplexed addresses are employed. The usual method is to employ one-half of the address pins that would be required without multiplexing and to use two extra pins $\overline{RAS}$ and $\overline{CAS}$. The $\overline{RAS}$ (row address strobe) pin when taken LOW indicates that the row address is being applied to the memory and the $\overline{CAS}$ (column address strobe) pin taken LOW indicates that the column address is applied.

### Example 6.8

A RAM has 12 address pins and one data input/output pin. Determine the organization of the RAM if multiplexing is (*a*) not used, and (*b*) used.

*Solution*
(*a*) Addressable locations = $2^{12}$ = 4096, so organization = 4k × 1.(*Ans.*)
(*b*) Addressable locations = $2^{24}$ = 16.78 × $10^6$, so organization = 16M × 1(*Ans.*)

Most DRAMs have a 1-bit word width although 4-bit width devices are also available; SRAMs are available in × 1, × 4 and × 8 versions and have a faster access time. The main advantage of DRAMs is the low cost per bit (also known as high density) but this is partly outweighed by the need for external control signals for the necessary refreshing and for address multiplexing. On the other hand, a SRAM requires no refreshing and does not use multiplexed addressing and so SRAM is advantageous for small memory systems.

Pseudo-static RAM (PSRAM) combines the advantages of DRAM and SRAM; it has the memory cells and refresh circuitry of DRAM, and the input/output circuitry of SRAM. A PSRAM appears to the other circuitry to be a DRAM and is cheaper than a DRAM; they are suitable for use in applications where high speed and low standby current are not important factors.

Recently, a new type of RAM has been developed that uses ferroelectric material to produce the storage capacitor in each memory cell. It offers an advantage in that the capacitance remains charged when the power supplies are switched off and hence it can provide permanent storage. A FRAM may be obtained in either DRAM or SRAM versions.

A form of non-volatile RAM can be obtained by using CMOS technology with a back-up battery. CMOS SRAM has a data retention mode in which the standby power dissipation is very low and the supply voltage can be reduced to just 2 V. This enables a battery to be used to provide a back-up system to retain the data when the power supply is off.

### Read-only Memory

The read-only memory (ROM) is a non-volatile device that stores information relatively long term. Long term may mean permanently with some types with which the stored data cannot be altered. These ROMs include those that are programmed by the manufacturer, masked ROM, and those that are programmed by the user, programmable ROM (PROM). The information may be read out of the memory but no new data can be written into it.

Other ROMs can have some, or all, of their stored data erased and new data programmed in whenever it is required to change the information. The oldest of these devices, known as *erasable PROM*, or EPROM or UVEPROM, has its stored data erased by exposing the device to ultraviolet light. A PROM has a transparent window in its package to allow the ultraviolet light to enter; care must be taken during use to ensure that the device is not exposed to daylight or the data will be lost over a period of time. All of the data stored in the memory must be erased at the same time. Once the unwanted data has been erased the EPROM can be reprogrammed, but since this may take 20 minutes or more to accomplish a PROM is not suitable to store frequently updated data. This reprogramming can be repeated many times.

The next development after the EPROM was the electrically erasable PROM, or EEPROM, which can have its stored data removed by applying a relatively high voltage (about 20 V) to the appropriate input. It is not necessary to erase and reprogram all the memory at the same time; if required selected bits only can be reprogrammed. The reprogramming process can take place in milliseconds. Recent devices automatically erase existing data as part of the WRITE cycle. They also contain some of the needed voltage-generation and pulse-shaping circuits on-chip. The latest EEPROMS have all the necessary voltage, shaping and timing functions on-chip. Their WRITE procedure is then the same as for a SRAM except that it takes much longer, typically about 10 ms. Unfortunately, EEPROMs are more expensive than EPROMs. Each time an EEPROM is wiped clean some damage is done to it and eventually (after some thousands of erasures) it may no longer be capable of storing information.

A NOVRAM is a memory that contains both SRAM and EEPROM on the same chip. Two additional pins, $\overline{\text{STORE}}$ and $\overline{\text{RECALL}}$, are used that control the movement of data between the SRAM and the EEPROM sections of the device. When the $\overline{\text{STORE}}$ pin is taken LOW the contents of the SRAM are transferred to the EEPROM. After a STORE operation has been carried out the same data is then held in both sections of the memory. When the $\overline{\text{RECALL}}$ pin is taken LOW the contents of the EEPROM are transferred to the SRAM and then both parts of the memory hold the data that was originally held in just the EEPROM. A NOVRAM has a lower density than EEPROM but a much shorter WRITE time. A *flash ROM* (FROM) is a form of ROM that can be erased electrically but the complete FROM must be erased in one go. The damage caused by reprogramming is much less than with a EEPROM so the device should have a much longer life. When the programming input pin $V_{pp}$ has no voltage applied to it the FROM acts just like an EPROM; when 12 V are applied to this terminal a register is enabled on to the address and the data buses to provide a high-level interface. This interface allows 'in-system write' reprogramming of non-volatile memory to be accomplished. Program updates can be carried fairly easily.

### Organization of a Memory

The capacity of a memory is the number of bits of information that it is able to store. The capacity may range from as little as 256 bits to (currently) 16 Megabits. The way in which the memory cells are connected together is known as the *organization* of the memory. Memories are organized so that 1, or 4, or 8 bits can be written into, or read out, simultaneously, i.e. the word width is 1, 4 or 8 bits. Thus a 16-kbit memory could be organized as 16k $\times$ 1, 4k $\times$ 4, or 2k $\times$ 8; in each case the total number of bits that can be stored is equal to 16 384. Often 8-bit organizations are referred to as *words*.

### Memory Technology

The majority of memories use CMOS technology because each cell then takes practically zero current from the power supply when it is not in the process of switching from one state to the other. Since a

**MOTOROLA**
# SEMICONDUCTOR
**TECHNICAL DATA**

## *Advance Information*

# 32K × 8 Bit Fast Static Random Access Memory

The MCM6206 is a 262,144 bit static random access memory organized as 32,768 words of 8 bits, fabricated using Motorola's high-performance silicon-gate CMOS technology. Static design eliminates the need for external clocks or timing strobes, while CMOS circuitry reduces power consumption and provides for greater reliability.

Chip enable ($\overline{E}$) controls the power-down feature. It is not a clock but rather a chip control that affects power consumption. In less than a cycle time after $\overline{E}$ goes high, the part automatically reduces its power requirements and remains in this low-power standby mode as long as $\overline{E}$ remains high. This feature provides significant system-level power savings. Another control feature, output enable ($\overline{G}$) allows access to the memory contents as fast as 12.5 ns (MCM6206-30).

The MCM6206 is packaged in a 300 or 600 mil, 28 pin plastic dual-in-line package or a 28 lead 300 or 400 mil plastic SOJ package with the JEDEC standard pinout.

- Single 5 V Supply, ±10%
- Fully Static—No Clock or Timing Strobes Necessary
- Fast Access Time—30, 35, or 45 ns (Maximum)
- Low Power Dissipation
- Two Chip Controls; $\overline{E}$ for Automatic Power Down
  $\overline{G}$ for Fast Access to Data
- Three State Outputs
- Fully TTL Compatible

### MCM6206

**NP PACKAGE**
**300 MIL PLASTIC**
**CASE 710B**

**P PACKAGE**
**600 MIL PLASTIC**
**CASE 710**

**J PACKAGE**
**400 MIL SOJ**
**CASE 810**

**NJ PACKAGE**
**300 MIL SOJ**
**CASE 810B**

**PIN ASSIGNMENT**

| | | | | |
|---|---|---|---|---|
| A14 | 1 | | 28 | V_CC |
| A12 | 2 | | 27 | $\overline{W}$ |
| A7 | 3 | | 26 | A13 |
| A6 | 4 | | 25 | A8 |
| A5 | 5 | | 24 | A9 |
| A4 | 6 | | 23 | A11 |
| A3 | 7 | | 22 | $\overline{G}$ |
| A2 | 8 | | 21 | A10 |
| A1 | 9 | | 20 | $\overline{E}$ |
| A0 | 10 | | 19 | DQ7 |
| DQ0 | 11 | | 18 | DQ6 |
| DQ1 | 12 | | 17 | DQ5 |
| DQ2 | 13 | | 16 | DQ4 |
| V_SS | 14 | | 15 | DQ3 |

**PIN NAMES**

| | |
|---|---|
| A0–A14 | Address |
| $\overline{W}$ | Write Enable |
| $\overline{E}$ | Chip Enable |
| $\overline{G}$ | Output Enable |
| DQ0–DQ7 | Data Input/Output |
| V_CC | +5 V Power Supply |
| V_SS | Ground |

**BLOCK DIAGRAM**

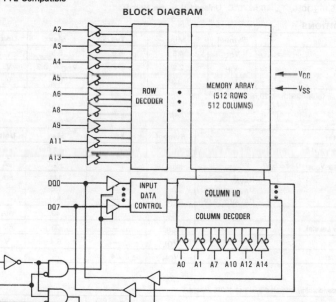

This document contains information on a new product. Specifications and information herein are subject to change without notice.

ADI1501R2

**Fig. 6.24** Data sheet for the MCM 6206 32k × 8 SRAM (*courtesy of Motorola*)

## TRUTH TABLE

| $\overline{E}$ | $\overline{G}$ | $\overline{W}$ | Mode | Supply Current | I/O Pin |
|---|---|---|---|---|---|
| H | X | X | Not Selected | $I_{SB}$ | High Z |
| L | H | H | Output Disabled | $I_{CC}$ | High Z |
| L | L | H | Read | $I_{CC}$ | $D_{out}$ |
| L | X | L | Write | $I_{CC}$ | $D_{in}$ |

X—Don't Care

This device contains circuitry to protect the inputs against damage due to high static voltages or electric fields; however, it is advised that normal precautions be taken to avoid application of any voltage higher than maximum rated voltages to this high-impedance circuit.

## ABSOLUTE MAXIMUM RATINGS (See Note)

| Rating | Symbol | Value | Unit |
|---|---|---|---|
| Power Supply Voltage | $V_{CC}$ | $-0.5$ to $+7.0$ | V |
| Voltage Relative to $V_{SS}$ for Any Pin Except $V_{CC}$ | $V_{in}, V_{out}$ | $-0.5$ to $V_{CC}+0.5$ | V |
| Output Current (per I/O) | $I_{out}$ | $\pm20$ | mA |
| Power Dissipation ($T_A=25°C$) | $P_D$ | 1.0 | W |
| Temperature Under Bias | $T_{bias}$ | $-10$ to $+85$ | °C |
| Operating Temperature | $T_A$ | 0 to $+70$ | °C |
| Storage Temperature—Plastic | $T_{stg}$ | $-55$ to $+125$ | °C |

NOTE: Permanent device damage may occur if ABSOLUTE MAXIMUM RATINGS are exceeded. Functional operation should be restricted to RECOMMENDED OPERATING CONDITIONS. Exposure to higher than recommended voltages for extended periods of time could affect device reliability.

## DC OPERATING CONDITIONS AND CHARACTERISTICS

($V_{CC}=5.0 \pm 10\%$, $T_A=0$ to 70°C, Unless Otherwise Noted)

### RECOMMENDED OPERATING CONDITIONS

| Parameter | Symbol | Min | Typ | Max | Unit |
|---|---|---|---|---|---|
| Supply Voltage (Operating Voltage Range) | $V_{CC}$ | 4.5 | 5.0 | 5.5 | V |
| Input High Voltage | $V_{IH}$ | 2.2 | — | $V_{CC}+0.3$ | V |
| Input Low Voltage | $V_{IL}$ | $-0.3*$ | — | 0.8 | V |

*$V_{IL}$ (min) = $-0.3$ V dc; $V_{IL}$ (min) = $-3.0$ V ac (pulse width ≤ 20 ns)

### DC CHARACTERISTICS

| Parameter | | Symbol | Min | Max | Unit |
|---|---|---|---|---|---|
| Input Leakage Current (All Inputs, $V_{in}=0$ to $V_{CC}$) | | $I_{lkg(I)}$ | — | $\pm1.0$ | $\mu A$ |
| Output Leakage Current ($\overline{E}=V_{IH}$, or $\overline{G}=V_{IH}$, $V_{out}=0$ to 5.5 V) | | $I_{lkg(O)}$ | — | $\pm1.0$ | $\mu A$ |
| Power Supply Current ($\overline{E}=V_{IL}$, $I_{out}=0$) | ($t_{AVAV}=30$ ns) | $I_{CC}$ | — | 145 | mA |
| | ($t_{AVAV}=35$ ns) | | — | 135 | |
| | ($t_{AVAV}=45$ ns) | | — | 130 | |
| Standby Current ($\overline{E}=V_{IH}$) (TTL Levels) | | $I_{SB1}$ | — | 40 | mA |
| Standby Current ($\overline{E} \geq V_{CC}-0.2$ V) (CMOS Levels) | | $I_{SB2}$ | — | 20 | mA |
| Output Low Voltage ($I_{OL}=8.0$ mA) | | $V_{OL}$ | — | 0.4 | V |
| Output High Voltage ($I_{OH}=-4.0$ mA) | | $V_{OH}$ | 2.4 | — | V |

### CAPACITANCE (f = 1.0 MHz, $T_A=25°C$, periodically sampled and not 100% tested.)

| Characteristic | | Symbol | Max | Unit |
|---|---|---|---|---|
| Input Capacitance | All Inputs Except $\overline{W}$ | $C_{in}$ | 6 | pF |
| | $\overline{W}$ | | 8 | |
| I/O Capacitance | | $C_{I/O}$ | 8 | pF |

## AC OPERATING CONDITIONS AND CHARACTERISTICS
($V_{CC}$=5 V ±10%, $T_A$=0 to 70°C, Unless Otherwise Noted)

Input Pulse Levels . . . . . . . . . . . . . . . . . . . . . . .0 to 3.0 V

Input Rise/Fall Time . . . . . . . . . . . . . . . . . . . . . . . .5 ns

Input Timing Measurement Reference Levels . . . . . . . . . 1.5 V

Output Timing Measurement Reference Levels . . . . . . . . 1.5 V

Output Load. . . . . . . . . . . . . . . . . . . . . . . .See Figure 1

### READ CYCLE 1 & 2 (See Note 1)

| Parameter | Symbol | Alt Symbol | MCM6206-30 | | MCM6206-35 | | MCM6206-45 | | Unit | Notes |
|---|---|---|---|---|---|---|---|---|---|---|
| | | | Min | Max | Min | Max | Min | Max | | |
| Read Cycle Time | $t_{AVAV}$ | $t_{RC}$ | 30 | — | 35 | — | 45 | — | ns | — |
| Address Access Time | $t_{AVQV}$ | $t_{AA}$ | — | 30 | — | 35 | — | 45 | ns | — |
| $\overline{E}$ Access Time | $t_{ELQV}$ | $t_{AC}$ | — | 30 | — | 35 | — | 45 | ns | — |
| $\overline{G}$ Access Time | $t_{GLQV}$ | $t_{OE}$ | — | 12.5 | — | 15 | — | 20 | ns | — |
| Enable Low to Enable High | $t_{ELEH}$ | $t_{CW}$ | 30 | — | 35 | — | 45 | — | ns | — |
| Output Hold from Address Change | $t_{AXQX}$ | $t_{OH}$ | 5 | — | 5 | — | 5 | — | ns | 2 |
| Chip Enable to Output Low-Z | $t_{ELQX}$ | $t_{CLZ}$ | 10 | — | 10 | — | 10 | — | ns | 2, 3 |
| Output Enable to Output Low-Z | $t_{GLQX}$ | $t_{OLZ}$ | 0 | — | 0 | — | 0 | — | ns | 2, 3 |
| Chip Enable to Output High-Z | $t_{EHQZ}$ | $t_{CHZ}$ | 0 | 20 | 0 | 20 | 0 | 20 | ns | 2, 3 |
| Output Enable to Output High-Z | $t_{GHQZ}$ | $t_{OHZ}$ | 0 | 17 | 0 | 17 | 0 | 17 | ns | 2, 3 |

NOTES:

1. $\overline{W}$ is high at all times for read cycles.
2. All high-Z and low-Z parameters are considered in a high or low impedance state when the output has made a 500 mV transition from the previous steady state voltage.
3. These parameters are periodically sampled and not 100% tested.

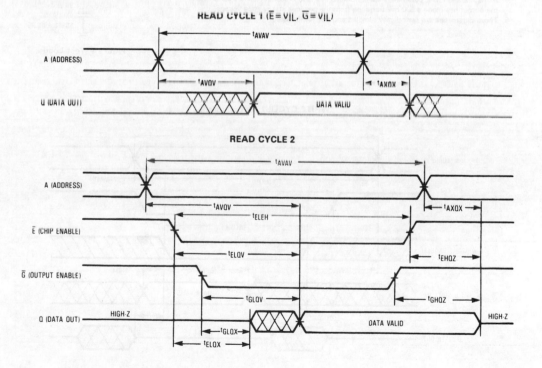

**READ CYCLE 1** ($\overline{E}$ = $V_{IL}$, $\overline{G}$ = $V_{IL}$)

**READ CYCLE 2**

## WRITE CYCLE 1 & 2 (See Note 1)

| Parameter | Symbol | Alt Symbol | MCM6206-30 | | MCM6206-35 | | MCM6206-45 | | Unit | Notes |
|---|---|---|---|---|---|---|---|---|---|---|
| | | | Min | Max | Min | Max | Min | Max | | |
| Write Cycle Time | $t_{AVAV}$ | $t_{WC}$ | 30 | — | 35 | — | 45 | — | ns | — |
| Address Setup to Write Low<br>Address Setup to Enable Low | $t_{AVWL}$<br>$t_{AVEL}$ | $t_{AS}$ | 0 | — | 0 | — | 0 | — | ns | 2 |
| Address Valid to Write High<br>Address Valid to Enable High | $t_{AVWH}$<br>$t_{AVEH}$ | $t_{AW}$ | 25 | — | 25 | — | 35 | — | ns | — |
| Data Valid to Write High<br>Data Valid to Enable High | $t_{DVWH}$<br>$t_{DVEH}$ | $t_{DW}$ | 15 | — | 15 | — | 20 | — | ns | — |
| Data Hold From Write High<br>Data Hold From Enable High | $t_{WHDX}$<br>$t_{EHDX}$ | $t_{DH}$ | 0 | — | 0 | — | 0 | — | ns | — |
| Write Recovery Time<br>Enable Recovery Time | $t_{WHAX}$<br>$t_{EHAX}$ | $t_{WR}$ | 0 | — | 0 | — | 0 | — | ns | 2 |
| Chip Enable to End of Write<br>Enable Low to Enable High | $t_{ELWH}$<br>$t_{ELEH}$ | $t_{CW}$ | 20 | — | 25 | — | 35 | — | ns | 1 |
| Write Pulse Width | $t_{WLWH}$ | $t_{WP}$ | 25 | — | 25 | — | 30 | — | ns | 3 |
| Write Low to Output High-Z | $t_{WLQZ}$ | $t_{WHZ}$ | 0 | 20 | 0 | 20 | 0 | 20 | ns | 4, 5 |
| Write High to Output Low-Z | $t_{WHQX}$ | $t_{WLZ}$ | 5 | — | 5 | — | 5 | — | ns | 4, 5 |

NOTES:
1. A write cycle starts at the latest transition of a low $\overline{E}$ or low $\overline{W}$. A write cycle ends at the earliest transition of a high $\overline{E}$ or high $\overline{W}$.
2. $\overline{W}$ must be high during all address transitions whenever $\overline{E}$ is low.
3. If $\overline{G}$ is enabled, allow an additional 15 ns $t_{WLWH}$ to avoid bus contention.
4. All high-Z and low-Z parameters are considered in a high or low impedance state when the output has made a 500 mV transition from the previous steady state voltage.
5. These parameters are periodically sampled and not 100% tested.

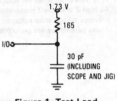

**Figure 1. Test Load**

## WRITE CYCLE 1 ($\overline{W}$ Controlled)

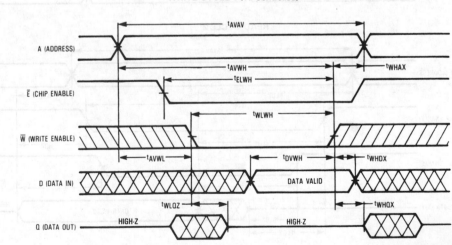

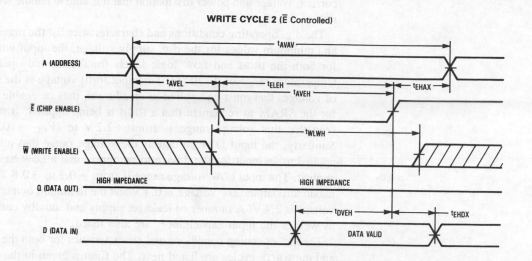

**WRITE CYCLE 2 ($\overline{E}$ Controlled)**

memory may contain many thousands of cells this can lead to a considerable saving in current and much smaller power dissipation than any other technology. Also employed, but less often than in the past, are NMOS memories; this technology is faster than CMOS but it has a greater power dissipation. Bipolar technology is about three times as fast as NMOS but its power dissipation is much greater. Typical speeds of the three technologies are: bipolar 5—50 ns, NMOS 50—400 ns and CMOS 60—500 ns.

## Memory Specification

The manufacturers of semiconductor memories provide data sheets for their devices. Figure 6.24 gives the data sheet for the MCM 6206 32k × 8 SRAM. The SRAM has an access time ranging from 30 to 45 ns and hence it is a high-speed CMOS device with TTL compatible input and output terminals. The data lines are labelled $DQ_0$ through to $DQ_7$; when data is written into the SRAM these lines are configured as inputs but when data is read out of the SRAM they are configured as outputs. The pin that controls the function of the data lines is labelled as write enable $\overline{W}$. A low level is applied to this input during a WRITE operation. There are 15 address lines, labelled $A_0$ through to $A_{14}$, allowing $2^{15} = 32\,768$, or 32k, 8-bit locations to be addressed. The total capacity of the memory is $32 \times 8 = 256$k or 262 144 bits. The IC also has two enable inputs; the chip enable $\overline{E}$ and the output enable $\overline{G}$. A LOW applied to the $\overline{E}$ input enables the SRAM and a LOW applied to the $\overline{G}$ input allows data to be accessed. The truth table and the timing diagrams show how these enable inputs are employed to control the WRITE operation.

The absolute maximum ratings for the device list the limits of current, voltage and power dissipation that it is able to handle without risk of damage.

The d.c. operating conditions and characteristics list the maximum and minimum values for the d.c. supply voltage, the input voltages for both the HIGH and LOW logic levels for the normal operating temperature range of 0–70 °C. The input HIGH voltage is the range of voltages that must be applied to an address, data or enable input for the SRAM to recognize that a HIGH is being applied. It can be seen that this voltage range is from +2.2 V to $(V_{CC} + 0.2)$ V. Similarly, the input LOW voltage is the voltage range that must be applied to an input for the device to recognize that a LOW has been applied. The input LOW voltage range is from −0.3 to +0.8 V. The maximum output LOW voltage is 0.4 V and the minimum output HIGH voltage is 2.4 V. A number of leakage, supply and standby currents, as well as the input capacitance, are also listed.

The a.c. operating conditions and characteristics for both the READ and the WRITE cycles are listed next. The figures given in the tables should be read in conjunction with the timing diagrams. Three access times are given. The address access time is the time that elapses between the application of a valid address, the $\bar{E}$ access time is the time from the chip enable input going LOW, and the $\bar{G}$ access time is the time between the $\bar{G}$ enable input going LOW, and valid data appearing at the input/output lines. The READ cycle time exceeds the access time to allow the internal circuitry time to get ready for the next READ cycle.

The speed with which data is written into the SRAM is controlled by the WRITE cycle time, and this can be seen to be equal to the READ cycle time. The WRITE cycle may be under the control of either the $\bar{E}$ input or the $\bar{W}$ input. Consider the $\bar{W}$-controlled WRITE cycle. The address lines are set to the address of the location to be written to. The enable input $\bar{E}$, and then the $\bar{W}$ input, are taken LOW. The $\bar{E}$ input must be kept LOW during the write cycle. Once the data has been written the $\bar{W}$ input, and then the $\bar{E}$ input, are taken HIGH and then a new address can be applied for the next READ or WRITE operation. It can be seen that some parameters are shown in the WRITE cycle as occupying a time of 0 ns. This is because some other SRAM chips take a certain time for these operations and the table indicates that this particular chip requires zero recovery time or zero data hold from write/enable high time.

The data sheet for the MCM 511000A 1M × 1 DRAM is given in Fig. 6.25. Generally, the information given in this data sheet follows the same headings as in the MCM 6206's data sheet. Additional parameters given include data concerning memory refresh and address multiplexing. Address multiplexing is employed to allow 1048 576 1-bit memory locations to be addressed using only 10 address lines. The data sheet includes a description of the operation of the DRAM including the refreshing arrangements.

MOTOROLA
# SEMICONDUCTOR
TECHNICAL DATA

## 1Mx1 CMOS Dynamic RAM
### Page Mode, Commercial and Industrial Temperature Range

The MCM511000A is a 1.0μ CMOS high-speed, dynamic random access memory. It is organized as 1,048,576 one-bit words and fabricated with CMOS silicon-gate process technology. Advanced circuit design and fine line processing provide high performance, improved reliability, and low cost.

The MCM511000A requires only ten address lines; row and column address inputs are multiplexed. The device is packaged in a standard 300-mil dual-in-line plastic package (DIP), a 300-mil SOJ plastic package, and a 100-mil zig-zag in-line package (ZIP).

- Two Temperature Ranges: Commercial — 0°C to 70°C
                           Industrial — –40°C to +85°C
- Three-State Data Output
- Common I/O with Early Write
- Fast Page Mode
- Test Mode
- TTL-Compatible Inputs and Output
- $\overline{RAS}$ Only Refresh
- $\overline{CAS}$ Before $\overline{RAS}$ Refresh
- Hidden Refresh
- 512 Cycle Refresh. MCM511000A = 8 ms
                      MCM51L1000A = 64 ms
- Unlatched Data Out at Cycle End Allows Two Dimensional Chip Selection
- Fast Access Time ($t_{RAC}$):
    MCM511000A-70 and MCM51L1000A-70 = 70 ns (Max)
    MCM511000A-80 and MCM51L1000A-80 = 80 ns (Max)
    MCM511000A-10 and MCM51L1000A-10 = 100 ns (Max)
- Low Active Power Dissipation:
    MCM511000A-70 and MCM51L1000A-70 = 440 mW (Max)
    MCM511000A-80 and MCM51L1000A-80 = 385 mW (Max)
    MCM511000A-10 and MCM51L1000A-10 = 330 mW (Max)
- Low Standby Power Dissipation:
    MCM511000A and MCM51L1000A = 11 mW (Max, TTL Levels)
    MCM511000A = 5.5 mW (Max, CMOS Levels)
    MCM51L1000A = 1.1 mW (Max, CMOS Levels)

## MCM511000A
## MCM51L1000A

P PACKAGE
300 MIL PLASTIC
CASE 707A

J PACKAGE
300 MIL SOJ
CASE 822

Z PACKAGE
PLASTIC
ZIG-ZAG IN-LINE
CASE 836

### PIN NAMES

| | |
|---|---|
| A0–A9 | Address Input |
| D | Data Input |
| Q | Data Output |
| W | Read/Write Enable |
| $\overline{RAS}$ | Row Address Strobe |
| $\overline{CAS}$ | Column Address Strobe |
| $V_{CC}$ | Power Supply (+5 V) |
| $V_{SS}$ | Ground |
| TF | Test Function Enable |
| NC | No Connection |

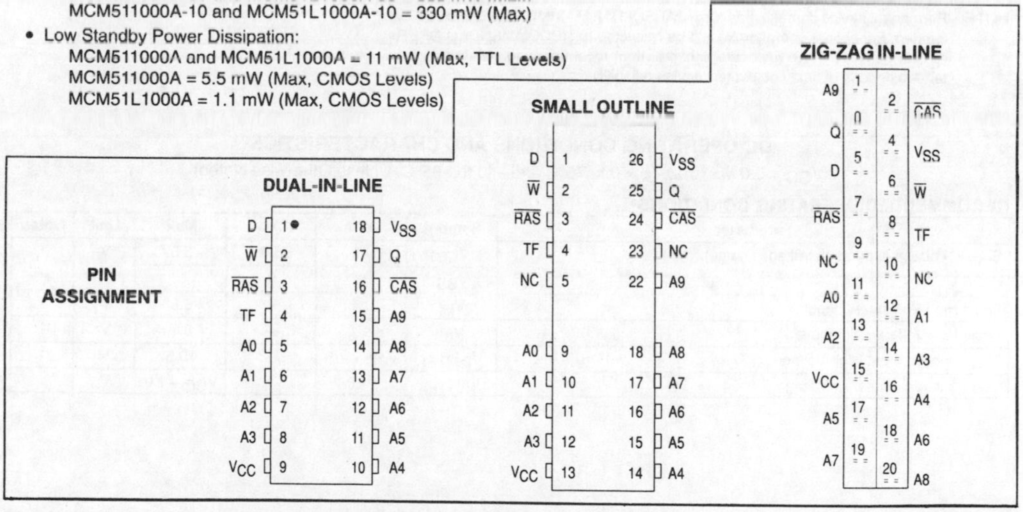

### DUAL-IN-LINE

PIN ASSIGNMENT

| | | | |
|---|---|---|---|
| D | 1 | 18 | $V_{SS}$ |
| $\overline{W}$ | 2 | 17 | Q |
| $\overline{RAS}$ | 3 | 16 | $\overline{CAS}$ |
| TF | 4 | 15 | A9 |
| A0 | 5 | 14 | A8 |
| A1 | 6 | 13 | A7 |
| A2 | 7 | 12 | A6 |
| A3 | 8 | 11 | A5 |
| $V_{CC}$ | 9 | 10 | A4 |

### SMALL OUTLINE

| | | | |
|---|---|---|---|
| D | 1 | 26 | $V_{SS}$ |
| $\overline{W}$ | 2 | 25 | Q |
| $\overline{RAS}$ | 3 | 24 | $\overline{CAS}$ |
| TF | 4 | 23 | NC |
| NC | 5 | 22 | A9 |
| A0 | 9 | 18 | A8 |
| A1 | 10 | 17 | A7 |
| A2 | 11 | 16 | A6 |
| A3 | 12 | 15 | A5 |
| $V_{CC}$ | 13 | 14 | A4 |

### ZIG-ZAG IN-LINE

| | | | |
|---|---|---|---|
| A9 | 1 | | |
| | | 2 | $\overline{CAS}$ |
| Q | 3 | | |
| | | 4 | $V_{SS}$ |
| D | 5 | | |
| | | 6 | $\overline{W}$ |
| $\overline{RAS}$ | 7 | | |
| | | 8 | TF |
| NC | 9 | | |
| | | 10 | NC |
| A0 | 11 | | |
| | | 12 | A1 |
| A2 | 13 | | |
| | | 14 | A3 |
| $V_{CC}$ | 15 | | |
| | | 16 | A4 |
| A5 | 17 | | |
| | | 18 | A6 |
| A7 | 19 | | |
| | | 20 | A8 |

**MOTOROLA**

**Fig. 6.25** Data sheet for the MCM 511000A 1M × 1 DRAM (*courtesy of Motorola*)

**BLOCK DIAGRAM**

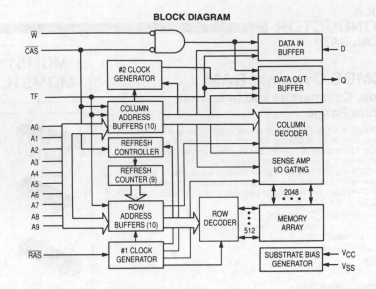

## ABSOLUTE MAXIMUM RATING (See Note)

| Rating | | Symbol | Value | Unit |
|---|---|---|---|---|
| Power Supply Voltage | | $V_{CC}$ | −1 to +7 | V |
| Voltage Relative to $V_{SS}$ for Any Pin Except $V_{CC}$ | | $V_{in}$, $V_{out}$ | −1 to +7 | V |
| Test Function Input Voltage | | $V_{in}$ (TF) | −1 to +10.5 | V |
| Data Out Current | | $I_{out}$ | 50 | mA |
| Power Dissipation | | $P_D$ | 600 | mW |
| Operating Temperature Range | Commercial | $T_A$ | 0 to +70 | °C |
| | Industrial | | −40 to +85 | |
| Storage Temperature Range | | $T_{stg}$ | −55 to +150 | °C |

NOTE: Permanent device damage may occur if ABSOLUTE MAXIMUM RATINGS are exceeded. Functional operation should be restricted to RECOMMENDED OPERATING CONDITIONS. Exposure to higher than recommended voltages for extended periods of time could affect device reliability.

This device contains circuitry to protect the inputs against damage due to high static voltages or electric fields; however, it is advised that normal precautions be taken to avoid application of any voltage higher than maximum rated voltages to this high-impedance circuit.

## DC OPERATING CONDITIONS AND CHARACTERISTICS

($V_{CC}$ = 5.0 V ±10%, $T_A$ = 0 to 70°C and −40 to +85°C, Unless Otherwise Noted)

### RECOMMENDED OPERATING CONDITIONS

| Parameter | Symbol | Min | Typ | Max | Unit | Notes |
|---|---|---|---|---|---|---|
| Supply Voltage (Operating Voltage Range) | $V_{CC}$ | 4.5 | 5.0 | 5.5 | V | 1 |
| | $V_{SS}$ | 0 | 0 | 0 | | |
| Logic High Voltage, All Inputs | $V_{IH}$ | 2.4 | — | 6.5 | V | 1 |
| Logic Low Voltage, All Inputs | $V_{IL}$ | −1.0 | — | 0.8 | V | 1 |
| Test Function Input High Voltage | $V_{IH}$ (TF) | $V_{CC}$ + 4.5 | — | 10.5 | V | 1 |
| Test Function Input Low Voltage | $V_{IL}$ (TF) | −1.0 | — | $V_{CC}$ + 1.0 | V | 1 |

MCM511000A • MCM51L1000A

## DC CHARACTERISTICS

| Characteristic | Symbol | Min | Max | Unit | Notes |
|---|---|---|---|---|---|
| $V_{CC}$ Power Supply Current | $I_{CC1}$ | | | mA | 3 |
| MCM511000A-70 and MCM51L1000A-70, $t_{RC}$ = 130 ns, $T_A$ = 0°C to 70°C | | — | 80 | | |
| MCM511000A-80 and MCM51L1000A-80, $t_{RC}$ = 150 ns, $T_A$ = 0°C to 70°C | | — | 70 | | |
| MCM511000A-10 and MCM51L1000A-10, $t_{RC}$ = 180 ns, $T_A$ = 0°C to 70°C | | — | 60 | | |
| MCM511000A-C70 and MCM51L1000A-C70, $t_{RC}$ = 130 ns, $T_A$ = −40°C to +85°C | | — | 85 | | |
| MCM511000A-C80 and MCM51L1000A-C80, $t_{RC}$ = 150 ns, $T_A$ = −40°C to +85°C | | — | 75 | | |
| MCM511000A-C10 and MCM51L1000A-C10, $t_{RC}$ = 180 ns, $T_A$ = −40°C to +85°C | | — | 65 | | |
| $V_{CC}$ Power Supply Current (Standby) ($\overline{RAS}=\overline{CAS}=V_{IH}$) | $I_{CC2}$ | | | mA | |
| MCM511000A- and MCM51L1000A-, $T_A$ = 0°C to 70°C | | — | 2 | | |
| MCM511000A-C and MCM51L1000A-C, $T_A$ = −40°C to +85°C | | — | 3 | | |
| $V_{CC}$ Power Supply Current During $\overline{RAS}$ Only Refresh Cycles ($\overline{CAS}=V_{IH}$) | $I_{CC3}$ | | | mA | 3 |
| MCM511000A-70 and MCM51L1000A-70, $t_{RC}$ = 130 ns, $T_A$ = 0°C to 70°C | | — | 80 | | |
| MCM511000A-80 and MCM51L1000A-80, $t_{RC}$ = 150 ns, $T_A$ = 0°C to 70°C | | — | 70 | | |
| MCM511000A-10 and MCM51L1000A-10, $t_{RC}$ = 180 ns, $T_A$ = 0°C to 70°C | | — | 60 | | |
| MCM511000A-C70 and MCM51L1000A-C70, $t_{RC}$ = 130 ns, $T_A$ = −40°C to +85°C | | — | 85 | | |
| MCM511000A-C80 and MCM51L1000A-C80, $t_{RC}$ = 150 ns, $T_A$ = −40°C to +85°C | | — | 75 | | |
| MCM511000A-C10 and MCM51L1000A-C10, $t_{RC}$ = 180 ns, $T_A$ = −40°C to +85°C | | — | 65 | | |
| $V_{CC}$ Power Supply Current During Fast Page Mode Cycle ($\overline{RAS} = V_{IL}$) | $I_{CC4}$ | | | mA | 3, 4 |
| MCM511000A-70 and MCM51L1000A-70, $t_{PC}$ = 40 ns, $T_A$ = 0°C to 70°C | | — | 60 | | |
| MCM511000A-80 and MCM51L1000A-80, $t_{PC}$ = 45 ns, $T_A$ = 0°C to 70°C | | — | 50 | | |
| MCM511000A-10 and MCM51L1000A-10, $t_{PC}$ = 55 ns, $T_A$ = 0°C to 70°C | | — | 40 | | |
| MCM511000A-C70 and MCM51L1000A-C70, $t_{PC}$ = 40 ns, $T_A$ = −40°C to +85°C | | — | 65 | | |
| MCM511000A-C80 and MCM51L1000A-C80, $t_{PC}$ = 45 ns, $T_A$ = −40°C to +85°C | | — | 55 | | |
| MCM511000A-C10 and MCM51L1000A-C10, $t_{PC}$ = 55 ns, $T_A$ = −40°C to +85°C | | — | 45 | | |
| $V_{CC}$ Power Supply Current (Standby) ($\overline{RAS}=\overline{CAS}=V_{CC}-0.2$ V) | $I_{CC5}$ | | | | |
| MCM511000A-, $T_A$ = 0°C to 70°C and MCM511000A-C, $T_A$ = −40°C to +85°C | | — | 1.0 | mA | |
| MCM51L1000A-, $T_A$ = 0°C to 70°C | | — | 200 | μA | |
| MCM51L1000A-C, $T_A$ = −40°C to +85°C | | — | 400 | μA | |
| $V_{CC}$ Power Supply Current During $\overline{CAS}$ Before $\overline{RAS}$ Refresh Cycle | $I_{CC6}$ | | | mA | 3 |
| MCM511000A-70 and MCM51L1000A-70, $t_{RC}$ = 130 ns, $T_A$ = 0°C to 70°C | | — | 80 | | |
| MCM511000A-80 and MCM51L1000A-80, $t_{RC}$ = 150 ns, $T_A$ = 0°C to 70°C | | — | 70 | | |
| MCM511000A-10 and MCM51L1000A-10, $t_{RC}$ = 180 ns, $T_A$ = 0°C to 70°C | | — | 60 | | |
| MCM511000A-C70 and MCM51L1000A-C70, $t_{RC}$ = 130 ns, $T_A$ = −40°C to +85°C | | — | 85 | | |
| MCM511000A-C80 and MCM51L1000A-C80, $t_{RC}$ = 150 ns, $T_A$ = −40°C to +85°C | | — | 75 | | |
| MCM511000A-C10 and MCM51L1000A-C10, $t_{RC}$ = 180 ns, $T_A$ = −40°C to +85°C | | — | 65 | | |
| $V_{CC}$ Power Supply Current, Battery Backup Mode ($t_{RC}$ = 125 μs, $t_{RAS}$ = 1 μs, $\overline{CAS}=\overline{CAS}$ Before $\overline{RAS}$ Cycle or 0.2 V, $\overline{W}$, D = $V_{CC}$ − 0.2 V or 0.2 V) | $I_{CC7}$ | | | μA | 3 |
| MCM51L1000A-, $T_A$ = 0°C to 70°C | | — | 300 | | |
| MCM51L1000A-C, $T_A$ = −40°C to +85°C | | — | 500 | | |
| Input Leakage Current (Except TF) (0 V ≤ $V_{in}$ ≤ 6.5 V) | $I_{lkg(I)}$ | −10 | 10 | μA | |
| Input Leakage Current (TF) (0 V ≤ $V_{in}$ (TF) ≤ $V_{CC}$ + 0.5 V) | $I_{lkg(I)}$ | −10 | 10 | μA | |
| Output Leakage Current ($\overline{CAS}$ = $V_{IH}$, 0 V ≤ $V_{out}$ ≤ 5.5 V) | $I_{lkg(O)}$ | −10 | 10 | μA | |
| Test Function Input Current ($V_{CC}$ + 4.5 V ≤ $V_{in}$ (TF) ≤ $V_{CC}$ ≤ 10.5 V) | $I_{in}$ (TF) | — | 1 | mA | |
| Output High Voltage ($I_{OH}$ = −5 mA) | $V_{OH}$ | 2.4 | — | V | |
| Output Low Voltage ($I_{OL}$ = 4.2 mA) | $V_{OL}$ | — | 0.4 | V | |

**CAPACITANCE** (f = 1.0 MHz, $T_A$ = 25°C, $V_{CC}$ = 5 V, Periodically Sampled Rather Than 100% Tested)

| Parameter | | Symbol | Max | Unit | Notes |
|---|---|---|---|---|---|
| Input Capacitance | D, A0–A9 | $C_{in}$ | 5 | pF | 4 |
| | $\overline{RAS}$, $\overline{CAS}$, $\overline{W}$, TF | | 7 | | |
| Output Capacitance ($\overline{CAS}$ = $V_{IH}$ to Disable Output) | Q | $C_{out}$ | 7 | pF | 4 |

NOTES:
1. All voltages referenced to $V_{SS}$.
2. Current is a function of cycle rate and output loading; maximum current is measured at the fastest cycle rate with the output open.
3. Measured with one address transition per page mode cycle.
4. Capacitance measured with a Boonton Meter or effective capacitance calculated from the equation: C = I$\Delta$t/$\Delta$V.

## AC OPERATING CONDITIONS AND CHARACTERISTICS

($V_{CC}$ = 5.0 V ±10%, $T_A$ = 0 to 70°C and –40 to +85°C, Unless Otherwise Noted)

**READ, WRITE, AND READ-WRITE CYCLES** (See Notes 1, 2, 3, 4, and 5)

| Parameter | Symbol | | MCM511000A-70 MCM51L1000A-70 | | MCM511000A-80 MCM51L1000A-80 | | MCM511000A-10 MCM51L1000A-10 | | Unit | Notes |
|---|---|---|---|---|---|---|---|---|---|---|
| | Std | Alt | Min | Max | Min | Max | Min | Max | | |
| Random Read or Write Cycle Time | $t_{RELREL}$ | $t_{RC}$ | 130 | — | 150 | — | 180 | — | ns | 6 |
| Read-Write Cycle Time | $t_{RELREL}$ | $t_{RWC}$ | 155 | — | 175 | — | 210 | — | ns | 6 |
| Page Mode Cycle Time | $t_{CELCEL}$ | $t_{PC}$ | 40 | — | 45 | — | 55 | — | ns | |
| Page Mode Read-Write Cycle Time | $t_{CELCEL}$ | $t_{PRWC}$ | 65 | — | 70 | — | 85 | — | ns | |
| Access Time from $\overline{RAS}$ | $t_{RELQV}$ | $t_{RAC}$ | — | 70 | — | 80 | — | 100 | ns | 7, 8 |
| Access Time from $\overline{CAS}$ | $t_{CELQV}$ | $t_{CAC}$ | — | 20 | — | 20 | — | 25 | ns | 7, 9 |
| Access Time from Column Address | $t_{AVQV}$ | $t_{AA}$ | — | 35 | — | 40 | — | 50 | ns | 7, 10 |
| Access Time from $\overline{CAS}$ Precharge | $t_{CEHQV}$ | $t_{CPA}$ | — | 35 | — | 40 | — | 50 | ns | 7 |
| $\overline{CAS}$ to Output in Low-Z | $t_{CELQX}$ | $t_{CLZ}$ | 0 | — | 0 | — | 0 | — | ns | 7 |
| Output Buffer and Turn-Off Delay | $t_{CEHQZ}$ | $t_{OFF}$ | 0 | 20 | 0 | 20 | 0 | 20 | ns | 11 |
| Transition Time (Rise and Fall) | $t_T$ | $t_T$ | 3 | 50 | 3 | 50 | 3 | 50 | ns | |
| $\overline{RAS}$ Precharge Time | $t_{REHREL}$ | $t_{RP}$ | 50 | — | 60 | — | 70 | — | ns | |
| $\overline{RAS}$ Pulse Width | $t_{RELREH}$ | $t_{RAS}$ | 70 | 10,000 | 80 | 10,000 | 100 | 10,000 | ns | |
| $\overline{RAS}$ Pulse Width (Fast Page Mode) | $t_{RELREH}$ | $t_{RASP}$ | 70 | 100,000 | 80 | 100,000 | 100 | 100,000 | ns | |
| $\overline{RAS}$ Hold Time | $t_{CELREH}$ | $t_{RSH}$ | 20 | — | 20 | — | 25 | — | ns | |
| $\overline{RAS}$ Hold Time from $\overline{CAS}$ Precharge (Page Mode Cycle Only) | $t_{CELREH}$ | $t_{RHCP}$ | 35 | — | 40 | — | 50 | — | ns | |
| $\overline{CAS}$ Hold Time | $t_{RELCEH}$ | $t_{CSH}$ | 70 | — | 80 | — | 100 | — | ns | |
| $\overline{CAS}$ Pulse Width | $t_{CELCEH}$ | $t_{CAS}$ | 20 | 10,000 | 20 | 10,000 | 25 | 10,000 | ns | |
| $\overline{RAS}$ to $\overline{CAS}$ Delay Time | $t_{RELCEL}$ | $t_{RCD}$ | 20 | 50 | 20 | 60 | 25 | 75 | ns | 12 |
| $\overline{RAS}$ to Column Address Delay Time | $t_{RELAV}$ | $t_{RAD}$ | 15 | 35 | 15 | 40 | 20 | 50 | ns | 13 |
| $\overline{CAS}$ to $\overline{RAS}$ Precharge Time | $t_{CEHREL}$ | $t_{CRP}$ | 5 | — | 5 | — | 5 | — | ns | |
| $\overline{CAS}$ Precharge Time (Page Mode Cycle Only) | $t_{CEHCEL}$ | $t_{CP}$ | 10 | — | 10 | — | 10 | — | ns | |
| Row Address Setup Time | $t_{AVREL}$ | $t_{ASR}$ | 0 | — | 0 | — | 0 | — | ns | |
| Row Address Hold Time | $t_{RELAX}$ | $t_{RAH}$ | 10 | — | 10 | — | 15 | — | ns | |
| Column Address Setup Time | $t_{AVCEL}$ | $t_{ASC}$ | 0 | — | 0 | — | 0 | — | ns | |
| Column Address Hold Time | $t_{CELAX}$ | $t_{CAH}$ | 15 | — | 15 | — | 20 | — | ns | |

(continued)

NOTES:
1. $V_{IH}$ min and $V_{IL}$ max are reference levels for measuring timing of input signals. Transition times are measured between $V_{IH}$ and $V_{IL}$.
2. An initial pause of 200 μs is required after power-up followed by 8 $\overline{RAS}$ cycles before proper device operation is guaranteed.
3. The transition time specification applies for all input signals. In addition to meeting the transition rate specification, all input signals must transition between $V_{IH}$ and $V_{IL}$ (or between $V_{IL}$ and $V_{IH}$) in a monotonic manner.
4. AC measurements $t_T$ = 5.0 ns.
5. TF pin must be at $V_{IL}$ or open if not used.
6. The specifications for $t_{RC}$ (min) and $t_{RWC}$ (min) are used only to indicate cycle time at which proper operation over the full temperature range (0°C ≤ $T_A$ ≤ 70°C and –40°C ≤ $T_A$ ≤ +85°C) is assured.
7. Measured with a current load equivalent to 2 TTL (–200 μA, +4 mA) loads and 100 pF with the data output trip points set at $V_{OH}$ = 2.0 V and $V_{OL}$ = 0.8 V.
8. Assumes that $t_{RCD}$ ≤ $t_{RCD}$ (max).
9. Assumes that $t_{RCD}$ ≥ $t_{RCD}$ (max).
10. Assumes that $t_{RAD}$ ≥ $t_{RAD}$ (max).
11. $t_{OFF}$ (max) defines the time at which the output achieves the open circuit condition and is not referenced to output voltage levels.
12. Operation within the $t_{RCD}$ (max) limit ensures that $t_{RAC}$ (max) can be met. $t_{RCD}$ (max) is specified as a reference point only; if $t_{RCD}$ is greater than the specified $t_{RCD}$ (max) limit, then access time is controlled exclusively by $t_{CAC}$.
13. Operation within the $t_{RAD}$ (max) limit ensures that $t_{RAC}$ (max) can be met. $t_{RAD}$ (max) is specified as a reference point only; if $t_{RAD}$ is greater than the specified $t_{RAD}$ (max) limit, then access time is controlled exclusively by $t_{AA}$.

## READ, WRITE, AND READ-WRITE CYCLES (Continued)

| Parameter | Symbol | | MCM511000A-70 MCM51L1000A-70 | | MCM511000A-80 MCM51L1000A-80 | | MCM511000A-10 MCM51L1000A-10 | | Unit | Notes |
|---|---|---|---|---|---|---|---|---|---|---|
| | Std | Alt | Min | Max | Min | Max | Min | Max | | |
| Column Address Hold Time Referenced to $\overline{RAS}$ | $t_{RELAX}$ | $t_{AR}$ | 55 | — | 60 | — | 75 | — | ns | |
| Column Address to $\overline{RAS}$ Lead Time | $t_{AVREH}$ | $t_{RAL}$ | 35 | — | 40 | — | 50 | — | ns | |
| Read Command Setup Time | $t_{WHCEL}$ | $t_{RCS}$ | 0 | — | 0 | — | 0 | — | ns | |
| Read Command Hold Time Referenced to $\overline{CAS}$ | $t_{CEHWX}$ | $t_{RCH}$ | 0 | — | 0 | — | 0 | — | ns | 14 |
| Read Command Hold Time Referenced to $\overline{RAS}$ | $t_{REHWX}$ | $t_{RRH}$ | 0 | — | 0 | — | 0 | — | ns | 14 |
| Write Command Hold Time Referenced to $\overline{CAS}$ | $t_{CELWH}$ | $t_{WCH}$ | 15 | — | 15 | — | 20 | — | ns | |
| Write Command Hold Time Referenced to $\overline{RAS}$ | $t_{RELWH}$ | $t_{WCR}$ | 55 | — | 60 | — | 75 | — | ns | |
| Write Command Pulse Width | $t_{WLWH}$ | $t_{WP}$ | 15 | — | 15 | — | 20 | — | ns | |
| Write Command to $\overline{RAS}$ Lead Time | $t_{WLREH}$ | $t_{RWL}$ | 20 | — | 20 | — | 25 | — | ns | |
| Write Command to $\overline{CAS}$ Lead Time | $t_{WLCEH}$ | $t_{CWL}$ | 20 | — | 20 | — | 25 | — | ns | |
| Data in Setup Time | $t_{DVCEL}$ | $t_{DS}$ | 0 | — | 0 | — | 0 | — | ns | 15 |
| Data in Hold Time | $t_{CELDX}$ | $t_{DH}$ | 15 | — | 15 | — | 20 | — | ns | 15 |
| Data in Hold Time Referenced to $\overline{RAS}$ | $t_{RELDX}$ | $t_{DHR}$ | 55 | — | 60 | — | 75 | — | ns | |
| Refresh Period          MCM511000A MCM51L1000A | $t_{HVRV}$ | $t_{RFSH}$ | —  — | 8 64 | —  — | 8 64 | —  — | 8 64 | ms | |
| Write Command Setup Time | $t_{WLCEL}$ | $t_{WCS}$ | 0 | — | 0 | — | 0 | — | ns | 16 |
| $\overline{CAS}$ to Write Delay | $t_{CELWL}$ | $t_{CWD}$ | 20 | — | 20 | — | 25 | — | ns | 16 |
| $\overline{RAS}$ to Write Delay | $t_{RELWL}$ | $t_{RWD}$ | 70 | — | 80 | — | 100 | — | ns | 16 |
| Column Address to Write Delay Time | $t_{AVWL}$ | $t_{AWD}$ | 35 | — | 40 | — | 50 | — | ns | 16 |
| $\overline{CAS}$ Precharge to Write Delay Time | $t_{CEHWL}$ | $t_{CPWD}$ | 35 | — | 40 | — | 50 | — | ns | 16 |
| $\overline{CAS}$ Setup Time for $\overline{CAS}$ Before $\overline{RAS}$ Refresh | $t_{RELCEL}$ | $t_{CSR}$ | 5 | — | 5 | — | 5 | — | ns | |
| $\overline{CAS}$ Hold Time for $\overline{CAS}$ Before $\overline{RAS}$ Refresh | $t_{RELCEH}$ | $t_{CHR}$ | 15 | — | 15 | — | 20 | — | ns | |
| $\overline{CAS}$ Precharge to $\overline{CAS}$ Active Time | $t_{REHCEL}$ | $t_{RPC}$ | 0 | — | 0 | — | 0 | — | ns | |
| $\overline{CAS}$ Precharge Time for $\overline{CAS}$ Before $\overline{RAS}$ Counter Test | $t_{CEHCEL}$ | $t_{CPT}$ | 40 | — | 40 | — | 50 | — | ns | |
| $\overline{CAS}$ Precharge Time | $t_{CEHCEL}$ | $t_{CPN}$ | 10 | — | 10 | — | 15 | — | ns | |
| Test Mode Enable Setup Time Referenced to $\overline{RAS}$ | $t_{TEHREL}$ | $t_{TES}$ | 0 | — | 0 | — | 0 | — | ns | |
| Test Mode Enable Hold Time Referenced to $\overline{RAS}$ | $t_{REHTEL}$ | $t_{TEHR}$ | 0 | — | 0 | — | 0 | — | ns | |
| Test Mode Enable Hold Time Referenced to $\overline{CAS}$ | $t_{CEHTEL}$ | $t_{TEHC}$ | 0 | — | 0 | — | 0 | — | ns | |

NOTES.
14. Either $t_{RRH}$ or $t_{RCH}$ must be satisfied for a read cycle.
15. These parameters are referenced to $\overline{CAS}$ leading edge in early write cycles and to $\overline{W}$ leading edge in delayed write or read-write cycles.
16. $t_{WCS}$, $t_{RWD}$, $t_{CWD}$, $t_{CPWD}$, and $t_{AWD}$ are not restrictive operating parameters. They are included in the data sheet as electrical characteristics only; if $t_{WCS} \geq t_{WCS}$ (min), the cycle is an early write cycle and the data out pin will remain open circuit (high impedance) throughout the entire cycle; if $t_{CWD} \geq t_{CWD}$ (min), $t_{RWD} \geq t_{RWD}$ (min), $t_{CPWD} \geq t_{CPWD}$ (min), and $t_{AWD} \geq t_{AWD}$ (min), the cycle is a read-write cycle and the data out will contain data read from the selected cell. If neither of these sets of conditions is satisfied, the condition of the data out (at access time) is indeterminate.

**READ CYCLE**

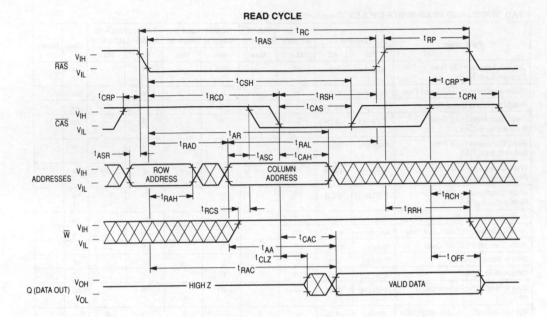

**EARLY WRITE CYCLE**

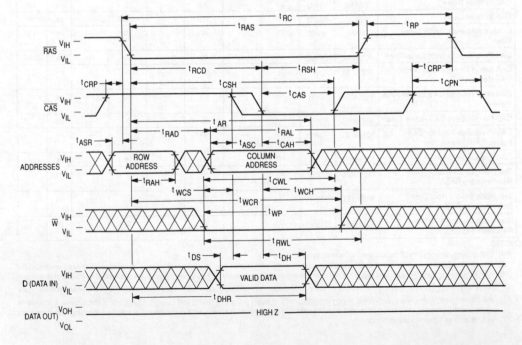

MCM511000A • MCM51L1000A

## READ-WRITE CYCLE

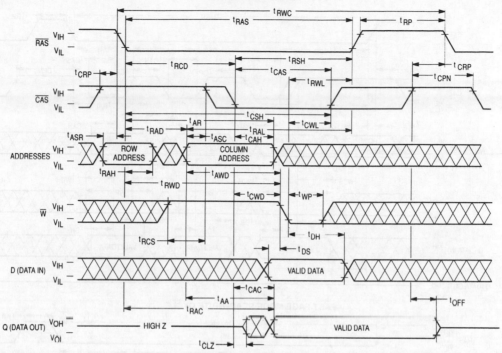

## FAST PAGE MODE READ CYCLE

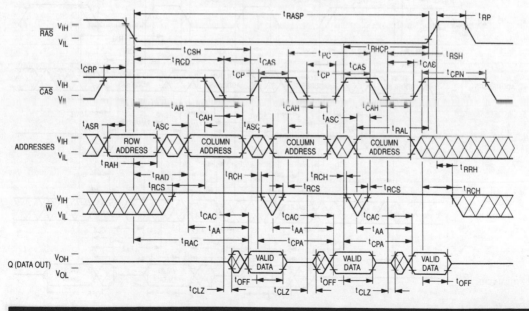

MCM511000A • MCM51L1000A

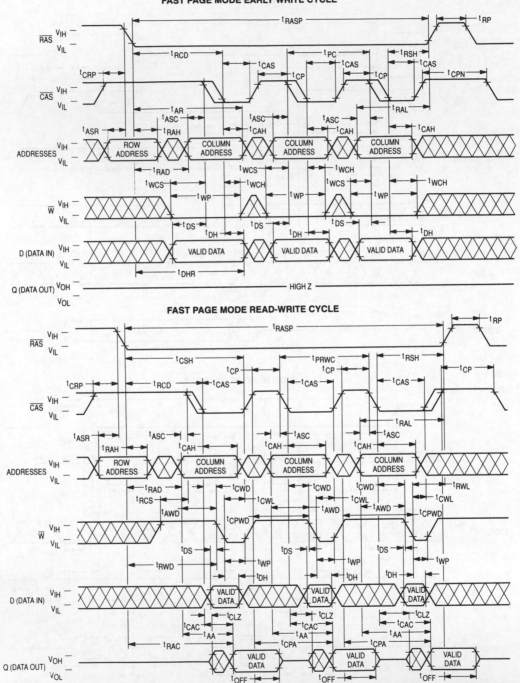

## FAST PAGE MODE EARLY WRITE CYCLE

## FAST PAGE MODE READ-WRITE CYCLE

**MCM511000A • MCM51L1000A**

## $\overline{\text{RAS}}$ ONLY REFRESH CYCLE
($\overline{\text{W}}$ and A9 are Don't Care)

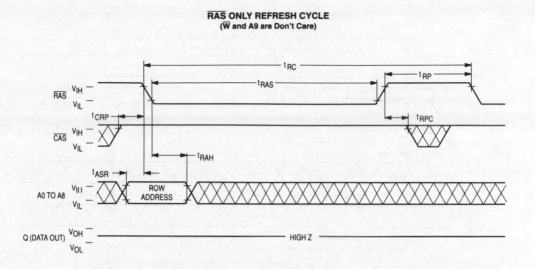

## $\overline{\text{CAS}}$ BEFORE $\overline{\text{RAS}}$ REFRESH CYCLE
(W and A0 to A9 are Don't Care)

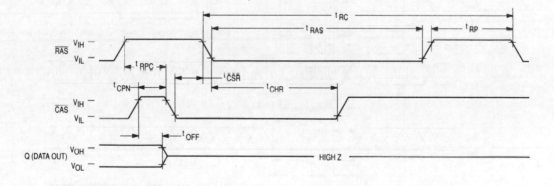

MCM511000A • MCM51L1000A

## HIDDEN REFRESH CYCLE (READ)

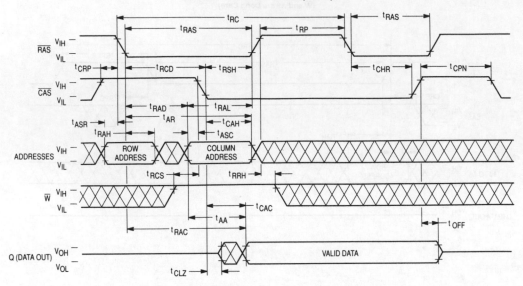

## HIDDEN REFRESH CYCLE (EARLY WRITE)

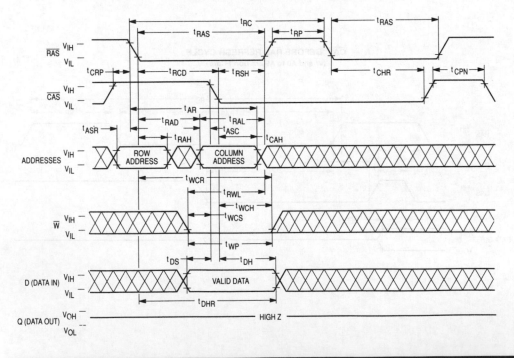

## CAS BEFORE RAS REFRESH COUNTER TEST CYCLE

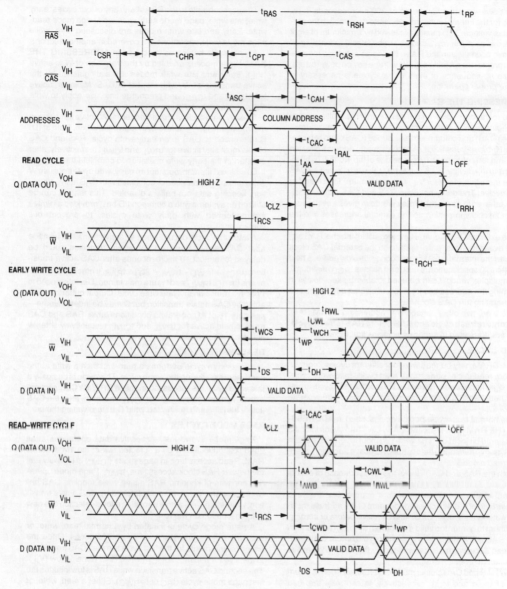

## DEVICE INITIALIZATION

On power-up an initial pause of 200 microseconds is required for the internal substrate generator to establish the correct bias voltage. This must be followed by a minimum of eight active cycles of the row address strobe (clock) to initialize all dynamic nodes within the RAM. During an extended inactive state (greater than 8 milliseconds with the device powered up), a wake up sequence of eight active cycles is necessary to ensure proper operation.

## ADDRESSING THE RAM

The ten address pins on the device are time multiplexed at the beginning of a memory cycle by two clocks, row address strobe ($\overline{RAS}$) and column address strobe ($\overline{CAS}$), into two separate 10-bit address fields. A total of twenty address bits, ten rows and ten columns, will decode one of the 1,048,576 bit locations in the device. $\overline{RAS}$ active transition is followed by $\overline{CAS}$ active transition (active = $V_{IL}$, $t_{RCD}$ minimum) for all read or write cycles. The delay between $\overline{RAS}$ and $\overline{CAS}$ active transitions, referred to as the **multiplex window**, gives a system designer flexibility in setting up the external addresses into the RAM.

The external $\overline{CAS}$ signal is ignored until an internal $\overline{RAS}$ signal is available. This "gate" feature on the external $\overline{CAS}$ clock enables the internal $\overline{CAS}$ line as soon as the row address hold time ($t_{RAH}$) specification is met (and defines $t_{RCD}$ minimum). The multiplex window can be used to absorb skew delays in switching the address bus from row to column addresses and in generating the $\overline{CAS}$ clock.

There are two other variations in addressing the 1M RAM: **$\overline{RAS}$ only refresh cycle** and **$\overline{CAS}$ before $\overline{RAS}$ refresh cycle**. Both are discussed in separate sections that follow.

## READ CYCLE

The DRAM may be read with four different cycles: "normal" random read cycle, page mode read cycle, read-write cycle, and page mode read-write cycle. The normal read cycle is outlined here, while the other cycles are discussed in separate sections.

The normal read cycle begins as described in **ADDRESSING THE RAM**, with $\overline{RAS}$ and $\overline{CAS}$ active transitions latching the desired bit location. The write ($\overline{W}$) input level must be high ($V_{IH}$), $t_{RCS}$ (minimum) before the $\overline{CAS}$ active transition, to enable read mode.

Both the $\overline{RAS}$ and $\overline{CAS}$ clocks trigger a sequence of events which are controlled by several delayed internal clocks. The internal clocks are linked in such a manner that the read access time of the device is independent of the address multiplex window. However, $\overline{CAS}$ must be active before or at $t_{RCD}$ maximum to guarantee valid data out (Q) at $t_{RAC}$ (access time from $\overline{RAS}$ active transition). If the $t_{RCD}$ maximum is exceeded, read access time is determined by the $\overline{CAS}$ clock active transition ($t_{CAC}$).

The $\overline{RAS}$ and $\overline{CAS}$ clocks must remain active for a minimum time of $t_{RAS}$ and $t_{CAS}$ respectively, to complete the read cycle. $\overline{W}$ must remain high throughout the cycle, and for time $t_{RRH}$ or $t_{RCH}$ after $\overline{RAS}$ or $\overline{CAS}$ inactive transition, respectively, to maintain the data at that bit location. Once $\overline{RAS}$ transitions to inactive, it must remain inactive for a minimum time of $t_{RP}$ to precharge the internal device circuitry for the next active cycle. Q is valid, but not latched, as long as the $\overline{CAS}$ clock is active. When the $\overline{CAS}$ clock transitions to inactive, the output will switch to High Z.

## WRITE CYCLE

The user can write to the DRAM with any of four cycles: early write, late write, page mode early write, and page mode read-write. Early and late write modes are discussed here, while page mode write operations are covered in another section.

A write cycle begins as described in **ADDRESSING THE RAM**. Write mode is enabled by the transition of $\overline{W}$ to active ($V_{IL}$). Early and late write modes are distinguished by the active transition of $\overline{W}$, with respect to $\overline{CAS}$. Minimum active time $t_{RAS}$ and $t_{CAS}$, and precharge time $t_{RP}$ apply to write mode, as in the read mode.

An early write cycle is characterized by $\overline{W}$ active transition at minimum time $t_{WCS}$ before $\overline{CAS}$ active transition. Data in (D) is referenced to $\overline{CAS}$ in an early write cycle. $\overline{RAS}$ and $\overline{CAS}$ clocks must stay active for $t_{RWL}$ and $t_{CWL}$, respectively, after the start of the early write operation to complete the cycle.

Q remains High Z throughout an early write cycle because $\overline{W}$ active transition precedes or coincides with $\overline{CAS}$ active transition, keeping data-out buffers disabled. This feature can be utilized on systems with a common I/O bus, provided all writes are performed with early write cycles, to prevent bus contention.

A late write cycle occurs when $\overline{W}$ active transition is made after $\overline{CAS}$ active transition. $\overline{W}$ active transition could be delayed for almost 10 microseconds after $\overline{CAS}$ active transition, ($t_{RCD} + t_{CWD} + t_{RWL} + 2t_T) \leq t_{RAS}$, if other timing minimums ($t_{RCD}$, $t_{RWL}$, and $t_T$) are maintained. D is referenced to $\overline{W}$ active transition in a late write cycle. Output buffers are enabled by $\overline{CAS}$ active transition but Q may be indeterminate—see note 16 of AC operating conditions table. $\overline{RAS}$ and $\overline{CAS}$ must remain active for $t_{RWL}$ and $t_{CWL}$, respectively, after $\overline{W}$ active transition to complete the write cycle.

## READ-WRITE CYCLE

A read-write cycle performs a read and then a write at the same address, during the same cycle. This cycle is basically a late write cycle, as discussed in the **WRITE CYCLE** section, except $\overline{W}$ must remain high for $t_{CWD}$ minimum after the $\overline{CAS}$ active transition, to guarantee valid Q before writing the bit.

## PAGE MODE CYCLES

Page mode allows fast successive data operations at all 2048 column locations on a selected row of the 1M dynamic RAM. Read access time in page mode ($t_{CAC}$) is typically half the regular $\overline{RAS}$ clock access time, $t_{RAC}$. Page mode operation consists of keeping $\overline{RAS}$ active while toggling $\overline{CAS}$ between $V_{IH}$ and $V_{IL}$. The row is latched by $\overline{RAS}$ active transition, while each $\overline{CAS}$ active transition allows selection of a new column location on the row.

A page mode cycle is initiated by a normal read, write, or read-write cycle, as described in prior sections. Once the timing requirements for the first cycle are met, $\overline{CAS}$ transitions to inactive for minimum of $t_{CP}$, while $\overline{RAS}$ remains low ($V_{IL}$). The second $\overline{CAS}$ active transition while $\overline{RAS}$ is low initiates the first page mode cycle ($t_{PC}$ or $t_{PRWC}$). Either a read, write, or read-write operation can be performed in a page mode cycle, subject to the same conditions as in normal operation (previously described). These operations can be intermixed in consecutive page mode cycles and performed in any order. The maximum number of consecutive page mode cycles is limited by $t_{RASP}$. Page mode operation is ended when $\overline{RAS}$ transitions to inactive, coincident with or following $\overline{CAS}$ inactive transition.

## REFRESH CYCLES

The dynamic RAM design is based on capacitor charge storage for each bit in the array. This charge degrades with time and temperature, thus each bit must be periodically **refreshed** (recharged) to maintain the correct bit state. Bits in the MCM511000A require refresh every 8 milliseconds while refresh time for the MCM51L1000A is 64 milliseconds..

Refresh is accomplished by cycling through the 512 row addresses in sequence within the specified refresh time. All the bits on a row are refreshed simultaneously when the row is addressed. Distributed refresh implies a row refresh every 15.6 microseconds for the MCM511000A and 124.8 microseconds for the MCM51L1000A. Burst refresh, a refresh of all 512 rows consecutively, must be performed every 8 milliseconds on the MCM511000A and 64 milliseconds on the MCM51L1000A.

A normal read, write, or read-write operation to the RAM will refresh all the bits (2048) associated with the particular row decoded. Three other methods of refresh, **$\overline{RAS}$-only refresh**, **$\overline{CAS}$ before $\overline{RAS}$ refresh**, and **Hidden refresh** are available on this device for greater system flexibility.

### $\overline{RAS}$-Only Refresh

$\overline{RAS}$-only refresh consists of $\overline{RAS}$ transition to active, latching the row address to be refreshed, while $\overline{CAS}$ remains high ($V_{IH}$) throughout the cycle. An external counter is employed to ensure all rows are refreshed within the specified limit.

### $\overline{CAS}$ Before RAS Refresh

$\overline{CAS}$ before $\overline{RAS}$ refresh is enabled by bringing $\overline{CAS}$ active before $\overline{RAS}$. This clock order actives an internal refresh counter that generates the row address to be refreshed. External address lines are ignored during the automatic retresh cycle. The output buffer remains at the same state it was in during the previous cycle (hidden refresh).

### Hidden Refresh

Hidden refresh allows refresh cycles to occur while maintaining valid data at the output pin. Holding $\overline{CAS}$ active at the end of a read or write cycle, while $\overline{RAS}$ cycles inactive for $t_{RP}$ and back to active, starts the hidden refresh. This is essentially the execution of a $\overline{CAS}$ before $\overline{RAS}$ refresh from a cycle in progress (see Figure 1).

### $\overline{CAS}$ BEFORE $\overline{RAS}$ REFRESH COUNTER TEST

The internal refresh counter of this device can be tested with a **$\overline{CAS}$ before $\overline{RAS}$ refresh counter test**. This test is performed with a read-write operation. During the test, the internal refresh counter generates the row address, while the external address supplies the column address. The entire array is refreshed after 512 cycles, as indicated by the check data written in each row. See $\overline{CAS}$ before $\overline{RAS}$ refresh counter test **cycle** timing diagram.

The test can be performed after a minimum of **eight $\overline{CAS}$ before $\overline{RAS}$** initialization cycles. Test procedure:

1. Write "0"s into all memory cells with normal write mode.
2. Select a column address, read "0" out and write "1" into the cell by performing the **$\overline{CAS}$ before $\overline{RAS}$ refresh counter test, read-write cycle**. Repeat this operation 512 times.
3. Read the "1"s which were written in step 2 in normal read mode.
4. Using the same starting column address as in step 2, read "1" out and write "0" into the cell by performing the **$\overline{CAS}$ before $\overline{RAS}$ refresh counter test, read-write cycle**. Repeat this operation 512 times.
5. Read "0"s which were written in step 4 in normal read mode.
6. Repeat steps 1 to 5 using complement data.

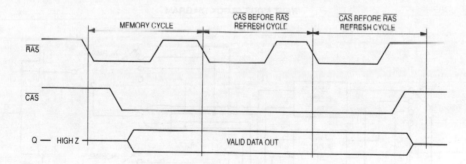

**Figure 1. Hidden Refresh Cycle**

## TEST MODE

Internal organization of this device (256K×4) allows it to be tested as if it were a 256K×1 DRAM. Only nine of the ten addresses (A0–A8) are used in test mode; A9 is internally disabled. A test mode write cycle writes data, D (data in), to a bit in each of the four 256K×1 blocks (B0–B3), in parallel. A test mode cycle reads a bit in each of the four blocks. If data is the same in all four bits, Q (data out) is the same as the data in each bit. If data is not the same in all four bits, Q is high Z. See truth table and test mode block diagram.

Test mode can be used in any timing cycle, including page mode cycles. The test mode function is enabled by holding the "TF" pin on "super voltage" for the specified period ($t_{TES}$, $t_{TEHR}$, $t_{TEHC}$; see **TEST MODE CYCLE**).

"Super voltage" = $V_{CC}$ + 4.5 V

where

4.5 V < $V_{CC}$ < ;5.5 V and maximum voltage = 10.5 V.

A9 is ignored in test mode. In normal operation, the "TF" pin must either be connected to $V_{IL}$, or left open.

### Test Mode Truth Table

| D | B0 | B1 | B2 | B3 | Q |
|---|----|----|----|----|---|
| 0 | 0 | 0 | 0 | 0 | 0 |
| 1 | 1 | 1 | 1 | 1 | 1 |
| — | | | Any Other | | High-Z |

## TEST MODE BLOCK DIAGRAM

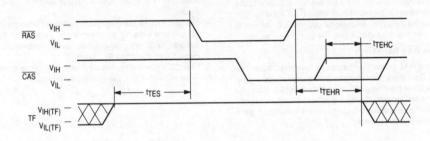

## TEST MODE BLOCK DIAGRAM

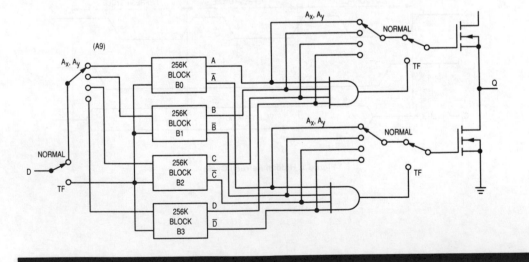

**J PACKAGE**
**300 MIL SOJ**
**CASE 822-03**

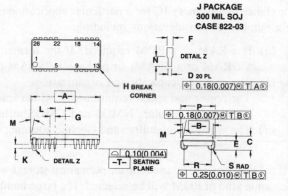

| DIM | MILLIMETERS | | INCHES | |
|-----|------|------|------|------|
| | MIN | MAX | MIN | MAX |
| A | 17.02 | 17.27 | 0.670 | 0.680 |
| B | 7.50 | 7.74 | 0.295 | 0.305 |
| C | 3.26 | 3.75 | 0.128 | 0.148 |
| D | 0.39 | 0.50 | 0.015 | 0.020 |
| E | 2.24 | 2.48 | 0.088 | 0.098 |
| F | 0.67 | 0.81 | 0.026 | 0.032 |
| G | 1.27 BSC | | 0.050 BSC | |
| H | — | 0.50 | — | 0.020 |
| K | 0.89 | 1.14 | 0.035 | 0.045 |
| L | 2.54 BSC | | 0.100 BSC | |
| M | 0° | 10° | 0° | 10° |
| N | 0.89 | 1.14 | 0.035 | 0.045 |
| P | 8.39 | 8.63 | 0.330 | 0.340 |
| R | 6.61 | 6.98 | 0.260 | 0.275 |
| S | 0.77 | 1.01 | 0.030 | 0.040 |

NOTES:
1. DIMENSIONING AND TOLERANCING PER ANSI Y14.5M, 1982.
2. CONTROLLING DIMENSION: INCH.
3. DIMENSION A & B DO NOT INCLUDE MOLD PROTRUSION. MOLD PROTRUSION SHALL NOT EXCEED 0.15(0.006) PER SIDE.
4. DIM R TO BE DETERMINED AT DATUM –T–.
5. FOR LEAD IDENTIFICATION PURPOSES, PIN POSITIONS 6,7,8,19,20, & 21 ARE NOT USED.
6. 822–01 AND –02 OBSOLETE, NEW STANDARD 822–03.

**Z PACKAGE**
**PLASTIC**
**ZIG-ZAG IN-LINE**
**CASE 836-02**

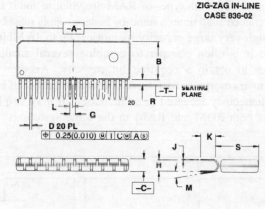

| DIM | MILLIMETERS | | INCHES | |
|-----|------|------|------|------|
| | MIN | MAX | MIN | MAX |
| A | 25.53 | 25.90 | 1.005 | 1.020 |
| B | 8.59 | 8.89 | 0.338 | 0.350 |
| C | 2.75 | 2.94 | 0.108 | 0.116 |
| D | 0.46 | 0.55 | 0.018 | 0.022 |
| G | 1.27 BSC | | 0.050 BSC | |
| H | 2.44 | 2.64 | 0.097 | 0.103 |
| J | 0.23 | 0.33 | 0.009 | 0.013 |
| K | 0.10 | 0.55 | 0.125 | 0.140 |
| L | 0.64 BSC | | 0.025 BSC | |
| M | 0° | 4° | 0° | 4° |
| R | 0.89 | 1.39 | 0.035 | 0.055 |
| S | 9.66 | 10.16 | 0.380 | 0.400 |

NOTES:
1. DIMENSIONING AND TOLERANCING PER ANSI Y14.5M, 1982.
2. CONTROLLING DIMENSION: INCH.
3. DIMENSION H TO CENTER OF LEAD WHEN FORMED PARALLEL.
4. DIMENSIONS A, B, AND S DO NOT INCLUDE MOLD PROTRUSION.
5. MOLD FLASH OR PROTRUSION SHALL NOT EXCEED 0.25(0.010).
6. 836–01 OBSOLETE, NEW STANDARD 836–02.

**Literature Distribution Centers:**
USA: Motorola Literature Distribution; P.O. Box 20912; Phoenix, Arizona 85036.
EUROPE: Motorola Ltd.; European Literature Center; 88 Tanners Drive, Blakelands, Milton Keynes, MK14 5BP, England.
JAPAN: Nippon Motorola Ltd.; 4-32-1, Nishi-Gotanda, Shinagawa-ku; Tokyo 141 Japan.
ASIA-PACIFIC: Motorola Semiconductors H.K. Ltd.; Silicon Harbour Center, No. 2 Dai King Street, Tai Po Industrial Estate, Tai Po, N.T., Hong Kong.

**MOTOROLA**

A23025-2  PRINTED IN USA  6/91  IMPERIAL LITHO  79814  12,000  MEM MOS YRAAAA

**MCM511000A • MCM51L1000A**

### Choice of a Memory

The choice of a memory IC for a particular application is determined by a number of considerations including:

(a) Is a RAM or a ROM required? If the former should it be a DRAM or a SRAM, or perhaps a PSRAM or a FRAM?
(b) The required capacity and organization.
(c) The power and speed requirements, which lead to a choice between the bipolar, NMOS and CMOS technologies.
(d) The cost, availability and second-sourcing of possible choices.

If the memory is to be used for the permanent storage of information then some kind of ROM will be selected. If a large number of ROMs are required it will prove cheaper to use mask-programed ROMs but for smaller numbers PROMs will prove to be more economical. If the stored information may on occasion need to be changed then either an EPROM or an EEPROM will be chosen. If the requirement is for a memory that can have data written into it as well as read out then a RAM must be selected. A small system will probably go for SRAMs since they are cheaper than DRAMs and space may not be quite as much a problem. Both types of RAM are volatile and if information is to be retained at all times a stand-by battery supply must be provided.

Although very large capacity memories, up to 16 Mbits, are now available it is often cheaper to combine several smaller-capacity memories to obtain a required bit capacity. Another reason for combining memories is that it may allow the number of bits that can be simultaneously accessed to be increased. It may also be possible to access both ROM and RAM in the same system.

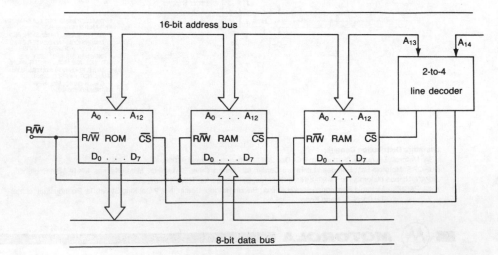

**Fig. 6.26** Memory decoding

In a large memory system the memory addresses are usually quoted using hexadecimal numbering. Consider, for example, a 32k × 1 RAM. There will be 15 address lines so that the lowest address will be 0H and the highest address will be 7FFH.

## Memory Decoding

Memory decoding consists of the use of a decoder to select one memory IC from a number that make up a memory system. Selection of one IC is achieved by taking its $\overline{CS}$ pin LOW while keeping the $\overline{CS}$ pins of all other ICs HIGH. The basic idea is illustrated by Fig. 6.26. The address bits $A_0$ to $A_{12}$ are used to select locations within the three memories and address bits $A_{13}$ and $A_{14}$ are used to select a particular memory.

### Example 6.9

A memory system includes one ROM and two RAMs all having the same capacity and × 8 word width. The ROM is located at addresses 0H through to 1FFFH. The two RAMs are located at addresses 2000H through to A000H. Determine the organization of the memory and draw the circuit.

### Solution

The ROM has an address size of 2000H = 4096 locations. Hence, it is organized as 4k × 8.

The RAM addresses go from 2000H to A000H so that each RAM has an address size of one-half 8000H = 4000H = 8192 locations. Hence each RAM is organized as 8k × 8. A 4k × 8 memory has 4096 locations so it must have 12 address lines $A_0$ through to $A_{11}$ and an 8k × 8 memory has 8192 locations so it must have 13 address lines $A_0$ through to $A_{12}$. Address lines $A_{13}$ and $A_{14}$ can be used to enable just one of three memory chips at a time. The address lines $A_0$ through to $A_{12}$ are connected together ($A_0$ to $A_0$, $A_1$ to $A_1$ etc.) as are the data lines $D_0$ through to $D_7$. The READ/WRITE inputs are also linked together. The arrangement of the three memory chips is shown by Fig. 6.26. An IC is selected by using the chip select CS input. The total number of addresses is 4096 + (2 × 8192) = 20 480 and this requires 15 address bits. The two most significant bits of the address are applied to the inputs of the 2-to-4 line decoder, e.g. the LS 74139, which has active LOW outputs. Three of the four decoder outputs are connected to the CS inputs of the memories so that any one of them can be enabled.

# 7 Design of Sequential Systems

A sequential system is one whose next output state depends not only on the input variables but also upon the present state of the output. This means that the output is fed back to one, or more, of the inputs to the circuit. A sequential circuit may be either synchronous or non-synchronous. In a synchronous sequential circuit all changes of state within the circuit take place under the control of a clock waveform. The application of each clock pulse to the circuit allows each stage in the circuit either to remain in its present state or to switch to its next state. The periodic time of the clock waveform must be greater than any propagation delays in the system. Synchronous sequential systems always include at least one flip-flop and they are easier to design than non-synchronous circuits. Commonly met examples of synchronous sequential circuits include the various counters and shift registers which were considered in the previous volume, *Digital Electron Technology*. A non-synchronous sequential circuit operates at its own speed; the output of one stage provides the input to the next and the circuit may be implemented using combinational logic only. Such a circuit cannot distinguish between consecutive inputs that are at the same logic level. The design of a non-synchronous circuit is more complicated than the design of a synchronous circuit because both stable and unstable states exist within the circuit.

An example of a simple non-synchronous sequential circuit is shown in Fig. 7.1. The circuit has two inputs, A and B, and a single output F. When the input variables change the next output state of the circuit is a function not only of those input variables but also of the present state of the output. It can be seen that the output fed back to the input has been labelled as f and not F; this is because there is always a short time delay between the output F changing state and this change appearing at the input. When analysing or designing a non-synchronous sequential circuit the next state F and the present state f must be considered separately.

The output of the circuit given in Fig. 7.1 is

$$F = \overline{\overline{A + B} + \overline{B + f}}$$
$$= (A + B).(B + f) = AB + Af + B + Bf$$
$$= Af + B$$

The design of both synchronous and non-synchronous sequential

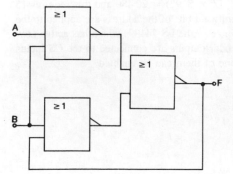

**Fig. 7.1** A non-synchronous sequential circuit

circuits makes use of the state diagram and the state table. However, while the state diagrams are the same for both kinds of circuit the state tables differ in some respects.

## State Diagrams

A *state diagram* is the means by which the required operation of a sequential circuit can be shown graphically. A state diagram uses circles to represent the states that may occur in the circuit. These circles are joined by lines that represent the transitions from one state to the next; the direction of a transition is shown by an arrowhead that points in the direction of the next state. The signal conditions that give rise to the transition are written alongside the transition lines.

In a synchronous system the present states are the internal states at the instant external data is applied. When a clock pulse occurs the next state is generated and this is fed back to the input to become the new present state. No further switching takes place until the next clock pulse occurs. During this time any errors due to unequal propagation delays in the logic disappear and so the internal states can be accurately defined and generated at the rate of one state per clock pulse. In a non-synchronous circuit the present states are the internal states at the instant external data is applied. The time taken to generate the next state depends only on the characteristics of the elements in the circuit and it is usually small. The next state is fed back to the input and it becomes the new present state.

Two different methods of labelling are employed:

(a) *Mealy*. Each of the circles is labelled with the state it represents, i.e. $S_0$, $S_1$, etc. Each transition path is labelled with the input(s) that cause that transition, a /, and the output that results from that transition. Thus, a label 01/1 indicates that the transition is caused by the input variables being A = 0, B = 1, with the resulting output being high. The state diagram of a J–K flip-flop, using this labelling, is shown by Fig. 7.2(a). State $S_0$ is the reset state of the flip-flop when the output Q = 0.

(i) If the inputs are J = 1, K = 0, or J = K = 1, the circuit will be set to have Q = 1. This is the state $S_1$.

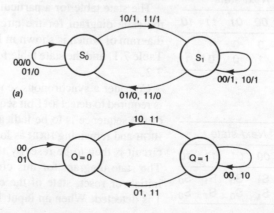

**Fig. 7.2** State diagrams of a J–K flip-flop: (a) Mealy; (b) Moore

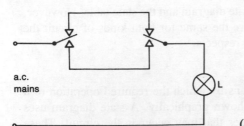

**Fig. 7.3** Simple logic circuit

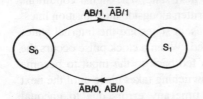

**Fig. 7.4** State diagram of Fig. 7.3

## Synchronous Sequential Circuits

**Table 7.1**

| Present state | | Next state | | | |
|---|---|---|---|---|---|
| $Q$ | $J-K$ | 00 | 01 | 11 | 10 |
| 0 | | 0 | 0 | 1 | 1 |
| 1 | | 1 | 0 | 0 | 1 |

**Table 7.2**

| Present state | | Next state | | | |
|---|---|---|---|---|---|
| | $AB$ | 00 | 01 | 11 | 10 |
| $S_0$ | | $S_1$ | $S_0$ | $S_1$ | $S_0$ |
| $S_1$ | | $S_1$ | $S_0$ | $S_1$ | $S_0$ |

(ii) If, instead, the inputs are $J = 0$, $K = 1$, or $J = K = 0$, the circuit will remain reset and it stays in its state $S_0$.

(iii) When the flip-flop is set, i.e. in state $S_1$, inputs $J = 0$, $K = 1$, or $J = K = 1$, will cause the circuit to reset when it goes back to its state $S_0$.

(iv) When in state $S_1$ the inputs are $J = K = 0$, or $J = 1$, $K = 0$, the circuit will remain in state $S_1$.

(b) *Moore*. The Moore state diagram of a J–K flip-flop is shown by Fig. 7.2(b). Each circle is labelled with the state that it represents, i.e. either $Q = 0$ or $Q = 1$, and the transition lines are marked only with the input variables leading to that state.

As an example of a state diagram consider the simple circuit shown in Fig. 7.3. This is the sort of arrangement employed in houses that enables the upstairs landing light to be turned ON and OFF by either one of two switches, one at the top, and the other at the bottom of the stairs. If the upward position of a switch is denoted by A, or B, and the downward position by $\bar{A}$, or $\bar{B}$, then clearly the light will be turned ON when both switches are UP, or when both switches are DOWN, i.e. $L = AB + \bar{A}\bar{B}$. The state diagram for this circuit is given in Fig. 7.4. The state $S_0$ represents the light being turned OFF and state $S_1$ represents the light being turned ON. Using the Mealy method of labelling the transition path from $S_0$ to $S_1$ is labelled $AB/1$ and $\bar{A}\bar{B}/1$. Similarly, the transition path from state $S_1$ to state $S_0$ is labelled $\bar{A}B/0$ and $A\bar{B}/0$.

## State Table

The operation of a sequential circuit can be listed by means of a *state table*. A state table has a number of rows equal to the number of possible states that the circuit is able to take up, and a number of columns equal to the number of possible combinations of the input variables. In each square of the state table the next state of the circuit is entered when it is in the state that is represented by the row label, and the input variables are those specified by the column heading.

The state table for a particular circuit is determined by looking at the state diagram for the circuit. For the J–K flip-flop, the state diagram of which is shown in Fig. 7.2(a), the state table is given by Table 7.1, and the state table for the lighting circuit is given in Table 7.2.

Consider a synchronous sequential circuit with a single input that is required to detect a 11 bit sequence in a digital waveform. Detection of the sequence is to be indicated by the output of the circuit going HIGH and remaining HIGH as long as consecutive 1s are received. The circuit is then to be reset by the next 0 pulse to arrive at the input. The state diagram for this circuit is shown by Fig. 7.5. $S_0$ is the initial, or reset, state of the circuit in which it rests until an input 1 is detected. When an input 1 is detected the circuit moves into its

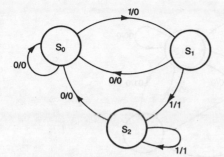

**Table 7.3**

| Present state | Next state | | Output | |
|---|---|---|---|---|
| Input x = | 0 | 1 | 0 | 1 |
| $S_0$ | $S_0$ | $S_1$ | 0 | 0 |
| $S_1$ | $S_0$ | $S_2$ | 0 | 1 |
| $S_2$ | $S_0$ | $S_2$ | 0 | 1 |

**Fig. 7.5** State diagram of circuit that detects an 11 bit sequence

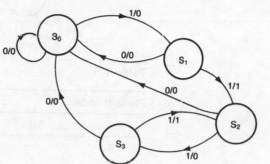

**Fig. 7.6** Modified version of Fig. 7.5

**Table 7.4**

| Present state | Next state | | Output | |
|---|---|---|---|---|
| Input x = | 0 | 1 | 0 | 1 |
| $S_0$ | $S_0$ | $S_1$ | 0 | 0 |
| $S_1$ | $S_0$ | $S_2$ | 0 | 1 |
| $S_2$ | $S_0$ | $S_3$ | 0 | 0 |
| $S_3$ | $S_0$ | $S_2$ | 0 | 1 |

state $S_1$; if, then, the next input bit is also a 1 the circuit will move into state $S_2$ and the output goes HIGH to indicate detection of the 11 bit sequence. If, however, the next input bit is a 0 the circuit reverts back to its initial state $S_0$. When the circuit is in state $S_2$ it will remain there until such time as a 0 is received when it will go into state $S_0$. The state table for this circuit is shown by Table 7.3.

If, now, the circuit is altered so that a third consecutive 1 will take the output LOW and a fourth 1 will take it HIGH again the state diagram must be modified as shown by Fig. 7.6. The state table for the amended circuit is given by Table 7.4.

*Example 7.1*

A circuit has two inputs A and B and a single output F. The output of the circuit is to go HIGH only if the inputs AB = 00, 01, 11 are received in that order. Once the output is HIGH it must remain HIGH as long as consecutive 11 bits are received. Obtain the state diagram and the state table for the circuit.

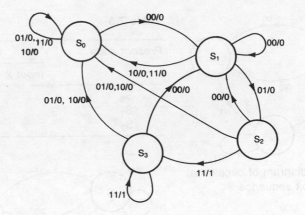

**Fig. 7.7**

**Table 7.5**

| Present state | | Next state | | | | Output F |
|---|---|---|---|---|---|---|
| | AB | 00 | 01 | 11 | 10 | |
| $S_0$ | | $S_1$ | $S_0$ | $S_0$ | $S_0$ | 0 |
| $S_1$ | | $S_1$ | $S_2$ | $S_0$ | $S_0$ | 0 |
| $S_2$ | | $S_1$ | $S_0$ | $S_3$ | $S_0$ | 1 |
| $S_3$ | | $S_1$ | $S_0$ | $S_3$ | $S_0$ | 0 |

*Solution*

The state diagram for the circuit is shown by Fig. 7.7 and the state table is given by Table 7.5.

*Example 7.2*

Draw the state diagram and the state table for a single-input circuit that will detect the occurrence of the bit sequence 010 (*a*) with no overlaps, i.e. 10101010 is two occurrences of the sequence, (*b*) with overlaps, i.e. 10101010 is three occurrences.

*Solution*

(*a*) The state diagram for the circuit is shown in Fig. 7.8(*a*), and the state table is given in Table 7.6.

(*b*) The state diagram is shown by Fig. 7.8(*b*) and the state table by Table 7.7.

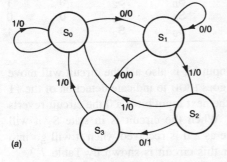

(*a*)

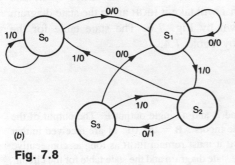

(*b*)

**Fig. 7.8**

**Table 7.6**

| Present state | | Next state | | Output |
|---|---|---|---|---|
| | Input x = | 0 | 1 | |
| $S_0$ | | $S_1$ | $S_0$ | 0 |
| $S_1$ | | $S_1$ | $S_2$ | 0 |
| $S_2$ | | $S_3$ | $S_0$ | 1 |
| $S_3$ | | $S_1$ | $S_0$ | 0 |

**Table 7.7**

| Present state | Next state | | Output |
|---|---|---|---|
| | Input x = 0 | 1 | |
| $S_0$ | $S_1$ | $S_0$ | 0 |
| $S_1$ | $S_1$ | $S_2$ | 0 |
| $S_2$ | $S_3$ | $S_0$ | 1 |
| $S_3$ | $S_1$ | $S_2$ | 0 |

## Excitation Maps

Synchronous sequential circuits are operated under the control of a system clock and they always employ at least one flip-flop. The flip-flops may be of either the J–K, D or T types. An *excitation map* is used to determine the necessary inputs to the flip-flops for the required next state to result when the next clock pulse arrives. The variables are the present states and the external inputs and the axes must be labelled using Gray code. The flip-flop input equations are obtained from the excitation map in the same way as combinational logic equations can be obtained from a Karnaugh map. The equations for the output F of the circuit and for the required flip flop inputs are derived separately. In writing down an excitation map it is necessary to have available the transition table for four types of flip-flop to be employed, and Table 7.8 gives the transition tables for the J–K, S–R, D and T flip-flops.

Consider Table 7.6 which has been rewritten in Table 7.9 to show

**Table 7.8**

| Present state | Next state | | | | | | | |
|---|---|---|---|---|---|---|---|---|
| Q | $Q^+$ | J–K | | D | T | S–R | |
| 0 | 0 | 0 | X | 0 | 0 | 0 | X |
| 0 | 1 | 1 | X | 1 | 1 | 1 | 0 |
| 1 | 0 | X | 1 | 0 | 1 | 0 | 1 |
| 1 | 1 | X | 0 | 1 | 0 | X | 0 |

**Table 7.9**

| Present state | | Next state | | | | Output |
|---|---|---|---|---|---|---|
| | | Input x = 0 | | 1 | | |
| | $Q_B$ $Q_A$ | $Q_B$ | $Q_A$ | $Q_B$ | $Q_A$ | |
| $S_0$ | 0  0 | 0 | 1 | 0 | 0 | 0 |
| $S_1$ | 0  1 | 0 | 1 | 1 | 1 | 0 |
| $S_2$ | 1  1 | 1 | 0 | 0 | 0 | 1 |
| $S_3$ | 1  0 | 0 | 1 | 0 | 0 | 0 |

the states of the $Q_A$ and $Q_B$ outputs of the two flip-flops that are necessary. Using Tables 7.8 and 7.9 the excitation maps for the circuit may be drawn. When the circuit is in its initial state $S_0$ and the input x to the circuit is at 0 the circuit should move into its state $S_1$. This means that it should change from $Q_A = Q_B = 0$ to $Q_A = 1$, $Q_B = 0$. From Table 7.8 it can be seen that this requires the following flip-flop inputs: $J_A = 1$, $K_A = X$, $J_B = 0$, $K_B = X$, $D_A = 1$, $D_B = 0$, $T_A = 1$ and $T_B = 0$. These values are entered in the $Q_BQ_A = 00$, $x = 0$ square of each excitation map. Similarly, when the input x $= 1$ the circuit should remain in state $S_0$, or $Q_A = Q_B = 0$ for both the present and the next states. Again from Table 7.8, this requires that $J_A = J_B = 0$, $K_A = K_B = X$, $D_A = D_B = 0$ and $T_A = T_B = 0$. These values are entered in the $Q_AQ_B = 00$, $x = 1$ square of each of the maps. The rest of the entries have been obtained in the same way.

(a) J–K flip-flops:

| $Q_BQ_A$ | | | | | $Q_BQ_A$ | | | | |
|---|---|---|---|---|---|---|---|---|---|
| x | 00 | 01 | 11 | 10 | x | 00 | 01 | 11 | 10 |
| 0 | 1 X | X 0 | X 1 | 1 X | 0 | 0 X | 0 X | X 0 | X 1 |
| 1 | 0 X | X 0 | X 1 | 0 X | 1 | 0 X | 1 X | X 1 | X 1 |

$J_A$, $K_A$            $J_B$, $K_B$

(b) D flip-flops:

| D | | | | | D | | | | |
|---|---|---|---|---|---|---|---|---|---|
| x | 00 | 01 | 11 | 10 | x | 00 | 01 | 11 | 10 |
| 0 | 1 | 1 | 0 | 1 | 0 | 0 | 0 | 1 | 0 |
| 1 | 0 | 1 | 0 | 0 | 1 | 0 | 1 | 0 | 0 |

(c) T flip-flops:

| T | | | | | T | | | | |
|---|---|---|---|---|---|---|---|---|---|
| x | 00 | 01 | 11 | 10 | x | 00 | 01 | 11 | 10 |
| 0 | 1 | 0 | 1 | 1 | 0 | 0 | 0 | 0 | 1 |
| 1 | 0 | 0 | 1 | 0 | 1 | 0 | 1 | 1 | 1 |

## Design of a Synchronous Sequential Circuit

The steps to be taken in the design of a symmetrical sequential circuit are as follows:

(a) Draw the state diagram of the required circuit.

(b) Determine the number of flip-flops needed and decide which kind of flip-flops are to be used.

(c) Use the state diagram to obtain the state table for the circuit.

(d) Draw the excitation maps.

(e) Use the excitation maps to obtain the equations for the flip-flop inputs.

(f) Determine an expression for the output F of the circuit.

(g) Using the equations obtained in (e) and (f) draw the circuit and decide which ICs are to be used.

As an example, design a circuit that will detect the bit sequence 1001 in a binary signal, including any overlaps. Detection of the sequence should be indicated by the output of the circuit going high.

The state diagram of the wanted circuit is shown in Fig. 7.9. Initially, the circuit is in state $S_0$ and here it will remain until a 1 bit is received. Then the circuit will move to state $S_1$ and will remain in this state until a 0 bit arrives at the input. The circuit then moves into its state $S_2$. Now a 0 bit followed by a 1 bit will move the circuit through state $S_3$ into state $S_4$ when the output will go HIGH to indicate that the bit sequence has been detected.

When the circuit is in state $S_2$ a 1 bit at the input means that the required sequence is not present but this 1 bit could be the start of a new wanted sequence and so the circuit must move back to state $S_1$.

When the circuit is in state $S_3$ a 0 bit at the input means that the sequence cannot now occur and the circuit should therefore go back to its initial state $S_0$.

When the circuit is in its final state $S_4$, which means that the wanted bit sequence has been detected, a 1 bit input means that the next sequence is not the wanted one and so the circuit reverts to state $S_0$. If, however, the next input bit is a 0 it could be the second bit of a following wanted sequence, i.e. 1001 001, and so the circuit moves into its state $S_2$.

There are five different states and hence three flip-flops are necessary. These may be either J–K, D or T flip-flops. The five states

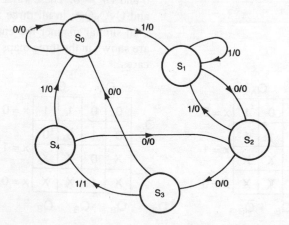

**Fig. 7.9** State diagram of circuit that detects a 1001 bit sequence

**Table 7.10**

| Present state | | | | Next state | | | | | | | | Output |
|---|---|---|---|---|---|---|---|---|---|---|---|---|
| $Q_C$ | $Q_B$ | $Q_A$ | Input x | $= 1$ | | | | $= 0$ | | | | F |
| | | | | $Q_C$ | $Q_B$ | $Q_A$ | | $Q_C$ | $Q_B$ | $Q_A$ | | |
| $S_0$  0 | 0 | 0 | $S_1$ | 0 | 0 | 1 | $S_0$ | 0 | 0 | 0 | | 0 |
| $S_1$  0 | 0 | 1 | $S_1$ | 0 | 0 | 1 | $S_2$ | 0 | 1 | 1 | | 0 |
| $S_2$  0 | 1 | 1 | $S_1$ | 0 | 0 | 1 | $S_3$ | 0 | 1 | 0 | | 0 |
| $S_3$  0 | 1 | 0 | $S_4$ | 1 | 1 | 0 | $S_0$ | 0 | 0 | 0 | | 0 |
| $S_4$  1 | 1 | 0 | $S_0$ | 0 | 0 | 0 | $S_2$ | 0 | 1 | 1 | | 1 |

are best numbered so that only one bit changes at a time in the present state table, and one possible numbering scheme is shown by Table 7.10.

### (a) D Flip-flops

When the circuit is in its initial state $S_0$ it should stay there if the input x = 0 or go into its state $S_1$ if x = 1. The first state requires that $D_A = 0$ and the second state requires that $D_A = 1$. Also $D_B = D_C = 0$ for both x = 0 and x = 1. Hence the excitation maps for the three flip-flops have these values entered in the $\overline{Q_A}\overline{Q_B}\overline{Q_C}\overline{x}$ and $\overline{Q_A}\overline{Q_B}\overline{Q_C}x$ squares.

Next, when the circuit is in state $S_1$, it should move to state $S_2$ if x = 0 but remain in state $S_1$ if x = 1. This requires that when x = 0, $Q_A$ stays at 1 and so $D_A = 1$, $Q_B$ changes from 0 to 1 which needs $D_B = 1$. $Q_C$ does not change so $D_C = 0$. Also, when x = 1 $Q_A$ remains at 1 so $D_A = 1$, while $Q_B$ and $Q_C$ both stay at 0 so $D_B = D_C = 0$. These values are entered into squares $Q_A\overline{Q_B}\overline{Q_C}\overline{x}$ and $Q_A\overline{Q_B}\overline{Q_C}x$ in all three maps.

Similarly, the circuit should go from state $S_2$ to state $S_3$ if x = 0, requiring $Q_A$ to change from 1 to 0 and so $D_A = 0$; $Q_B$ stays at 1 and $Q_C$ stays at 0 so that $D_B = 1$ and $D_C = 0$. If x = 1 the state of the circuit should change from $S_2$ to $S_1$ so that $D_A = 1$, $D_B = 0$ and $D_C = 0$. These values are entered into the squares $Q_AQ_B\overline{Q_C}\overline{x}$ and $Q_AQ_B\overline{Q_C}x$ in all three maps.

In similar manner the entries for $S_3$ and $S_4$ have been made and are shown in the three maps. The remainder of the squares are 'don't cares'.

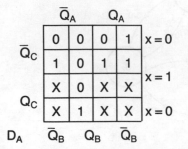

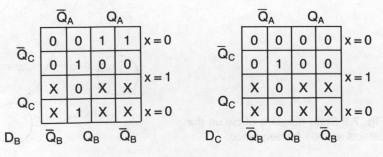

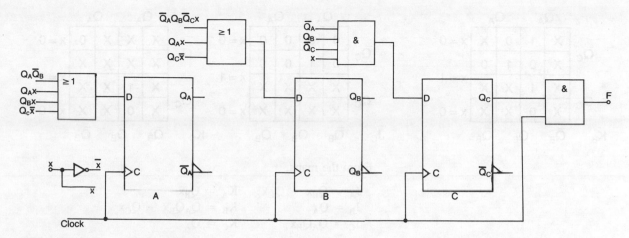

**Fig. 7.10** D flip-flop implementation of Fig. 7.9

From the excitation maps,

$$D_A = Q_A\overline{Q}_B + Q_A\overline{x} + Q_Bx + Q_C\overline{x},$$
$$D_B = Q_A\overline{x} + \overline{Q}_AQ_B\overline{Q}_Cx + Q_C\overline{x} \quad \text{and} \quad D_C = \overline{Q}_AQ_B\overline{Q}_Cx$$

The circuit is shown in Fig. 7.10.

### (b) J—K Flip-flops

The same procedure as in (a) is followed but now it is necessary to determine the required states of both the J and the K inputs of each flip-flop for a required change in the state of the circuit to take place. As an example, consider the circuit when it is state $S_0$. When $x = 0$ the circuit must stay in state $S_0$ and so $Q_A$, $Q_B$ and $Q_C$ must all remain at 0, hence the required inputs are $J = 0$ and $K = X$. These values are entered in the $\overline{Q}_A\overline{Q}_B\overline{Q}_C\overline{x}$ square of each map. If $x = 1$ the circuit must move to state $S_1$ in which $Q_A$ alone changes from 0 to 1. Hence the necessary input to flip-flop A is $J_A = 1$ and $K_A = X$. These values are entered in the $\overline{Q}_A\overline{Q}_B\overline{Q}_Cx$ squares of the $J_A$ and $K_A$ maps. The corresponding squares in the other maps have the same values as for $\overline{x}$ entered. Continuing in this way gives the excitation maps shown.

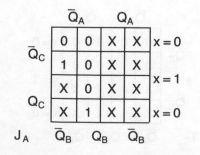

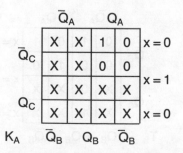

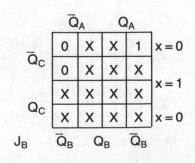

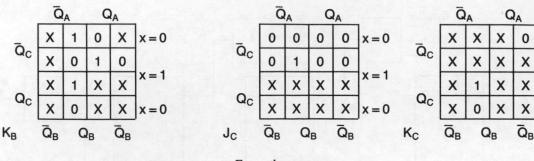

From the maps,

$$J_A = \overline{Q}_B x + Q_C \overline{x} \qquad K_A = Q_B \overline{x}$$
$$J_B = Q_A \qquad\qquad K_B = \overline{Q}_A \overline{Q}_B \overline{x} + Q_C x$$
$$J_C = \overline{Q}_A Q_B x \qquad K_C = \overline{Q}_A$$

The circuit is shown in Fig. 7.11. It can be seen that the extra flexibility due to the J–K flip-flop having two input terminals leads to simpler combinational logic.

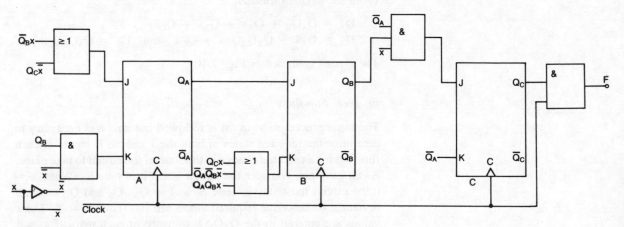

**Fig. 7.11** J–K flip-flop implementation of Fig. 7.9

*(c) The T Flip-flop*

The excitation maps obtained using the transition table for a T flip-flop are as follows:

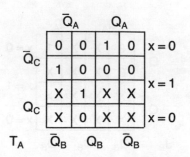

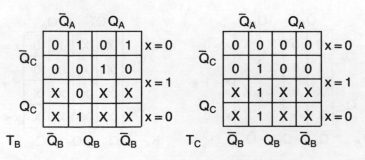

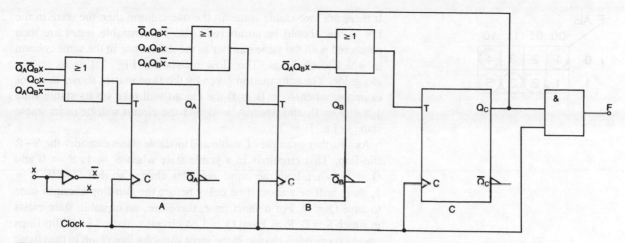

**Fig. 7.12** T flip-flop implementation of Fig. 7.9

From the maps,

$$T_A = \overline{Q}_A\overline{Q}_Bx + Q_Cx + Q_AQ_B\overline{x}$$
$$T_B = Q_AQ_B\overline{x} + Q_AQ_Bx + Q_AQ_B\overline{x}$$
$$T_C = Q_C + \overline{Q}_AQ_Bx$$

The circuit is shown in Fig. 7.12.

## Non-synchronous Sequential Circuits

A non-synchronous sequential circuit does not have a clock to regulate the events that take place within the circuit. The control inputs to a non-synchronous sequential circuit are pulses on one or more inputs. The circuit responds immediately to any input change instead of waiting until the next clock pulse arrives, as does a synchronous circuit. If an unchanging sequence of bits, say 1111, is applied to a synchronous circuit this will be taken as a series of 1 inputs at each clock pulse, but if the same sequence is applied to a non-synchronous circuit it would be taken as a single 1 input. Timing may therefore be difficult to design unless it can be arranged that the input variables never change state simultaneously.

An example of a non-synchronous sequential circuit is shown by Fig. 7.1 for which the output was found to be F = Af + B. This equation has been plotted on excitation map (a).

A stable state is one for which the entry for the output F of the circuit and the condition of f for that row are the same. The stable states in the circuit are usually ringed (see map (b)). An unstable state exists when the circuit is in the process of changing from one stable state to another; for an unstable state the F entry differs from the condition for f for that row. Unstable states are not ringed.

The stable and unstable states for this circuit are shown in the next map. If, now, the stable states are numbered, in the sequence AB = 00, 01, 11 and 10, the *flow matrix* for the circuit will be obtained.

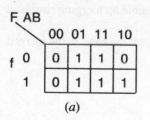

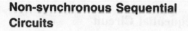

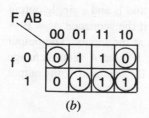

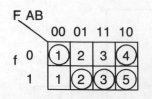

F AB

|   | 00 | 01 | 11 | 10 |
|---|----|----|----|----|
| f 0 | ①  | 2  | 3  | ④  |
| 1   | 1  | ②  | ③  | ⑤  |

If there are two stable states in the one column then the state in the
f = 0 row should be numbered first. The unstable states are then
numbered with the same number as the stable state in the same column
to which each leads. The flow matrix for Fig. 7.1 is as shown
alongside. The information given by the flow matrix shows that, for
example, when A = B = 0 the circuit will take up its stable state
1, i.e. F = 0, and when A = B = 1 the circuit will be in its stable
state 3, i.e. F = 1.

As a further example of stable and unstable states consider the S−R
flip-flop. This circuit is in a stable state when S = 1, R = 0 and
Q = 1. When both the input variables change so that S = 0, R =
1, there will be a short time delay before the flip-flop changes state
to have Q = 0. For a short time, therefore, an unstable state exists
in which S = 0, R = 1 and Q = 1. A circuit with two S−R flip-flops
could experience a change in the input variables that result in the circuit
changing state from $Q_A = Q_B = 1$ to $Q_A = Q_B = 0$. This change
in state could, possibly, occur simultaneously but it is much more
likely that there will be an unstable state, with either $Q_A = 1$ and
$Q_B = 0$, or $Q_A = 0$ and $Q_B = 1$, because of the different switching
speeds of the two flip-flops. If the outputs $Q_A$ and $Q_B$ are applied
as inputs to other circuitry incorrect system operation is very likely
to occur. To avoid such problems non-synchronous sequential circuits
are usually designed so that only one input variable is allowed to
change at any particular instant in time.

### Design of a Non-synchronous Sequential Circuit

The design procedure for a non-synchronous sequential system is as
follows:

(a) Draw the state diagram.
(b) Obtain the *primitive state (flow) table*.
(c) Simplify the primitive state table if possible by merging rows and
    obtain the merged state table.
(d) Allocate secondary variables to identify all the rows in the merged
    state table.
(e) Draw the flow matrix.
(f) Draw the excitation map(s).
(g) Derive the logic equations.
(h) Draw the designed circuit.

Consider a circuit that has two inputs A and B and a single output
F. The output of the circuit is to go HIGH if the sequence A = B =
0, A = 0 and B = 1, A = B = 1 occurs at the inputs. The output
should then remain high as long as A = 1. If input A should change
from 1 to 0 and then back to 1, while input B remains at 1, the output
should go HIGH.

## (a) State Diagram

The initial state $S_0$ of the circuit is when $A = B = 0$. If B changes to 1 the circuit moves to its state $S_1$. When the circuit is in state $S_1$ a change in input B from 1 to 0 moves the circuit back to state $S_0$ but if input A changes from 0 to 1 (when $A = B = 1$) the circuit goes into its state $S_2$ and output goes HIGH.

Now a change in A from 1 to 0 (AB = 01) will move the circuit into state $S_1$ and the output of the circuit will go LOW; but a change in B from 1 to 0 (AB = 10) moves the circuit to state $S_3$ and the output remains HIGH. When the circuit is in state $S_1$ *after* being in state $S_2$ a move to state $S_2$ will also take the output HIGH. If, in either state $S_1$ or $S_3$, the next input change gives $A = B = 0$ the circuit will revert to its initial state $S_0$. The circuit will go from state $S_3$ to state $S_2$ if B changes from 0 to 1 and the output will then remain HIGH.

If, when in state $S_0$, A changes from 0 to 1 before B has changed from 0 to 1 a sequence of states is entered none of which will lead to the output of the circuit going HIGH.

The state diagram is shown in Fig. 7.13.

## (b) Primitive State (Flow) Table

The primitive state, or flow, table can be obtained from the state diagram. This table must have one row for each stable state and so for this circuit there will be seven rows. An entry must be made in the table for each unstable state; these entries indicate the next stable state that will be taken up by the circuit for the particular input variables AB for that column in the table. If a particular condition cannot occur — in this circuit because simultaneous changes in A and B are not permitted — it is represented in the table by a dash.

The initial stable state of the circuit is $S_0$ and this is represented

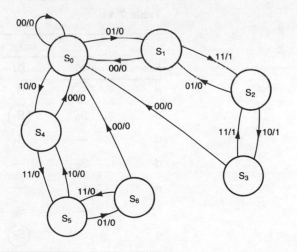

**Fig. 7.13** State diagram of a non-synchronous sequential circuit

| Row | AB | | | | Output |
|-----|----|----|----|----|--------|
| | 00 | 01 | 11 | 10 | F |
| a | $(S_0)$ | $S_1$ | — | $S_4$ | 0 |

in the first row of the table under the column heading AB = 00. Since this is a stable state it is ringed. If, then, A = 0 and B = 1 the circuit will go into an unstable state $S_1$ until it moves into stable state $S_1$. Thus, unstable state $S_1$ is entered in the first row under the column heading AB = 01. If, instead, AB = 10 the circuit enters unstable state $S_4$ before it goes to its stable state $S_4$. Hence $S_4$ is entered in the first row under the column heading 10. Since A and B cannot change simultaneously the condition A = B = 1 does not happen and a dash is entered at the intersection of the first row and the 11 column. The first row entries in the primitive state table, therefore, are as shown alongside.

The output of the circuit is 0 since the conditions for the output to go HIGH have not yet occurred. The rest of the entries into the primitive state table are determined in a similar manner leading to Table 7.11.

### (c) Row Merging

Once the primitive state table has been obtained it will often contain more stable states than are necessary and so the next step should be to attempt to reduce the number of rows in the table. Two or more rows may be merged together if they have the same numbered states, stable or unstable, in each column of the table. Dashed entries are treated as 'don't cares' and may be assumed to have any state number. Each row in the primitive state table may only appear in *one* merged row. The output $F$ of the circuit is ignored; it can later be obtained from the primitive state table.

Thus, rows a and b in Table 7.11 can be merged to give the merged row

$(S_0)$  $(S_1)$  $S_2$  $S_4$

**Table 7.11**

| Row | Input variables AB | | | | Output |
|-----|----|----|----|----|--------|
| | 00 | 01 | 11 | 10 | F |
| a | $(S_0)$ | $S_1$ | — | $S_4$ | 0 |
| b | $S_0$ | $(S_1)$ | $S_2$ | — | 0 |
| c | — | $S_1$ | $(S_2)$ | $S_3$ | 1 |
| d | $S_0$ | — | $S_2$ | $(S_3)$ | 1 |
| e | $S_0$ | — | $S_5$ | $(S_4)$ | 0 |
| f | — | $S_6$ | $(S_5)$ | $S_4$ | 0 |
| g | $S_0$ | $(S_6)$ | $S_5$ | — | 0 |

**Table 7.12**

| Rows | AB | 00 | 01 | 11 | 10 |
|------|-----|-----|-----|-----|-----|
| a/b | | (S₀) | (S₁) | S₂ | S₄ |
| c/d | | S₀ | S₁ | (S₂) | (S₃) |
| e/f/g | | S₀ | (S₆) | (S₅) | (S₄) |

**Table 7.13**

| Rows | AB | 00 | 01 | 11 | 10 |
|------|-----|-----|-----|-----|-----|
| a | | (S₀) | S₁ | — | S₄ |
| b/c/d | | S₀ | (S₁) | (S₂) | (S₃) |
| e/f/g | | S₀ | (S₆) | (S₅) | (S₄) |

and rows c and d can be merged to give the merged row

$$S_0 \quad S_1 \quad \text{(}S_2\text{)} \quad \text{(}S_3\text{)}$$

The merged state table is given by Table 7.12.

Alternatively, rows b, c and d could be merged together leaving row a on its own. The merged state table would then be as shown by Table 7.13.

Other possibilities also exist, e.g. rows a and e could be merged together.

### (d) Secondary Variables

Since there are only three rows — in either of the merged state tables — only two secondary variables x and y will be required to identify them. These are variables that are provided by the circuit itself (see Fig. 7.14). The rows should be allotted x and y values that place adjacently rows that contain similar numbered stable and unstable states. In Table 7.12 there is movement from row a/b to row e/f/g [$S_4$ to $S_4$ and $S_0$ to $S_0$] and also between rows a/b and c/d, but none between rows c/d and e/f/g. Hence a suitable numbering would be

row c/d     $xy = 00$
row a/b     $xy = 01$
row e/f/g     $xy = 11$

The xy combination 10 is then not used.

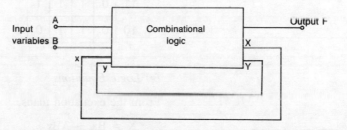

**Fig. 7.14** Secondary variables

### (e) Flow Matrix

The flow matrix maps the stable and unstable states with 'don't cares' indicated by a dash. The flow matrix for Table 7.12, using the suggested xy row allocations, is shown alongside.

The flow matrix is to be minimized using the Karnaugh map looping technique. Hence the next step is to allocate suitable state numbers to all of the 'don't cares'. A possible choice is illustrated by the next flow matrix, shown on page 250.

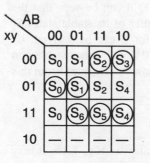

### (f) Excitation Map

The excitation map for the circuit is drawn by writing down the values of X and Y for each stable state that agrees with the xy entry in that

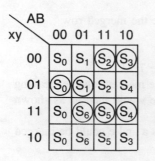

row. Thus, in the position occupied by stable states $S_2$ and $S_3$ in the top row of the flow matrix 00 is entered in the excitation map. Similarly, $S_0$ and $S_1$ in the second row are represented by 01. Next, the entries for the stable states are copied for the unstable states of the same number. This is shown by the next maps:

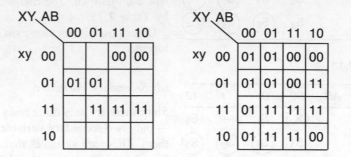

Lastly, separate excitation maps can be drawn for both X and Y as shown:

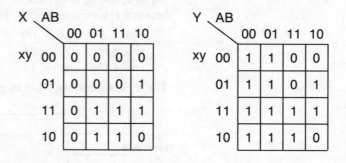

### (g) Logic Equations

From the excitation maps,

$$X = Bx + A\bar{B}y$$
$$Y = \bar{A} + \bar{B}y + Bx$$

From the primitive state table (Table 7.11) it can be seen that the output F is HIGH when the circuit is in either of its stable states $S_2$ or $S_3$. From the flow matrix a mapping for F can be derived, by writing 1 for each $S_2$ and $S_3$ entry and a 0 for all other stable and unstable states. Thus, the output map is shown alongside. From this map $F = AB\bar{x} + AB\bar{y}$

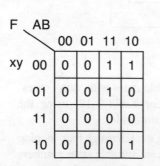

### (h) The Circuit

The circuit is shown, using AND and OR gates, in Fig. 7.15. It may be implemented using random logic, probably using either NAND gates or NOR gates only, or, more likely, one of the other methods that are available such as a multiplexer or a ROM.

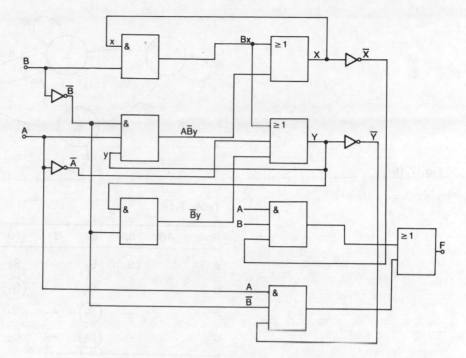

**Fig. 7.15**

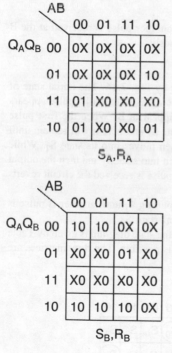

$S_A, R_A$

$S_B, R_B$

The design of a non-synchronous sequential circuit may also be carried out using flip-flops. The merged state table given in Table 7.12 has three rows so that two flip-flops are required. Suppose that S–R flip-flops are employed. Two excitation maps, labelled $Q_A^+Q_B^+$, AB and $Q_AQ_B$, are required, one for $S_AR_A$ and one for $S_BR_B$ as shown. Referring to the excitation map for the circuit (on p. 250) and to Table 7.8: for the present states $Q_AQ_B$ of the flip-flop outputs to become the required next states $Q_A^+Q_B^+$ it is necessary to compare the row assignments and the table entries to determine the conditions necessary to set and reset each S–R flip-flop. For example, in the top left-hand square of the $S_AR_A$ map the present and next states of $Q_A$ are both 0 and hence, from Table 7.8, $S_A = 0$ and $R_A = X$. Similarly, for the top left-hand square of the $S_BR_B$ map, $Q_B$ is to change from 0 to 1 and so $S_B = 1$ and $R_B = 0$. The remainder of the entries in the map have been obtained in the same way.

From the maps

$$S_A = A\overline{B}Q_B$$
$$R_A = \overline{A}B + \overline{B}Q_B$$
$$S_B = \overline{A}\,\overline{B} + \overline{A}BQ_A + BQ_A\overline{Q_B}$$
$$R_B = ABQ_A$$

*Example 7.3*

Design a circuit that has two inputs A and B and a single output F. The output is to go high whenever the sequence A = 1, B = 11 in that order occurs.

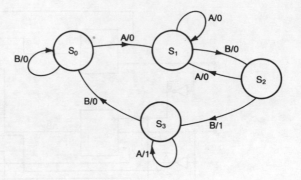

**Fig. 7.16**

**Table 7.14**

| Row | AB | 00 | 01 | 11 | 10 | Output F |
|-----|----|----|----|----|----|----------|
| a | | (S₀) | S₀ | — | S₁ | 0 |
| b | | — | S₁ | — | (S₁) | 0 |
| c | | — | (S₂) | — | S₁ | 0 |
| d | | — | (S₃) | — | S₃ | 1 |

The output should then remain high until another pulse is received at the B input. A and B never go high simultaneously.

*Solution*

The state diagram for the circuit is shown by Fig. 7.16. The initial state of the circuit is $S_0$ and the circuit will remain in this state until a pulse appears at input A, no matter how many pulses appear at B. When the first pulse appears at input A the circuit moves to state $S_1$ and here it will remain until a pulse appears at input B. The circuit then moves into its state $S_2$. While in state $S_2$ another B pulse moves the circuit into state $S_3$ and then the output goes HIGH. If, when in state $S_2$, another A pulse is received the circuit reverts to state $S_1$.

The circuit will remain in state $S_3$, outputting 1, until another B pulse is received when it will go back to its initial state $S_0$.

The primitive state table, obtained from the state diagram is given by Table 7.14. It can be seen that no row merging is possible. The flow matrices are as follows:

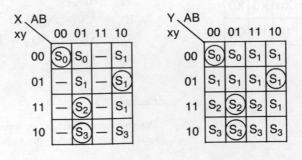

The excitation maps are:

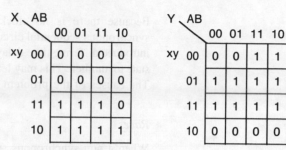

From the maps,

$$X = \bar{A}x + Bx + x\bar{y}$$
$$Y = A\bar{x} + y$$

The output F of the circuit is high only when the circuit is in stable state $S_3$ and from the flow matrix

$$F = x\bar{y}$$

The circuit is shown in Fig. 7.17.

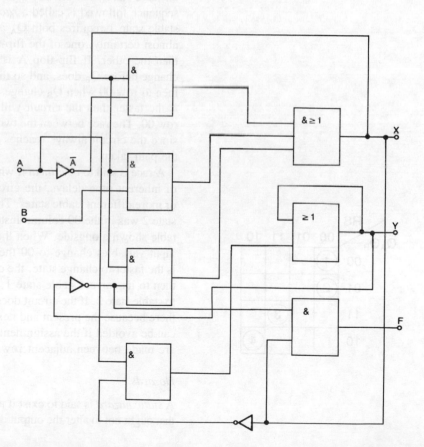

**Fig. 7.17**

## Hazards and Races

Because there is no clock to regulate the operation of a non-synchronous sequential circuit, different ICs, gates or flip-flops may introduce differing time delays when a circuit changes from one stable state to another. This may lead to the occurrence of incorrect outputs. The causes of the problem are known as *hazards* and as *races*.

### Races

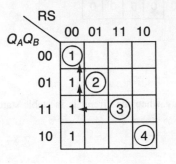

When a non-synchronous sequential circuit moves from one stable state to another it may go through one or more unstable states. If there is only one possible path between the two states the sequence that is followed is known as a *cycle*. The flow table alongside is for a circuit that has four stable states and uses flip-flops. If the circuit is in its stable state 3 when the input variables change from 11 to 00 the circuit will immediately move to its unstable state 1 in row 11 and column 00. It then moves to unstable state 1 in row 01 and, finally, to stable state 1 in row 00. The cycle is indicated by the arrows.

When there are two or more possible paths for the sequence of unstable states, but the circuit always gets into the correct state the sequence followed is called a *race*. The move from stable state 3 to stable state 1 requires both $Q_A$ and $Q_B$ to change from 1 to 0 but, almost certainly, one of the flip-flops will be faster to change state than the other. If flip-flop A is the faster to switch then $Q_A$ will change before $Q_B$ does, and so the circuit moves to row 01 first and then to row 00 when $Q_B$ changes. If, on the other hand, flip-flop B is the faster then the circuit will move to row 10 first and then to row 00. The race between the two flip-flops is said to be *non-critical* since the circuit always reaches its correct next state regardless of the path taken.

A race is said to be *critical* if when different paths are taken, because of inherent time delays, the circuit may end up in any one of two or more different stable states. This situation could occur if the stable state 2 was in the 00 column instead of the 01 column as in the flow table shown alongside. When the circuit is in stable state 3 and the input variables change to 00 then, depending upon which flip-flop is the faster to change state, the circuit could move (i) to row 10 and then to its correct stable state 1, or (ii) move to row 01 and thence to stable state 2. If the circuit does get into stable state 2 it will remain there because the present and next states are the same. Critical races can be avoided if the assignment of rows is made so that transitions are made between adjacent rows only (see p. 249).

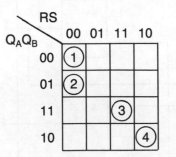

### Hazards

A *static hazard* is said to exist if a change in the input variables occurs that ought not to alter the output does in fact cause the output to change

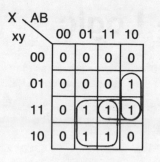

momentarily. A static hazard may arise if any two 1 entries in an excitation map are not looped together. There is, for example, a static hazard in the excitation map which is shown alongside.

The minimal solution is $X = Bx + A\bar{B}y$. To remove the static hazard the adjacent squares AB = xy = 11, and AB = 10 xy = 11, must also be looped together. The equation for the circuit is then

$$X = Bx + A\bar{B}y + Axy$$

# 8 Programmable Logic Devices

The need to design more compact digital systems than is possible using SSI/MSI/LSI devices has led to the use of much more complex ICs. An *application-specific integrated circuit* (ASIC) is an IC that has been designed for a specific application. A single ASIC can replace a large number of standard SSI/MSI/LSI devices and in some cases a complete board of ICs may be replaced by a single chip. A *full-custom* ASIC is very costly and time consuming to get into its final design stage, although once this point has been reached production costs are relatively low. The use of a full-custom ASIC can only be economically justified when high-volume production is anticipated, generally in excess of about 200 000 items. A cheaper and quicker alternative is the *semi-custom* ASIC which employs a cell library; these devices are either *gate array* ASICs or *standard cell* ASICs.

A gate array consists of a large number of gates that are formed within the silicon but that are left completely unconnected. A required circuit is produced by the manufacturer depositing a metal interconnection network to the customer's specification. Further metal connects appropriate points in the circuit to the IC package pins. The production of a gate array ASIC is subject to considerable wastage because of the difficulties experienced in designing the gate interconnections even though a *computer-aided design* (CAD) package is employed. As a result many of the gates always remain unused. In addition, the various interconnection routeings which have to be chosen lead to increased propagation delays so that the performance of the final circuit is not usually as good as could be obtained from a full-custom design. Gate arrays may employ both bipolar, TTL, $I^2L$, ECL, NMOS and CMOS technology, but the majority of new designs are CMOS devices.

The alternative approach to semi-custom ASIC design is to use standard cells. A standard cell ASIC is fabricated from a number of predefined building blocks that are listed by the manufacturer in a *cell library*. A designer selects the combination of standard cells required for his design and specifies, using a CAD package, how they are to be interconnected to produce the wanted circuit. The information is then sent to the manufacturer who completes the design using the same masking techniques that are employed for a full-custom design. This method of developing an ASIC has the advantage that, because each standard cell has been previously tested, it is very likely that the circuit will work correctly. The standard cell ASIC combines the best features of using standard parts and the inherent flexibility of relatively low-cost design.

Gate array ASICs are somewhat cheaper than standard cell ASICs and they are economical to use for production runs of between 5000 and 50 000 items. Standard cell ASICs do not become economical until the required production run is in the region of 50 000 to 200 000 items.

Any practical electronic system must be able to interface with the real analogue world and this means that circuits such as op-amps, filters, analogue-to-digital converters, and digital-to-analogue converters must be employed. With the introduction of BiCMOS technology it has become possible to mix analogue and digital circuitry on the same chip. All-analogue ASICs are also now possible. However, the fabrication process cannot produce components with tightly specified parameters and so differential op-amps are employed wherever possible. Filters use active RC circuits that employ the switched-capacitor technique (p. 71). The CMOS sections of an ASIC are used for the digital circuits and the bipolar sections are used for analogue circuitry and for interfacing. A library of cells gives the designer a (relatively) easy way of producing a system; the library will include such circuits as (a) digital — gates, flip-flops, counters; (b) analogue — voltage regulators, op-amps, oscillators, timers.

## Programmable Logic Devices

Programmable logic provides designers with an alternative to both standard IC devices and full-custom or semi-custom ASICs. A *programmable logic device* (PLD) allows the user to program a chip to perform a particular logic function. Some PLDs can only be programmed once but others may be erased and reprogrammed as often as required. They are economical to use in relatively small numbers, i.e. 1 to about 5000, and they are used extensively in modern digital systems to replace large numbers of SSI/MSI circuits.

There are several different kinds of PLD available and these include the following: (a) programmable logic array (PLA); (b) programmable array logic (PAL); (c) programmable gate array (PGA). In addition, the programmable read only memory (PROM) is sometimes regarded as a PLD rather than a memory device. Both bipolar and MOS technologies are used for PLDs. A PLD may be either a mask or a field programmable device. Field programmable types (FPLDs) may employ either fusible links or CMOS erasable cells. In a fusible FPLD the device is supplied with all its links intact and user must disconnect, by blowing a fuse, all those connections that are not required for that particular design.

A PLD consists of a combination of an AND array that feeds into an OR array, or a number of OR gates, inverters and, sometimes, flip-flops. The PLD can be programmed to implement a wanted logic

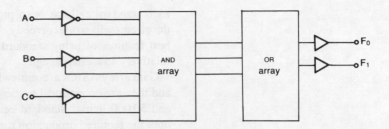

**Fig. 8.1** Programmable logic device

function, to provide flip-flops, counters or registers, or almost any other digital circuit. The basic block diagram of a PLD is shown by Fig. 8.1. The input variables A, B and C, are applied to the input buffer amplifiers which generate the complement of each variable. The input variables and their complements are applied to an AND array which forms the products of the input variables. These products are then passed on to an OR array, or to a number of OR gates, in which the sums of the products are formed. Lastly, these sums of product terms are applied to an output inverter and amplified before appearing at the output terminals of the circuit.

The dimensions of a PLD are $m \times n \times p$, where $m$ is the number of inputs (sometimes known as the width of the circuit), $n$ is the number of product terms generated, and $p$ is the number of outputs. Thus, a $7 \times 4 \times 2$ PLD has seven inputs, four product terms and two outputs.

The four kinds of PLD differ from one another in that:

(a) A programmable read only memory has a fixed AND array and a programmable OR array.
(b) A programmable array logic device has a programmable AND array and a fixed OR array.
(c) A programmable logic array device has both its AND array and its OR array programmable.
(d) A programmable gate array is a gate array that can be programmed, and later reprogrammed, an unlimited number of times.

Some of these devices can be programmed in the field and then they are likely to be known as a field programmable gate array (FPGA), a field programmable logic array (FPLA), etc. An FPGA is an array of programmable AND gates and has no OR gates at all. If a device is not field programmable the user submits the required design to the manufacturer and the IC is programmed using a mask technique.

Both the AND array and the OR array consist of a matrix of rows and columns. A connection between a row and a column is 'hard-wired' in a fixed array and made by a diode or a transistor, bipolar or MOS, in a programmable array. Permanent connections are indicated by a blob, programmed connections by a X. This is shown by Fig. 8.2 which shows (a) no connection, (b) a fixed connection and (c) a programmed connection.

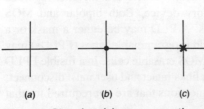

**Fig. 8.2** Showing (a) no connection, (b) a permanent connection and (c) a programmed connection in a PLD array

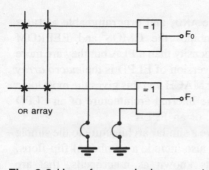

**Fig. 8.3** Use of an exclusive OR gate to obtain either an active-high or active-low output

In some cases an output exclusive OR gate is used instead of an inverter as shown by Fig. 8.3. If the fuse is left intact an output will go HIGH when the input to the gate is HIGH but if the fuse is blown the output will go HIGH when the gate input is LOW.

The AND array of a PLD determines the maximum number of input variables allowed, and the OR array determines the maximum number of sum-of-product terms. If the logic equation to be implemented is too large for the PLD then it must be reduced using a Karnaugh map, or by some other method. No minimization need be attempted if the number of input variables/product terms does not exceed the number which the PLD is able to accommodate.

Each row and column in a field programmable array is linked by a fused diode, or transistor, as shown by Figs 8.4(*a*) and (*b*). Programming an array is carried out by blowing the fuse at each row/column intersection that is to be non-connected. If a row/column connection is wanted then the fuse at that point is not blown. Since a blown fuse can neither be restored, nor replaced, a fuse-programmable PLD is not reprogrammable. However, CMOS PLDs are programmed in a similar way to an EEPROM and can be reprogrammed when required. The most common form of CMOS PLD is the *generic array logic* (GAL) device. This has a programmable AND array, a fixed OR array, and an output stage that can be configured to perform different functions. *Programmable electrically erasable logic* (PEEL) and *erasable PLD* (ELPD) devices allow a user to move up to more complex circuits without any need to change either technology or technique. The PEEL family includes

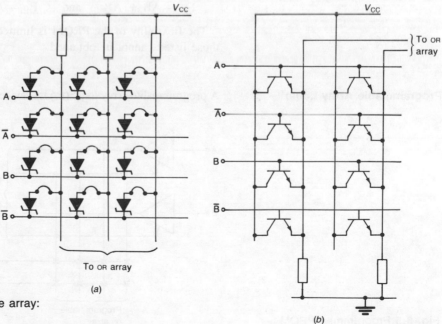

**Fig. 8.4** Field programmable array: (*a*) diode; (*b*) transistor

devices with both programmable AND, and programmable AND/OR arrays. An ELPD device combines the CMOS and EEPROM technologies to provide a higher density than a PLA but they are more expensive and slower. A recent version of ELPD is the *macro array CMOS high-performance PLD* or MACH and this gives the most logic per pin of any kind of PLD. The internal architecture of an ELPD is the same as that of an FPAL or an FPLA.

Some more complex PLDs have a similar architecture to the simple PLD shown in Fig. 8.1 but they also include a number of flip-flops, or even more complex circuits known as macrocells, that are associated with the output pins.

### Programmable Read-only Memory

A programmable read-only memory (PROM) is often regarded as a PLD rather than as a memory device. A PROM has a fixed AND array followed by a programmable OR array. Because of its fixed AND array, which acts like a fixed address decoder, a PROM must utilize all the possible combinations of the input variables. This means that the number of addressed locations is equal to $2^n$, where $n$ is the number of input variables. The size of the address decoder, as well as of the AND array, doubles for each additional input variable. This, in turn, leads to inefficient usage of the ROM when only a few addresses are to be used. Figure 8.5 shows a simple PROM circuit. The outputs of the fixed AND array are the products AB, $\overline{A}$B, A$\overline{B}$ and $\overline{A}\overline{B}$. The OR array may be fuse-programmed to output any combination(s) of the sums of these products. For example, the outputs could be

$$F_0 = AB + \overline{A}\overline{B} \quad \text{and} \quad F_1 = AB + A\overline{B}.$$

The flexibility of the PROM is limited since products other than those listed cannot be obtained.

### Programmable Array Logic

A programmable array logic (PAL) device consists of a programmable

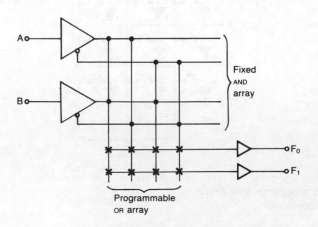

**Fig. 8.5** Programmed PROM

AND array followed by a fixed OR array and it is designed to perform AND/OR logic. The use of a programmable AND array allows input addresses to be specified as reduced product terms, i.e. any subset of product terms can be chosen from $2^n$ locations in the AND array. It is not necessary, as with a PROM, to provide all $2^n$ locations in the AND array. For example, a 16 × 48 × 8 PAL (or a PLA) must have an AND array large enough to store 48 words. A PROM with the same number of input variables would have to be large enough to decode the addresses of $2^{16} = 65\,536$ words.

Figure 8.6 shows the arrangement of a small 2 × 4 × 2 PAL. The product terms generated by the AND array depend upon its programming. Before a logic function is implemented by a PAL the equation must be expressed in its sum-of-products form. Suppose, for example, that the PAL is to implement the functions

$$F_0 = A\bar{B} + \bar{A}B \quad \text{and} \quad F_1 = A\bar{B} + AB.$$

The programming of the AND array should then be as shown by Fig. 8.7.

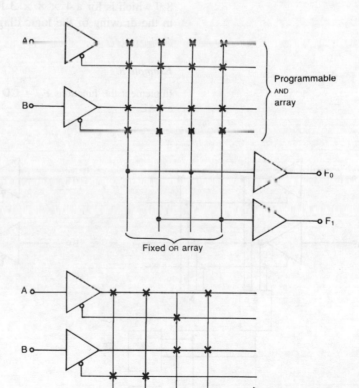

Fig. 8.6 2 × 4 × 2 PAL

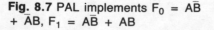

Fig. 8.7 PAL implements $F_0 = A\bar{B} + \bar{A}B$, $F_1 = A\bar{B} + AB$

## Representation of a PAL

The drawing of a PAL (or a PLA) rapidly becomes complicated as its dimensions increase and a 'shorthand' method of drawing is usually employed. The method is illustrated by redrawing the $2 \times 4 \times 2$ PAL as shown by Fig. 8.8(a). The programmed and fixed connections are shown in the usual way and each one is connected to an input of a four-input AND gate. The output of each AND gate is connected to an input of both the OR gates. In general, each AND gate has an input for each input variable *and* its complement and this means that the number of inputs per AND gate is equal to twice the number of input variables. An alternative method of drawing is sometimes used and this is shown by Fig. 8.8(b); here the programming fuses are shown as being connected in series with each input to a multiple-input AND gate.

The diagrams will rapidly become very confusing as the number of input variables is increased and hence it is customary to show only one input to each gate. This method of drawing is illustrated by Fig. 8.9 which is for a $4 \times 8 \times 3$ PAL. The same method is also used in the drawing of the logic diagram of a PLA.

### Example 8.1

Implement the functions $F_0 = \bar{C}D + C\bar{D}$, $F_1 = \bar{C}\bar{D} + AB + \bar{A}D$ using a PAL.

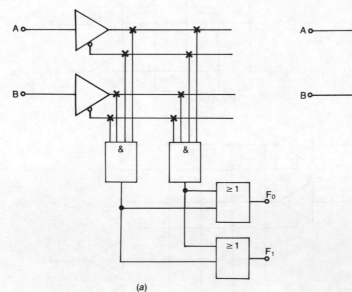

(a)

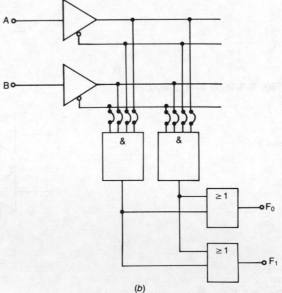

(b)

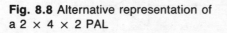

**Fig. 8.8** Alternative representation of a $2 \times 4 \times 2$ PAL

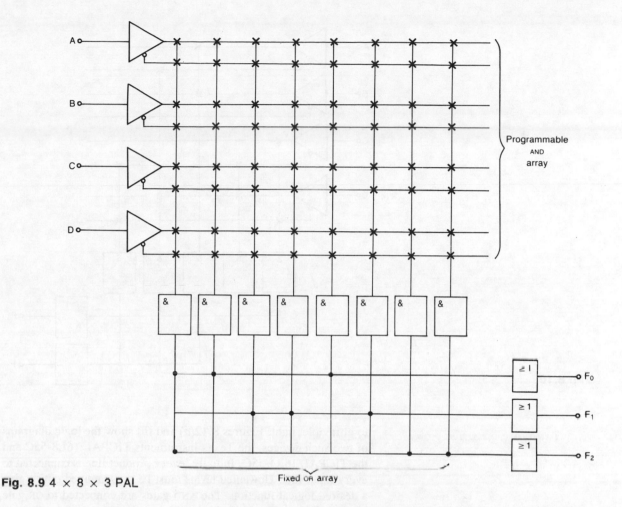

**Fig. 8.9** 4 × 8 × 3 PAL

*Solution*
The required circuit is shown in Fig. 8.10.

The lack of flexibility of a PAL is made clear by this example. The fixed OR array needs to have the connections shown in Fig. 8.10 or the required functions could not be implemented. The PAL devices are available in a wide variety of dimensions and specifying the OR connections becomes a task of device selection rather than one of programming. Some examples of available PALs are given in Fig. 8.11 using the customary symbols.

Many PALs may have some outputs fed back to the input and/or include D flip-flops associated with one or more of the outputs.

The CMOS PALs are compatible with HC logic and are easy to program since they employ a programming cell consisting of a floating-gate device similar to that used in an EPROM. Programming can be done manually but usually it is accomplished using programming equipment. The programming is erased by exposing the PAL

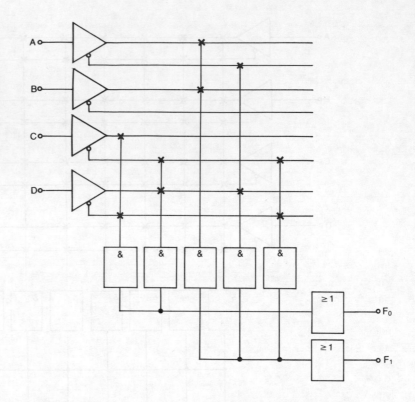

**Fig. 8.10**

to ultraviolet light. Figures 8.12(a) and (b) show the logic diagrams of two such devices, the Texas Instruments TICPAL 16L8-55C and the TICPAL 16R4-55C. Initially, every product line is connected to every input line. Unwanted terms must be programmed out to give a desired logical function. The AND gates are connected to OR gate inputs as shown. The vertical lines numbered at the top of the diagram represent the input, and inverted input, lines. The horizontal lines numbered on the left of the diagram are the product terms and represent AND gates shown in a simplified form. Each intersection of a horizontal line with a vertical line represents a separate input to the AND gate. Each intersection may be programmed to either keep the connection or to disconnect it. A dot indicates each point where a permanent connection is made.

The user of a PAL places an X at each intersection that is to be maintained when the device is programmed. Figure 8.13(a) shows a single programmed line and Fig. 8.13(b) shows its equivalent circuit. If all the product terms are left intact the output of the line will be of the form $A\bar{A}B\bar{B}C\bar{C}D\bar{D}$ etc. and hence it will be at the logical 0 voltage level. To simplify the drawings this state is usually indicated by inserting a X inside the AND gate symbol as shown by Fig. 8.14.

The 16L8 PAL (see Fig. 8.12) can be programmed to give various gates and three examples follow.

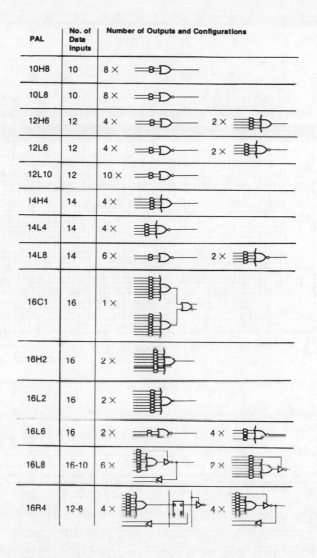

| PAL | No. of Data Inputs | Number of Outputs and Configurations | | |
|---|---|---|---|---|
| 10H8 | 10 | 8 × | | |
| 10L8 | 10 | 8 × | | |
| 12H6 | 12 | 4 × | 2 × | |
| 12L6 | 12 | 4 × | 2 × | |
| 12L10 | 12 | 10 × | | |
| 14H4 | 14 | 4 × | | |
| 14L4 | 14 | 4 × | | |
| 14L8 | 14 | 6 × | 2 × | |
| 16C1 | 16 | 1 × | | |
| 16H2 | 16 | 2 × | | |
| 16L2 | 16 | 2 × | | |
| 16L6 | 16 | 2 × | 4 × | |
| 16L8 | 16–10 | 6 × | 2 × | |
| 16R4 | 12–8 | 4 × | 4 × | |

**Fig. 8.11** Some available PALs

### (a) Inverter

Suppose that input terminal 2 is used as the input to the inverter and terminal 19 as the output. Then, see Fig. 8.15, the connections are kept on the intersection of the non-inverting line of input 2 and product line 0. All other connections on product line 0 are broken. The other seven inputs to the OR gate associated with output terminal 19 must all be enabled with the 0 logical level and so all the connections on product lines 1 to 7 are left intact.

### (b) Two-input OR Gate

If inputs 3 and 4 are allocated as the two inputs to the AND gate and terminal 18 is allotted as the output then the AND gate is implemented

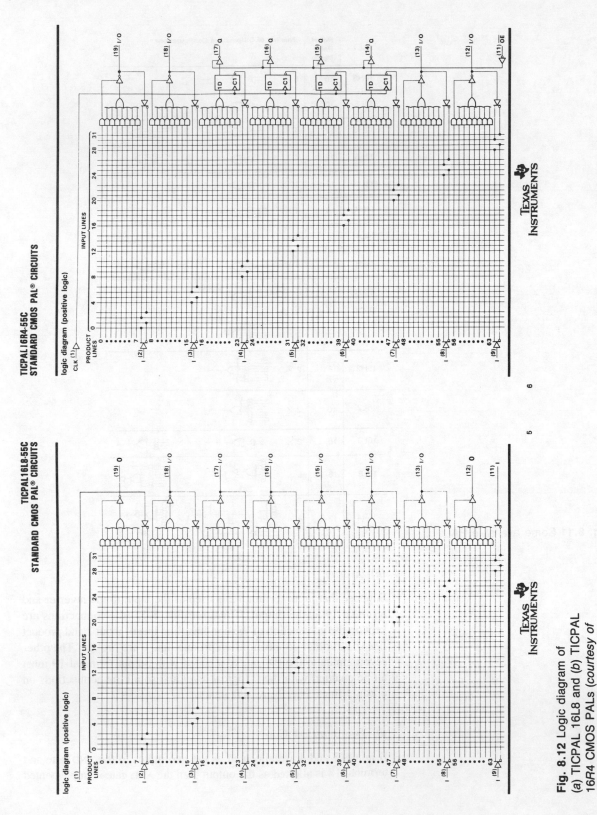

**Fig. 8.12** Logic diagram of (a) TICPAL 16L8 and (b) TICPAL 16R4 CMOS PALs (*courtesy of Texas instruments*)

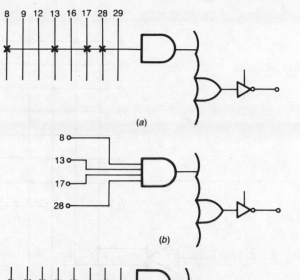

**Fig. 8.13** (*a*) A single programmed line in a PAL; (*b*) its equivalent circuit

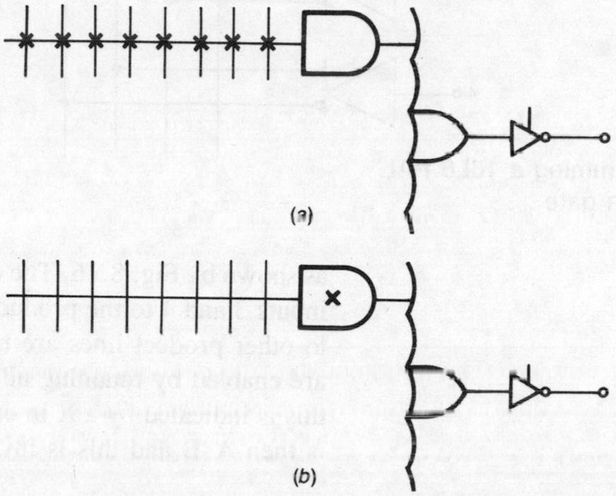

**Fig. 8.14** (*a*) All intersections in a PAL are programmed in; (*b*) 'shorthand' representation of (*a*)

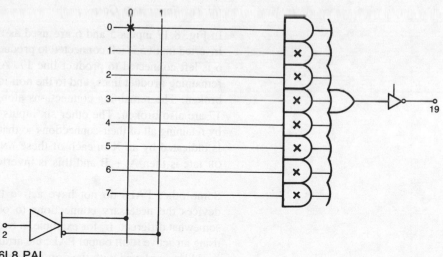

**Fig. 8.15** Programming a 16L8 PAL as an inverter

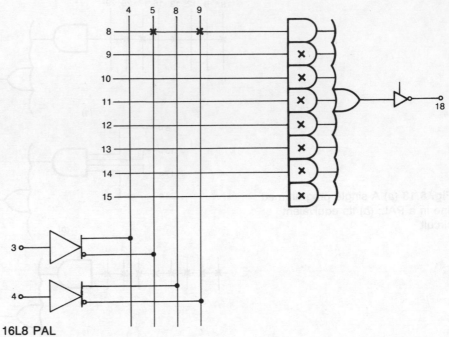

**Fig. 8.16** Programming a 16L8 PAL as a two-input OR gate

as shown by Fig. 8.16. The connections of the inverting lines of both inputs 3 and 4 to the product line 8 are retained and all connections to other product lines are broken. The other inputs to the OR gate are enabled by retaining all their connections to the input lines and this is indicated by a X in each AND gate. The output of the OR gate is then $\overline{A} . \overline{B}$ and this is inverted to give $\overline{\overline{A} . \overline{B}} = A + B$.

### (c) Two-input AND Gate

In Fig. 8.17 inputs 5 and 6 are used as the inputs to the AND gate. Inverted input 5 is left connected to product line 16 and inverted input 6 is left connected to product line 17. All other connections to the remaining product lines, and to the non-inverting inputs 5 and 6, are broken. The remaining connections along the product lines 16 and 17 are also broken. The other six inputs to the OR gate are enabled by retaining all of their connections so that these inputs are LOW. This is indicated by an X in each of these AND gates. The output of the OR fate is then $\overline{A} + \overline{B}$ and this is inverted to give $\overline{\overline{A} + \overline{B}} = A.B$.

Some other PALs do not have active LOW outputs and for these devices the necessary connections to obtain a particular gate are somewhat different. If, for example, a two-input AND gate is required using an active HIGH output PAL, the arrangement shown in Fig. 8.16 could be employed with the non-inverting inputs used instead of the inverting inputs.

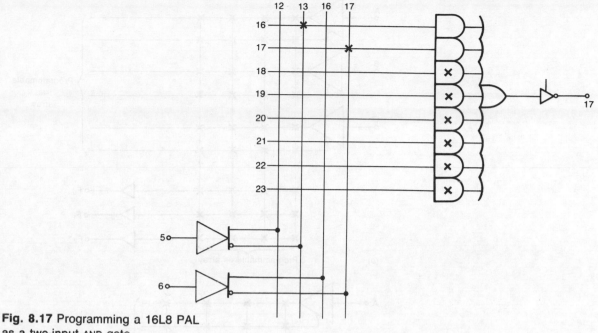

**Fig. 8.17** Programming a 16L8 PAL as a two-input AND gate

**Programmable Logic Array**

A programmable logic array (PLA) is more flexible than the PROM and the PAL since both its AND array and its OR array are fully programmable. All possible product terms can be generated by the AND array and any, or all, of these terms can be added by the OR array to form sum-of-product terms. Because of this inherent flexibility a PLA is able to implement many functions that a PAL or PROM could not tackle. When a PLA includes flip-flops it is often known as a *field programmable logic sequencer* (FPLS). The basic circuit of a PLA and the 'shorthand representation' of it is shown by Figs 8.18(*a*) and (*b*) respectively.

*Example 8.2*

Use a PLA to implement the Boolean expressions $F_0 = ABCD + A\bar{B}\bar{C}D + \bar{A}B\bar{C}\bar{D}$, $F_1 = A + BC$, and $F_2 = \bar{B}C + B\bar{C}$.

*Solution*
The programmed PLA is shown by Fig. 8.19.

If the equation to be implemented will fit into the dimensions of the PLA there will be no need to attempt to minimize the equation. Otherwise, it will be necessary to start minimizing until the total number of product terms required is no longer larger than the number the PLA can provide. Often the number of product terms needed can be reduced by the use of an external inverter. Consider the function

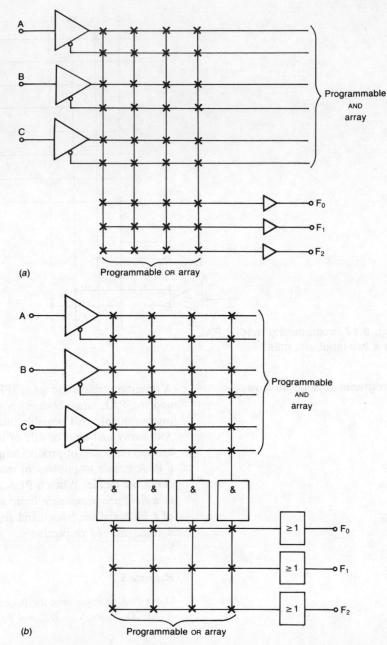

**Fig. 8.18** Programmable logic array:
(a) logic circuit; (b) 'shorthand'
representation

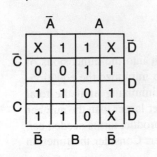

$$F_0 = B\bar{C}\bar{D} + AB\bar{C}D + A\bar{B}\bar{C}D + \bar{B}CD + \bar{A}C\bar{D} + \bar{A}BC$$

with 'don't cares' $\bar{A}\bar{B}\bar{C}\bar{D}$, $\bar{A}\bar{B}CD$ and $A\bar{B}C\bar{D}$. The Karnaugh mapping of the function is shown alongside. Looping the 1 squares gives

$$F_0 = A\bar{C} + A\bar{B} + \bar{A}C + \bar{C}D$$

This equation involves four product terms. If, instead, the 0 squares are looped $\bar{F} = \bar{A}\bar{C}D + ABC$ and this equation only has two product terms.

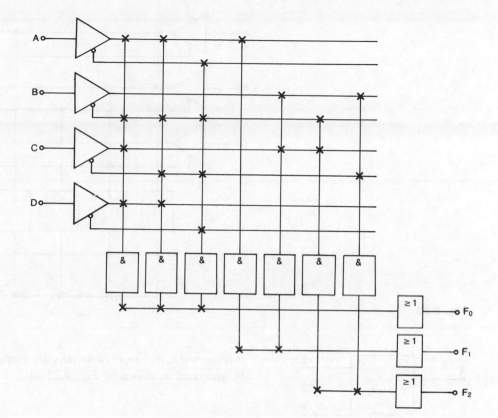

**Fig. 8.19**

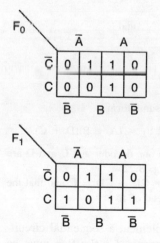

### Example 8.3

Implement the functions

$$F_0 = \overline{A}B\overline{C} + AB\overline{C} + ABC \quad \text{and}$$
$$F_1 = \overline{A}BC + AB\overline{C} + ABC + A\overline{B}C$$

using a 3 × 4 × 2 PLA.

### Solution

If an attempt is made to implement the two equations as they stand a 3 × 8 × 2 PLA would be required so some minimization is necessary. The Karnaugh mappings of the equations are shown alongside. From the maps $F_0 = BC + AB$ and $F_1 = \overline{B}C + B\overline{C} + AB$ and these functions are shown implemented by a 3 × 4 × 2 PLA in Fig. 8.20

An alternative way of reducing the number of product terms required involves the decoding of the input variables before they are applied to the PLA. The binary functions, rather than input variables, are ANDed together and this gives fewer product terms. Usually, the input variables are partitioned into twos and applied to 2-to-4 line decoders. The equation to be implemented must first be put into its product-of-sums form. If the inputs to the 2-to-4 line decoder are A and B then the functions generated are as shown by Fig. 8.21(a). If an inverter

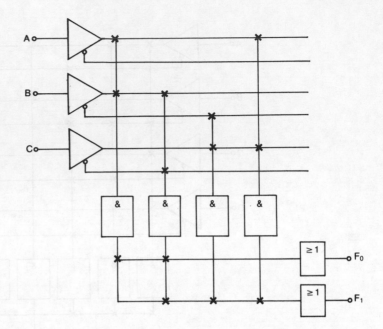

**Fig. 8.20**

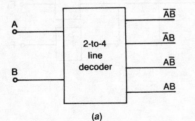

(a)

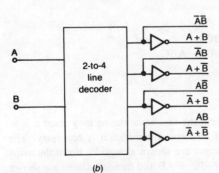

(b)

**Fig. 8.21** Use of a decoder to produce input variables

is connected to the output of the decoder another four functions could be generated as shown by Fig. 8.21(b).

*Example 8.4*

Implement the equations

$$F_0 = ABC + ABD + \bar{A}\bar{C} + B\bar{C} \qquad \text{and}$$
$$F_1 = A\bar{D} + A\bar{C} + B\bar{D} + B\bar{C}$$

using a PLA and a 2-to-4 line decoder.

*Solution*

Putting the equations into their product-of-sums form

$$F_0 = AB(D + C) + (\bar{A} + \bar{B})\bar{C} \quad \text{and } F_1 = (A + B)(\bar{D} + \bar{C})$$

The input variables A and B are applied to one decoder and C and D are applied to another decoder as shown by Fig. 8.22.

Note that $\bar{C} = (\bar{C} + D)(\bar{C} + \bar{D}) = \bar{C} + \bar{C}D + \bar{C}\bar{D} = \bar{C}$ and that the decoded PLA has three product terms instead of five.

A PLA can also be employed to implement a sequential circuit. If a synchronous sequential circuit is required a flip-flop must be inserted in the feedback path. For a non-synchronous sequential circuit a flip-flop is not necessary but the system will then be susceptible to hazards caused by unequal propagation delays. The state table of the required circuit should be rewritten in terms of the necessary product terms, $P_1$, $P_2$, etc. and the present, A and B, and the next, $A^+$ and $B^+$, values of the fed-back variables, and the input variable

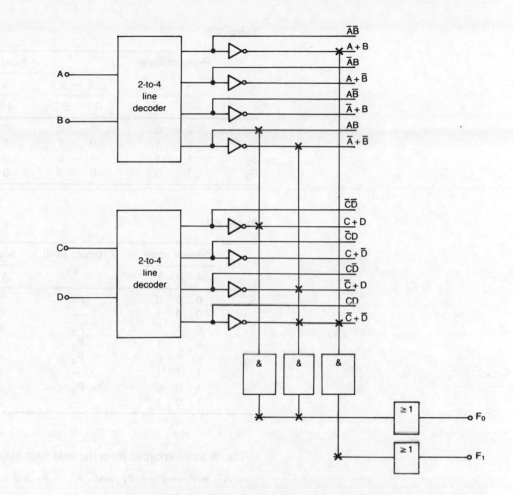

**Fig. 8.22**

x. The required product terms from the AND array can then be read from the left-hand side of the modified state table. Product terms are *not* wanted for any rows in the table for which there is no 1 entry for any one of $A^+$, $B^+$, or the output F. The required sum-of-product terms from the OR array are read from the right-hand side of the table. Lastly, the output F of the circuit is read from the row(s) in the table for which the output is HIGH.

### Example 8.5

Implement the circuit in Example 7.2 using a PLA.

*Solution*
The state table of the required circuit is given by Table 7.9 which is reproduced here, headings slightly modified. The table is rewritten in slightly modified form in Table 8.1.

The AND terms are read from the present state values of A, B and x. Thus

$$P_1 = \bar{A}\bar{B}\bar{x} \quad P_2 = \bar{A}\bar{B}x \quad P_3 = \bar{A}Bx \quad P_4 = A\bar{B}\bar{x} \quad \text{and} \quad P_5 = A\bar{B}x$$

**Table 7.9**

| | Present state | | Next state | | | | Output |
|---|---|---|---|---|---|---|---|
| | A | B | x = 0 | | x = 1 | | |
| | | | $A^+$ | $B^+$ | $A^+$ | $B^+$ | |
| $S_0$ | 0 | 0 | 1 | 0 | 0 | 0 | 0 |
| $S_1$ | 0 | 1 | 0 | 1 | 1 | 1 | 0 |
| $S_2$ | 1 | 1 | 1 | 0 | 0 | 0 | 1 |
| $S_3$ | 1 | 0 | 0 | 1 | 0 | 0 | 0 |

**Table 8.1**

| | Present state | | | Product term | Next state | | Output |
|---|---|---|---|---|---|---|---|
| | A | B | x | | $A^+$ | $B^+$ | |
| $S_0$ | 0 | 0 | 0 | $P_1$ | 1 | 0 | 0 |
| | 0 | 0 | 1 | — | 0 | 0 | 0 |
| $S_1$ | 0 | 1 | 0 | $P_2$ | 0 | 1 | 0 |
| | 0 | 1 | 1 | $P_3$ | 1 | 1 | 0 |
| $S_2$ | 1 | 1 | 0 | $P_4$ | 1 | 0 | 1 |
| | 1 | 1 | 1 | — | 0 | 0 | 0 |
| $S_3$ | 1 | 0 | 0 | $P_5$ | 0 | 1 | 0 |
| | 1 | 0 | 1 | — | 0 | 0 | 0 |

The OR terms are read from the next state values of A and B. Thus

$$A^+ = P_1 + P_3 + P_4 \quad \text{and} \quad B^+ = P_2 + P_3 + P_5$$

The output F of the circuit is HIGH in only one row in the table and hence $F = P_4$.

The PLA implementation of the circuit is shown by Fig. 8.23.

Normally PLAs are available in the following sizes: $16 \times 48 \times 8$, $18 \times 42 \times 10$, $16 \times 45 \times 12$, $22 \times 42 \times 10$ and $20 \times 45 \times 12$.

## Output Macrocells

Some PLDs are provided with output circuits known as *macrocells*. A macrocell replaces simple gates and flip-flops with circuits that can be programmed to function in any one of a number of different ways. Macrocells can be regarded as the basic building blocks available to the user. *Macrofunctions* are circuits of greater complexity and are composed of a number of macrocells. Simple macrofunctions can, in turn, be combined to build higher-level macrofunctions until the required system is obtained. Examples of macrofunctions include adders, comparators, registers, counters and decoders.

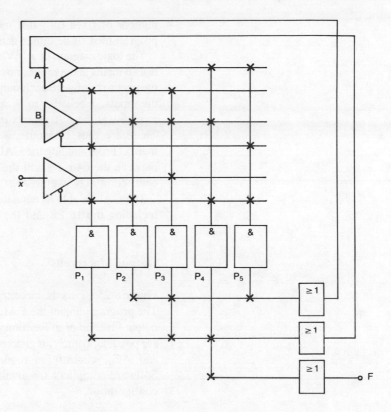

**Fig. 8.23**

**Field Programmable Gate Array**

A field programmable gate array (FPGA) is a high-density ASIC that can be configured by the user. Some FPGAs are of similar architecture to gate arrays while others have more in common with a PLD. An FPGA can be designed and produced in a few days and it may have its programming changed in a few hours. The device consists of a number of programmable logic elements and a programmable interconnection matrix. The design of an FPGA is carried out with the aid of a CAD package with the appropriate software. One form of FPGA is the *electrically reconfigurable array* (ERA) which uses an on-chip RAM to control the routeing of signals between logic elements. Simply by loading the RAM with the correct data the array can be programmed to provide the wanted circuit.

**Generic Array Logic**

*Generic array logic* (GAL)† devices are electrically erasable CMOS ICs that combine reconfigurable logic, CMOS low-power dissipation and TTL high-speed performance. A GAL can be reprogrammed as often as required. The internal design of a GAL is very similar to that of a PAL and it has a programmable AND matrix to which a

---

†Generic array logic (GAL) is the registered trademark of Lattice Semiconductor Corporation.

number of fixed OR gates may be connected. The outputs may be programmed to be either active HIGH or active LOW.

The logic diagram of a 16V8 GAL is shown in Fig. 8.24. The device has 16 inputs and 8 programmable *output logic macrocells* which allow the user to configure each output in a number of different ways. Both the inputs, labelled 1 to 9, and the outputs, labelled 12 to 19, are available at the AND matrix in both inverted and non-inverted form. There are thus 16 input rows and 16 output columns in the AND matrix. Programming the GAL consists of making chosen connections between the 64 rows and the 32 columns of the matrix. Each made connection acts like an AND gate.

The GAL is able to emulate many of the common PAL structures including the 16 R8 and the 16 L8.

## Output Macrocells

Figure 8.25 shows the circuitry of an output logic macrocell (OLMC). The programming of the GAL sets the OLMC to perform a particular output function or to function as an input. Three global configurations are possible: namely, registered, complex, and simple. Two bits, SYN and $AC_0$, control the mode configuration for all the OLMCs. Software compilers are available to set each OLMC to the desired configuration.

### Registered Mode

In the registered mode an OLMC can be configured as a dedicated registered output or to provide an input/output (I/O) function. The two configurations are shown in Figs 8.26(a) and (b). Registered outputs have eight product terms per output and I/Os have seven product terms per output.

### Complex Mode

In the complex mode an OLMC may be configured as either an output only or as an I/O circuit. The two configurations are similar to those of a 16 L8 or 16 P8 PAL. The two configurations are shown in Figs 8.26(c) and (d). All OLMCs have seven product terms per output since one term is used for programmable output enable control.

### Simple Mode

Three simple mode configurations are shown in Figs 8.26 (e), (f) and (g). Each OLMC is configured as either a dedicated input or as a dedicated active combinational output.

The polarity of the output, i.e. active HIGH or LOW, is implemented by the exclusive-OR gate that follows the OR gate from the AND array.

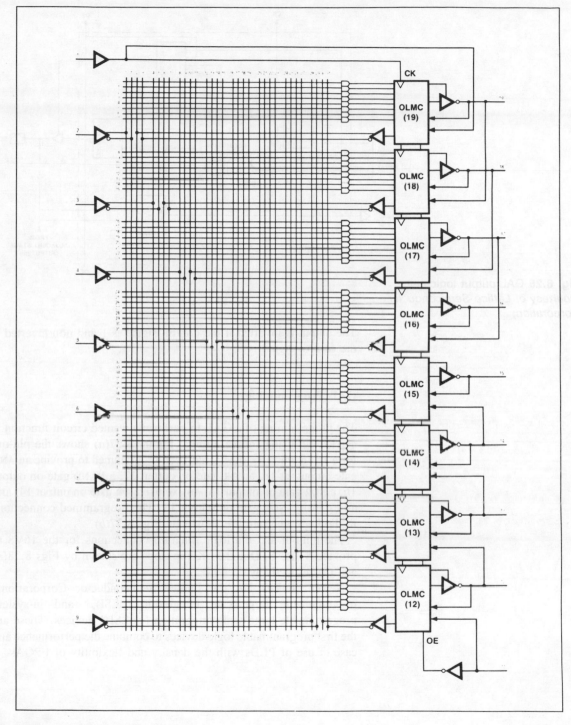

**Fig. 8.24** Logic diagram of a 16V8
GAL (*courtesy of Lattice
Semiconductor Corporation*)

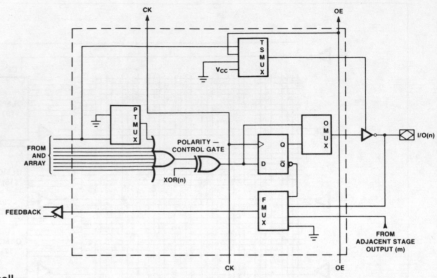

**Fig. 8.25** GAL output logic cell
(*courtesy of Lattice Semiconductor
Corporation*)

The output is inverted if the control signal is 1 and non-inverted if the control signal is 0.

## GAL Programming

The programming of a GAL to perform a wanted circuit function is carried out using software tools. Figure 8.27(*a*) shows the pin-out for a 16V8 when the device has been programmed to provide an AND gate on output 18, an OR gate on output 17, a NAND gate on output 16, a NOR gate on output 15, an exclusive OR gate on output 14, and an exclusive NOR gate on output 13. The programmed connections in the logic diagram are shown by Fig. 8.27(*b*).

The pin-out and the logic diagram connections for the 16V8 to provide S−R, T, D and J−K flip-flops are shown by Figs 8.28(*a*) and (*b*) respectively.

Newer devices are the Lattice Semiconductor Corporation's programmable large-scale integration (pLSI),† and in-system programmable large-scale integration (ispLSI)† devices. These are the first programmable logic devices to combine the performance and ease of use of PLDs with the density and flexibility of FPGAs.

---

†Trademarks of the Lattice Semiconductor Corporation.

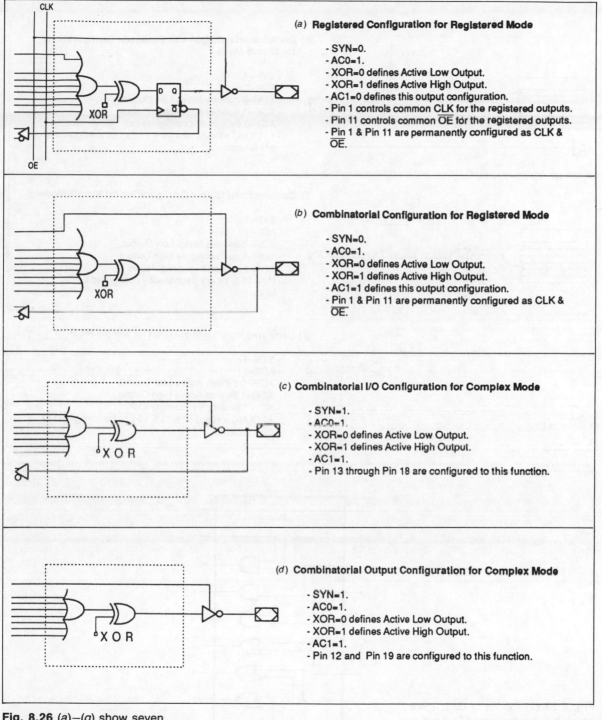

(a) **Registered Configuration for Registered Mode**

- SYN=0.
- AC0=1.
- XOR=0 defines Active Low Output.
- XOR=1 defines Active High Output.
- AC1=0 defines this output configuration.
- Pin 1 controls common CLK for the registered outputs.
- Pin 11 controls common $\overline{OE}$ for the registered outputs.
- Pin 1 & Pin 11 are permanently configured as CLK & $\overline{OE}$.

(b) **Combinatorial Configuration for Registered Mode**

- SYN=0.
- AC0=1.
- XOR=0 defines Active Low Output.
- XOR=1 defines Active High Output.
- AC1=1 defines this output configuration.
- Pin 1 & Pin 11 are permanently configured as CLK & $\overline{OE}$.

(c) **Combinatorial I/O Configuration for Complex Mode**

- SYN=1.
- AC0=1.
- XOR=0 defines Active Low Output.
- XOR=1 defines Active High Output.
- AC1=1.
- Pin 13 through Pin 18 are configured to this function.

(d) **Combinatorial Output Configuration for Complex Mode**

- SYN=1.
- AC0=1.
- XOR=0 defines Active Low Output.
- XOR=1 defines Active High Output.
- AC1=1.
- Pin 12 and Pin 19 are configured to this function.

**Fig. 8.26** (a)–(g) show seven possible output configurations of a GAL OLMC (*courtesy of Lattice Semiconductor Corporation*)

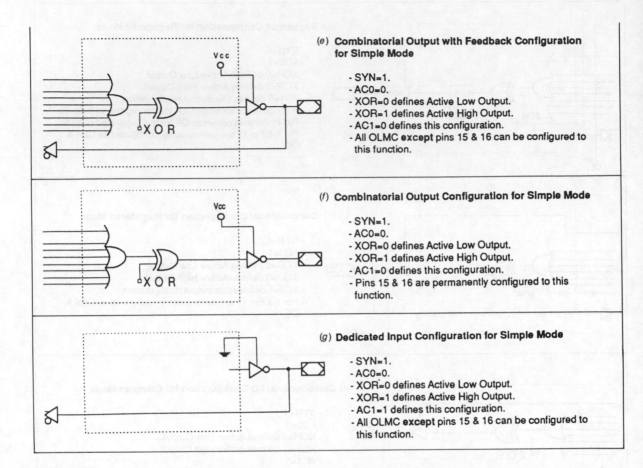

(e) **Combinatorial Output with Feedback Configuration for Simple Mode**

- SYN=1.
- AC0=0.
- XOR=0 defines Active Low Output.
- XOR=1 defines Active High Output.
- AC1=0 defines this configuration.
- All OLMC **except** pins 15 & 16 can be configured to this function.

(f) **Combinatorial Output Configuration for Simple Mode**

- SYN=1.
- AC0=0.
- XOR=0 defines Active Low Output.
- XOR=1 defines Active High Output.
- AC1=0 defines this configuration.
- Pins 15 & 16 are permanently configured to this function.

(g) **Dedicated Input Configuration for Simple Mode**

- SYN=1.
- AC0=0.
- XOR=0 defines Active Low Output.
- XOR=1 defines Active High Output.
- AC1=1 defines this configuration.
- All OLMC **except** pins 15 & 16 can be configured to this function.

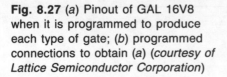

**Fig. 8.27** (a) Pinout of GAL 16V8 when it is programmed to produce each type of gate; (b) programmed connections to obtain (a) (*courtesy of Lattice Semiconductor Corporation*)

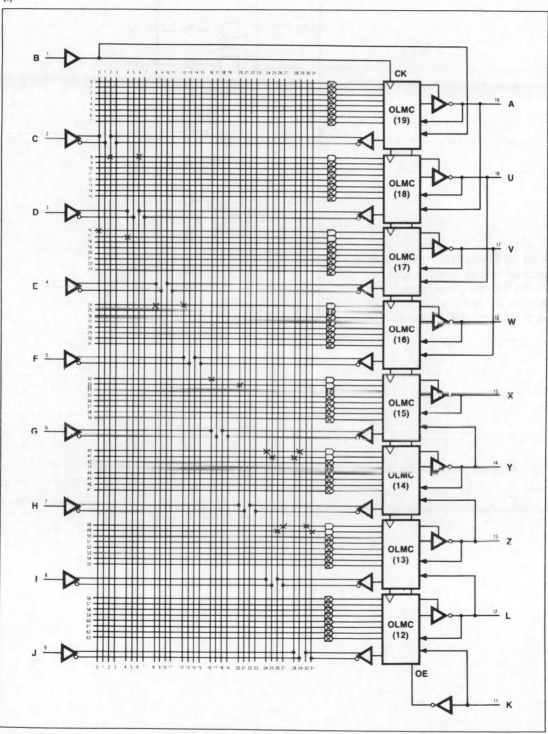

(a)

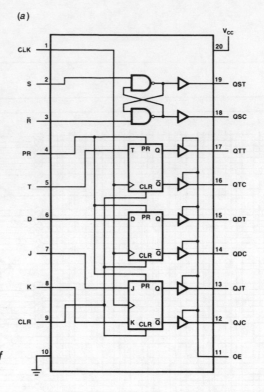

**Fig. 8.28** (a) Pinout GAL 16V8 when it is programmed to produce each type of flip-flop; (b) programmed connections to obtain (a) (*Courtesy of Lattice Semiconductor Corporation*)

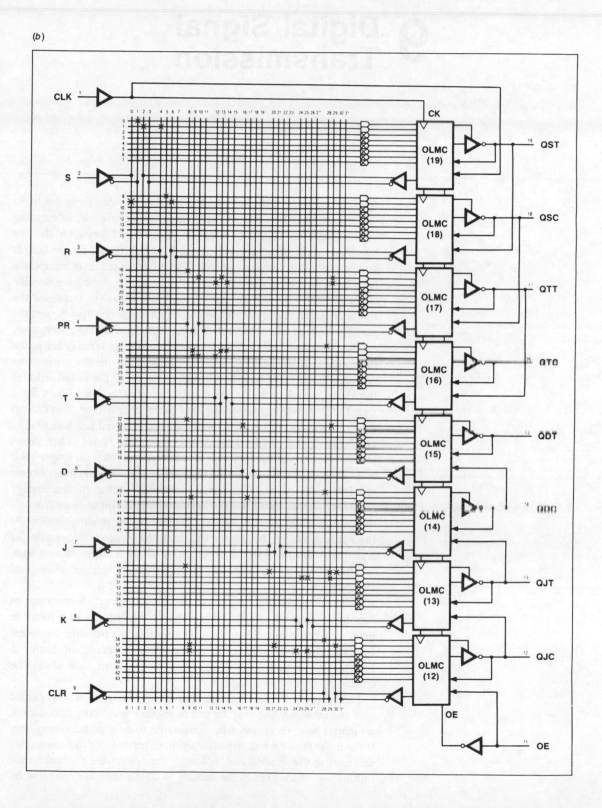

# 9 Digital Signal Transmission

Modern high-speed digital systems employ devices from the high-speed logic families and this introduces problems when designing interconnections. Signals must be transmitted between devices mounted on the same printed circuit board, between circuit boards and between different equipments. The different pieces of equipment may be very near to one another or they may be some considerable distance apart. As the bit rate of a digital signal is increased the fundamental frequency of the digital waveform increases also and the point is reached where the wavelength of the fundamental frequency — and, of course, all of its harmonics — is of the same order as the dimensions of the circuit board. The lengths of the conductors connecting different boards will then also be of the same order as the signal wavelength. At these high bit rates the time taken for a signal to change from one logical level to the other will be short enough for the transition to be completed before the signal has had time to reach the driven or receiving device, or circuit board. This means that the lumped-circuit approach to interconnections is no longer valid and all interconnections must be regarded as transmission lines. Unless the interconnection is *correctly terminated* some of the energy contained in the signal will be reflected by the input impedance of the receiving device and will be returned to the sending device. At the sending end of the circuit the reflected signal will degrade the signal waveform. A signal propagating along a 'transmission line' is also likely to be affected by external sources of noise and interference.

Different equipments may also be interconnected. Some form of cabling is generally employed which may be only about a metre in length, as is the case for a printer connected to a personal computer for example, or may be some tens, or even hundreds, of metres in length. Unless the bit rate is low these connections must always be treated as transmission lines.

The criterion for deciding whether a connection must be regarded as a transmission line may be put in another way. Any connection, no matter how short, must be considered to be a transmission line, if the time taken by a voltage transition to travel over the connection and back to the sending end, is longer than either the risetime or the falltime — which ever is the shorter — of the transition. At low bit rates, with a signal risetime greater than 4 ns, the assumption of a lumped-element connection circuit will usually give accurate results. For high-speed circuits, however, where the risetimes are probably shorter than 2 ns, unwanted transmission line effects, such as ringing, may occur on connections as short as 15 cm.

### Example 9.1

A 30 m length of cable is used to transmit (*a*) a 10 kb/s, (*b*) a 10 Mb/s digital signals between two points. Should the connection be treated as a transmission line?

### Solution

(*a*) Wavelength $\lambda = (3 \times 10^8)/(10 \times 10^3) = 30$ km. The cable is only $0.001\lambda$ in length and so transmission line effects will be negligible.
    (*Ans.*)

(*b*) Wavelength $\lambda = (3 \times 10^8)/(10 \times 10^6) = 30$ m. The cable is now 1 wavelength long and so the connection must be regarded as a transmission line.    (*Ans.*)

### Example 9.2

A PCB track is 10 cm long and is to carry a signal which has a risetime of 2 ns and a falltime of 3 ns. Determine whether the track should be considered as a transmission line.

### Solution

Propagation delay $= 1/(3 \times 10^8)$ s/m. The time for the signal to travel up and down the track $= (2 \times 0.01)/(3 \times 10^8) = 67$ ps. This is a much shorter time than 2 ns so the track need not be treated as a transmission line.    (*Ans.*)

## Transmission Lines

Any transmission line consists of two conductors that provide a path over which current and voltage waves are able to propagate. The two conductors may consist of a twisted pair, a pair in a multi-pair cable, a coaxial pair, a PCB track and an earth (or ground) plane, or even a single conductor with an earth return. In all cases the two conductors are separated from one another by a dielectric that may be air, some form of plastic, or with PCB epoxy glass. Figure 9.1 shows the twisted pair, coaxial pair, surface microstrip, and stripline versions of transmission lines. An earth plane consists of a continuous conductive plane over one surface (usually the component side), of the circuit board. It provides a return path for signal currents that is directly underneath the signal-carrying conductors. The only breaks in the earth plane are for necessary current paths. Because the plane is flat it provides a low-inductance earth from any point on the board; hence it reduces the effects of stray capacitances by referring them to earth and so avoiding unwanted feedback paths. An earth plane should always be used with high-speed circuitry.

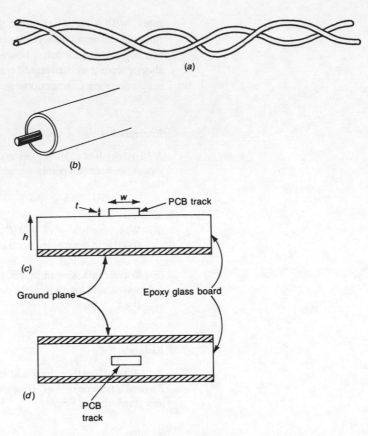

**Fig. 9.1** Types of transmission line: (a) twisted pair; (b) coaxial pair; (c) surface microstrip; (d) stripline

Any transmission line has series resistance $R$, series inductance $L$, shunt capacitance $C$ and shunt conductance $G$, evenly distributed along the length of the line. Very often the two conductors are of different physical dimensions and/or made from different materials and so have different values of resistance and inductance. Most backplanes and PCB tracks have very small resistances and inductances and hence their losses are negligibly small. The other types of transmission line are generally of longer length and their losses may not be negligible, particularly at the higher bit rates. However, all lines will have a negligible loss if their length is short enough when they are said to be *lossfree*. The longer a line is the more energy it will lose by radiation and this is another source of losses.

The performance of a digital transmission line is described in terms of both its *secondary coefficients* and its capacitance. The most important of the secondary coefficients for a line carrying digital signals are the characteristic impedance and the propagation delay.

The *characteristic impedance* $Z_0$ of a transmission line is determined by its physical dimensions and by the relative permittivity of the dielectric material. The concept of characteristic impedance is illustrated by Fig. 9.2; it is the input impedance of a line that is terminated in the characteristic impedance. At all points along the

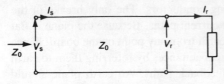

**Fig. 9.2** Characteristic impedance

length of the line the characteristic impedance is equal to the ratio voltage/current; thus at the sending end of the line $Z_0 = V_s/I_s$ and at the receiving end $Z_0 = V_r/I_r$. A line that is terminated in its characteristic impedance is said to be *correctly terminated*. At the higher bit rates when connections must be treated as transmission lines, $\omega L \gg R$ and $\omega C \gg G$ and then $Z_0$ is purely resistive and is given by

$$Z_0 = \sqrt{(L/C)} \tag{9.1}$$

For a microstrip line,

$$Z_0 = [87/(\sqrt{\epsilon_r} + 1.41)]\log_e[(5.98h)/(0.8w + t)] \tag{9.2}$$

Typically, $\epsilon_r = 5$ for glass-epoxy.

Typically, the characteristic impedances of the different kinds of transmission lines are: (*a*) twisted pair $50-125\ \Omega$; (*b*) multiple pair cable $62-100\ \Omega$; (*c*) coaxial pair $50-75\ \Omega$; (*d*) ribbon coaxial 75 and 93 Ω; (*e*) PCB track: (i) surface microstrip 70 Ω typical, (ii) stripline 50 Ω typical.

Figure 9.3 (*a*) shows a simple circuit in which an OR gate is connected to another OR gate by a connection which is long enough to be regarded as a transmission line. The input impedance of a gate is much higher than the characteristic impedance $Z_0$ of the line and this *mismatch* at the receiving end of the circuit will cause *reflections* to occur. This means that some, or all, of the incident voltage will be reflected by the receiving OR gate and returned along the line back

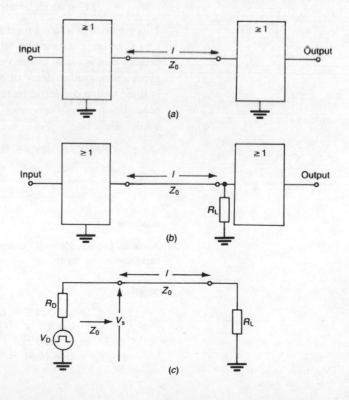

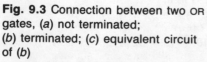

**Fig. 9.3** Connection between two OR gates, (*a*) not terminated; (*b*) terminated; (*c*) equivalent circuit of (*b*)

to the sending gate. To prevent reflections taking place the line must be correctly terminated. A load resistor $R_L$ can be connected between the gate input terminal and earth as shown by Fig. 9.3(b). The value of $R_L$ should be equal to $Z_0$. The input impedance of the line is then also equal to $Z_0$ and the circuit may be redrawn as shown by Fig. 9.3(c). When the driving gate changes state from 1 to 0, or from 0 to 1, the output voltage of the gate changes by $\Delta V_D$ and the voltage transition applied to the line is

$$\Delta V_s = \Delta V_D Z_0/(R_D + Z_0) \tag{9.3}$$

For many digital circuits the output resistance $R_D$ of the driving circuit is small compared to $Z_0$ and then $\Delta V_s \simeq \Delta V_D$. This voltage step will propagate along the line and appear across the load resistance after a time delay of $T_D$ seconds. If the line losses are negligibly small the received voltage transition will be equal to the transmitted voltage transition. The time $T_D$ taken for the voltage transition to travel over the line from one end to the other is known as the *propagation delay*.

The velocity with which a digital signal travels along a line is known as the *velocity of propagation* $V_p$ and it is given by

$$V_p = 1/\sqrt{LC} \text{ metres/seconds} \tag{9.4}$$

The propagation delay per metre is equal to the reciprocal of the velocity of propagation, i.e.

$$t_D = \sqrt{LC} \text{ seconds/metre} \tag{9.5}$$

This means that a line $l$ metres in length has a propagation delay of $l\sqrt{LC}$ seconds. For a line with air as the dielectric the velocity of propagation is equal to the velocity of light, i.e $3 \times 10^8$ m/s. This gives a propagation delay of 3.33 ns/m. If the inter-conductor space is filled with a dielectric material of relative permittivity $\epsilon_r$ then the propagation delay becomes $t_D' = 3.33\sqrt{\epsilon_r}$ nanoseconds/metre. For a microstrip line

$$t_d' = 3.33\sqrt{(0.48\epsilon_r + 0.67)} = 5.8 \text{ ns/m}$$

if $\epsilon_r = 5$.

### Example 9.3

A backplane has $Z_0 = 75\,\Omega$ and $\epsilon_r = 4.3$. Calculate its inductance and capacitance per metre.

*Solution*

$$t_D' = 3.33\sqrt{4.3} = 6.91 \text{ ns/m}$$
$$Z_0 \times t_D' = \sqrt{(L/C)} \times \sqrt{LC} = L = 75 \times 6.91 \times 10^{-9}$$
$$= 0.52\ \mu\text{H/m}. \quad (Ans.)$$
$$t_D'/Z_0 = C = (6.91 \times 10^{-9})/75 = 92 \text{ pF/m} \quad (Ans.)$$

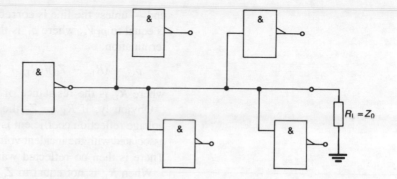

**Fig. 9.4** Line with n devices connected

Figure 9.4 shows a line of length $l$ metres that has $n$ devices connected to it at various points along the line. Each device has a high input resistance that does not shunt the line and an input capacitance of $C_{in}$ picofarads. The total capacitance $C_A$ added to the line is therefore $C_A = nC_{in}l$ picofarads/metre.

The effective characteristic impedance of the line will be reduced by this added capacitance to

$$Z_0' = Z_0 / \sqrt{(1 + C_A/C)} \tag{9.6}$$

and the propagation delay of the line will be increased to

$$t_D'' = t_D' \sqrt{(1 + C_A/C)} \tag{9.7}$$

Increasing the capacitance $C$ of a line has several adverse effects upon an interface:

(a) More current must be supplied by the driving circuit so that a higher average current is taken from the power supply with consequent increased power dissipation.

(b) The output impedance of the driving circuit forms a voltage divider with the characteristic impedance of the line. Increasing $C$ decreases $Z_0$ and this, in turn, reduces the signal amplitude at the receiver.

(c) The slew rate of the driving circuit will be reduced which will increase the risetime and the falltime of the voltage transitions and so limit the maximum data rate possible. The maximum allowable length of line can be determined by dividing the maximum line capacitance by the capacitance/metre of the cable used. The maximum line capacitance is specified by EIA 232 D/E (p. 305) as 2500 pF.

## Reflections

When a digital signal has been propagated along a transmission line and has arrived at the load presented by the driven device at the far end, it will be *reflected* and returned down the line towards the sending end — unless the line is correctly terminated. The reflected voltage is equal to $\rho_L V_s$, where $\rho_L$ is the *voltage reflection coefficient* at the termination.

$$\rho_L = (R_L - Z_0)/(R_L + Z_0) \tag{9.8}$$

where $R_L$ is the resistance of the load terminating the line.

Obviously, if $R_L = Z_0$ the line is correctly terminated and the voltage reflection coefficient is zero and this means that all the power associated with the incident voltage wave will be absorbed by the load. There is then no reflected wave.

When $R_L$ is not equal to $Z_0$ the reflected wave propagates along the line, in the opposite direction to before, towards the sending end. When it arrives at the sending end it will be reflected again, unless the sending end is correctly terminated, this time with a voltage reflection coefficient of $\rho_D$.

$$\rho_D = (R_D - Z_0)/(R_D + Z_0) \tag{9.9}$$

The reflected voltage $\rho_D \rho_L V_s$ will now propagate along the line towards the driven circuit to be reflected again and so on. The reflected waves continually propagate backwards and forwards along the line with a diminishing amplitude, because the line always has some losses and the two voltage reflection coefficients are not (usually) equal to unity. The most common result of not taking into account the transmission line nature of a high-speed digital connection is *ringing*. Ringing is caused by multiple reflections from the two ends of an un-terminated line; it may cause a single pulse to be interpreted by the receiver as being three, or even more, pulses.

Figure 9.5 shows two digital circuits connected together by a length of line. The driving circuit has an output resistance $R_D$ of approximately $0\,\Omega$ so that the voltage reflection coefficient at the sending end of the line is $-1$. The driven circuit has a very high input resistance, $R_L$, so that the voltage reflection coefficient at this end of the line is $+1$.

Suppose that initially there is a steady state d.c. voltage of $2.5\,\text{V}$ on the line as shown by Fig. 9.6(a) and that at time $t = 0$ this is suddenly reduced to $0.5\,\text{V}$. The voltage transition is a $-2\,\text{V}$ step and this step is propagated along the line towards the driven circuit (see Fig. 9.6(b)). As the $-2\,\text{V}$ step travels along the line it reduces the

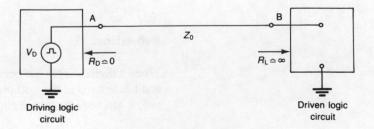

**Fig. 9.5** Digital circuit

Driving logic circuit

Driven logic circuit

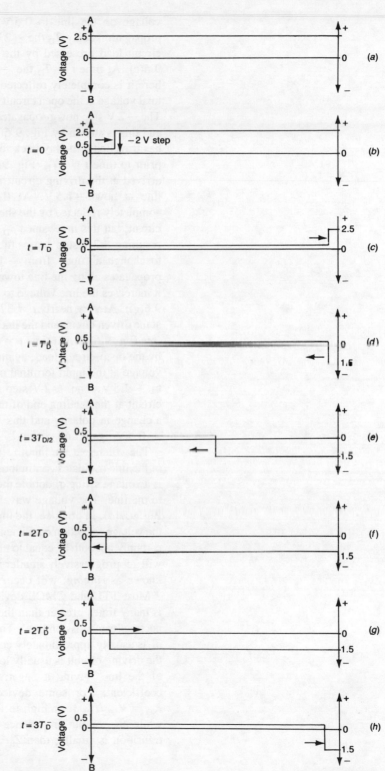

**Fig. 9.6** Reflected waves on a digital circuit

voltage on the line to 0.5 V. After a time slightly less than the propagation delay $T_D$ the $-2$ V step has almost arrived at the open-circuit load presented by the driven circuit; this is shown by Fig. 9.6(c). At time $t = T_D$ the $-2$ V step reaches the open circuit and here it is completely reflected with zero change in its polarity. The total voltage at the open circuit is then equal to $+0.5 - 2 = -1.5$ V. The $-2$ V step now propagates back towards the driving logic circuit and this is shown by Fig. 9.6(d). At time $t = 3T_D/2$ the $-2$ V step has travelled half-way back to the driving circuit (Fig. 9.6(e)). Just prior to time $t = 2T_D$ (Fig. 9.6(f)), the $-2$ V step has very nearly arrived at the driving circuit and the voltage at all points along the line is now $-1.5$ V. At time $t = 2T_D$ the returned wave is completely reflected by the short-circuit load presented by the driving circuit, but this time, since $\rho_D = -1$, with a change in polarity. The resulting $+2$ V step causes the voltage at the sending end of the line to change abruptly from $-1.5$ to $+0.5$ V. The $+2$ V step now propagates along the line towards the receiving end and as it travels it increases the line voltage to its original value of $+0.5$ V (see Fig. 9.6(g)). At very nearly $t = 3T_D$ the $+2$ V step is just about to arrive at the driven circuit and the line voltage is everywhere equal to $+0.5$ V (see Fig. 9.6(h)). At time $t = 3T_D$ the $+2$ V voltage step is reflected by the open-circuit load, again with zero change in polarity, and the voltage at the input terminal of the driven circuit abruptly increases to $+2.5$ V. The $+2$ V step now travels back towards the driving circuit at the sending end of the line where it is again reflected with a change in polarity and this $-2$ V step reduces the line voltage to $+0.5$ V again and so on.

The voltage at the input to the driven circuit at the receiving end of the line oscillates continuously between $+2.5$ and $-1.5$ V, which is a voltage swing of double the original change in the voltage applied to the line. The voltage waveform at the driven circuit is shown by Fig. 9.7(a). In practice, the line will inevitably have some losses and the voltage reflection coefficients at either end of the circuit will almost certainly not both be equal to unity. Consequently, the reflected waves will get progressively smaller in amplitude and a damped oscillation, known as *ringing*, will occur (see Fig. 9.7(b)).

Most TTL and CMOS devices have a high input impedance that is many times greater than the characteristic impedance of the connecting 'transmission line'. Therefore, the load reflection coefficient $\rho_L$ is usually approximately equal to 1. The output impedance $R_D$ of the driving circuit is usually lower than the characteristic impedance of the line, giving a negative, but not unity, voltage reflection coefficient. For some devices, however, the output impedance $R_D = V_{OL}/I_{OL}$ for a high-to-low transition is much higher than $Z_0$, while the output impedance $R_D = V_{OH}/I_{OH}$ for a low-to-high transition is smaller than $Z_0$.

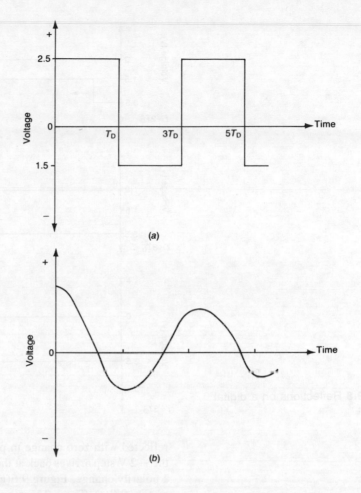

**Fig. 9.7** Received voltage at end of a non-terminated line

### Example 9.4

An LS 7400 NAND gate has $V_{OH} = 3.4\,\text{V}$, $V_{OL} = 0.2\,\text{V}$, $I_{OH} = 0.4\,\text{mA}$ and $I_{OL} = 16\,\text{mA}$. Two such gates are linked by a transmission line of characteristic impedance 70 Ω. Calculate the voltage reflection coefficient at each end of the line.

*Solution*

For the driving gate $R_D = 3.4/(0.4 \times 10^{-3}) = 8500\,\Omega$ for a 1-to-0 transition and $R_D = 0.2/(16 \times 10^{-3}) = 12.5\,\Omega$ for a 0-to-1 transition. Hence

$$\rho_D = (8500 - 70)/(8500 + 70) = 0.984 \quad (Ans.)$$

or

$$\rho_D = (12.5 - 70)/(12.5 + 70) = -0.697 \quad (Ans.)$$
$$\rho_L = +1 \quad (Ans.)$$

When the driving impedance is greater than the characteristic impedance of the line the reflected wave at the sending end will be

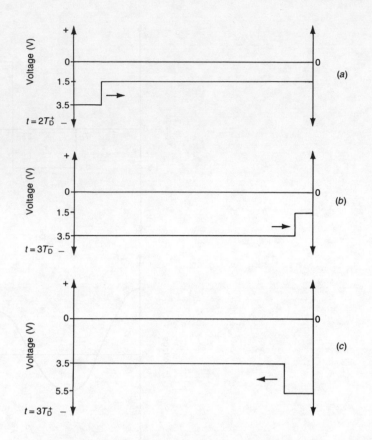

**Fig. 9.8** Reflections on a digital circuit

reflected with zero change in polarity. Referring to Fig. 9.6, when the $-2\,$V step arrives back at the driving device it is reflected without a polarity change. Figure 9.6($g$) must therefore be altered as shown by Fig. 9.8($a$). The $-2\,$V step travels back towards the driven circuit increasing the line voltage to $-3.5\,$V as it propagates. After time $t = 3T_D^-$ (Fig. 9.8($b$)) the line voltage is at all points equal to $-3.5\,$V. At time $t = 3T_D$ the $-2\,$V step is reflected, again with zero change in polarity, to increase the voltage at this end of the line to $-5.5\,$V. The $-2\,$V step now propagates along the line towards the sending end increasing the line voltage to $-5.5\,$V as it travels (Fig. 9.8($c$)). Multiple reflections at both ends of the line now cause the voltage at the input to the driven circuit to 'stair-step' in $-4\,$V steps. In practice, line losses and non-unity voltage reflection coefficients mean that the voltage steps are of progressively smaller amplitude. However, false operation of the driven circuit may occur and the signal waveshape will be distorted.

## The Lattice Diagram

Consideration of the effects of multiple reflections on a digital transmission line is made easier if a *lattice diagram* is employed. A

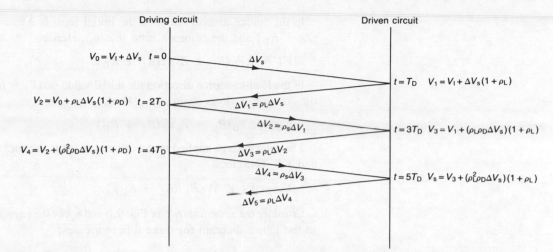

**Fig. 9.9** Lattice diagram

lattice diagram has axes of line length in the horizontal direction and of time in the vertical downwards direction. It enables the progress of a voltage transition over a mismatched line to be indicated by a zigzag plot in which the gradient of the zigzag path represents the velocity of propagation on the line.

An example of a lattice diagram is shown by Fig. 9.9. The initial steady d.c. voltage on the line is $V_1$. When a voltage transition $\Delta V_D$ occurs at the sending end of the line the voltage applied to the line is

$$V_0 = V_1 + \Delta V_D Z_0/(R_D + Z_0) = V_1 + \Delta V_s$$

The voltage transition $\Delta V_s$ propagates along the line and arrives at the other end after a time $t = T_D$. Some of the incident voltage is now reflected so that the total voltage at this end of the line is

$$V_1 = V_1 + \Delta V_s(1 + \rho_L)$$

The reflected voltage $\Delta V_1 = \rho_L \Delta V_s$ propagates along the line towards the sending end where some of it, $\rho_D \Delta V_1$, is reflected. The total voltage at the sending end of the line is now

$$V_2 = V_0 + \rho_L \Delta V_s(1 + \rho_D).$$

A fraction of this voltage, $\Delta V_2 = \rho_D \Delta V_1$, now propagates towards the driven circuit where it will be partially reflected and so on. The amplitudes of the voltages at each end of the line after each reflection are given in the lattice diagram. The multiple reflections continue to propagate back and forth on the line until their amplitudes become negligibly small.

The final value of the voltage applied to the driven device is equal to the initial voltage plus the sum to infinity of the incident and reflected voltages at the receiving end of the line. This is a geometric series whose sum to infinity is given by

$$\text{Sum to infinity} = (\text{initial term})/(1 - \text{common ratio}) \quad (9.10)$$

In the source-to-load direction the initial term is $\Delta V_s = \Delta V_D Z_0 / (Z_0 + R_D)$ and the common ratio is $\rho_L \rho_D$. Hence,

$$V'_L = \Delta V_D (R_L + Z_0)/2(R_L + R_D).$$

In the load-to-source direction the initial value is $\Delta V_1 = \rho_L \Delta V_s$ and the common ratio is again $\rho_L \rho_D$. Hence

$$V''_L = \Delta V_D (R_L - Z_0)/2(R_L + R_D)$$

The total voltage applied to the driven circuit is the sum of $V_I$, $V'_L$ and $V''_L$. Therefore

$$V_L = V_I + \Delta V_D R_L/(R_L + R_D) \qquad (9.11)$$

Consider the system shown in Fig. 9.6 and apply the equations given in the lattice diagram for three different cases.

*(a) $\rho_L = +1$, $\rho_D = -1$*

$$V_1 = 2.5\,\text{V} \qquad \Delta V_s = -2\,\text{V}$$
$$V_0 = 2.5 - 2 = 0.5\,\text{V}$$
$$V_1 = 2.5 - 2(1 + 1) = -1.5\,\text{V}$$
$$V_2 = 0.5 + (1 \times -2)(1 - 1) = 0.5\,\text{V}$$
$$V_3 = -1.5 + (1 \times -1 \times -2)(1 + 1) = 2.5\,\text{V}$$
$$V_4 = 0.5 + (1^2 \times -1 \times -2)(1 - 1) = 0.5\,\text{V}$$
$$V_5 = 2.5 + (1^2 \times (-1)^2 \times -2)(1 + 1) = -1.5\,\text{V etc.}$$

as was obtained before in an alternative manner.

*(b) $\rho_L = \rho_D = +1$*

Assume that a $-2\,\text{V}$ transition is applied to the line.

$$V_1 = 2.5\,\text{V} \qquad \Delta V_s = -2\,\text{V}$$
$$V_0 = 2.5 - 2 = 0.5\,\text{V}$$
$$V_1 = 2.5 - 2(1 + 1) = -1.5\,\text{V}$$
$$V_2 = 0.5 + (1 \times -2)(1 + 1) = -3.5\,\text{V}$$
$$V_3 = -1.5 + (1 \times 1 \times -2)(1 + 1) = -5.5\,\text{V}$$
$$V_4 = -3.5 + (1^2 \times 1 \times -2)(1 + 1) = -7.5\,\text{V}$$
$$V_5 = -5.5 + (1^2 \times 1^2 \times -2)(1 + 1) = -9.5\,\text{V etc.}$$

as before.

*(c) $\rho_L = +1$, $\rho_D = -0.3$*

Since $R_D$ is no longer approximately equal to $0\,\Omega$ not all of the $-2\,\text{V}$ transition will be applied to the line.

$$(R_D - Z_0)/(R_D + Z_0) = -0.3$$

and hence $R_D = 0.54 Z_0$. Now

$$\Delta V_s = -2 Z_0/1.54 Z_0 = -1.3\,\text{V}$$

$$V_0 = 2.5 - 1.3 = 1.2\,\text{V}$$
$$V_1 = 2.5 - 1.3(1 + 1) = -0.1\,\text{V}$$
$$V_2 = 1.2 + (1 \times -1.3)(1 - 0.3) = 0.29\,\text{V}$$
$$V_3 = -0.1 + (1 \times -0.3 \times -1.3)(1 + 1) = 0.68\,\text{V}$$
$$V_4 = 0.29 + (1^2 \times -0.3 \times -1.3)(1 - 0.3) = 0.563\,\text{V}$$
$$V_5 = 0.68 + (1^2 \times (-0.3)^2 \times -1.3)(1 + 1) = 0.446\,\text{V}$$

etc.

### Risetime

If the length of the line is short the risetime of the input voltage transition will be greater than twice the propagation delay. The voltage wave reflected at the receiving end of the line will then arrive back at the sending end while the driving circuit is still in the process of changing from one output state to the other. The reflected wave is then partly overridden by the rising edge of the input transition and so ringing is reduced. This means that the longer the risetime of the input voltage transition the longer will be the length of line that may be employed before ringing effects must be taken into account.

The maximum permissible length of line that may be considered as a lumped circuit is given by equation (9.12), i.e.

$$l_{(max)} = t_r/2t_D \tag{9.12}$$

where $t_r$ is the risetime of input transition and $t_D$ the propagation delay per metre.

### Example 9.5

An LS circuit is to send signals over the pair of conductors that link two circuit boards. The propagation delay is 4.92 ns/m and the minimum transition time is 1.6 ns. Calculate the maximum line length that may be used without reflections being taken into account.

### Solution

$$v = 1/(4.92 \times 10^{-9}) = 2.032 \times 10^8\,\text{m/s}.$$

Hence

$$l_{max} = 2.023 \times 10^8 \times (1.6 \times 10^{-9})/2 = 0.162\,\text{m}. \qquad (Ans.)$$

### Line Terminations

To minimize the occurrence of ringing and/or stair-stepping on a transmission line carrying digital signals the line should be correctly terminated at one end, or at both ends if it is either a bidirectional line or a party line.

A resistor $R$ can be connected in series with the output terminal of the driving device as shown by Fig. 9.10. The value of the resistor

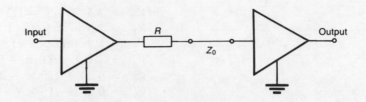

**Fig. 9.10** Series termination

should be equal to the difference between the characteristic impedance of the line and the output resistance $R_D$ of the driving device, i.e. $R = Z_0 - R_D$. The best terminating resistors are carbon or metal film types with the shortest possible leads. Clearly, the method can only be employed when $R_D < Z_0$. The voltage transition applied to the line will only be equal to one-half of the voltage change developed by the driving device when it changes state. If, however, the driven device has a high input impedance the voltage transition at its input terminal will, because of the 'zero-polarity change' reflection, be equal to the generated transition. This is then the same voltage as would be obtained if the line had been correctly terminated at the far end when a series resistor would not be required. The series termination will cause a small increase, typically 2 ns, in the propagation delay over the line and will also increase the transition time. The series termination is preferred when the load is lumped at the end of the line since it gives a good noise margin, but it is not suitable for a line that is supplying several distributed loads.

When the driving device has a differential output, like many line drivers, it will be necessary to connect a resistor in series with both of its outputs as shown by Fig. 9.11. The main disadvantage of the single-ended line is that it has a poor noise immunity. There are two reasons for this:

(a) Any noise induced into the earth return will adversely affect the received data.
(b) A shift in the earth potential at the receiver end of the line may give an apparent reduction in the switching threshold of the receiver.

The single-ended circuit is also prone to cross-talk from adjacent circuits. These disadvantages are overcome by the use of a differential connection between the driving and receiving devices.

The parallel termination resistor method is shown by Fig. 9.12. The parallel resistor should have the same value as the characteristic impedance of the line. The resistor is not always connected to earth as shown in Fig. 9.12; it may instead be connected to either $+ V_{CC}$

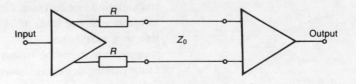

**Fig. 9.11** Series termination for a differential circuit

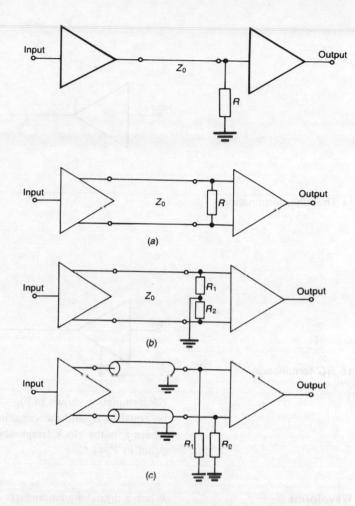

**Fig. 9.12** Parallel termination

**Fig. 9.13** Different methods of terminating a differential circuit

or to + 3 V. This method of terminating a line does not increase the propagation delay of the circuit as much as does the series resistor method but the d.c. power dissipated in the parallel resistor may present a problem.

A differential line driver/line receiver circuit may connect the terminating resistor between the two input terminals of the receiver (see Fig. 9.13(a)). This circuit, however, has a poor noise immunity and better arrangements are shown by Figs 9.13(b) and (c). The value of each resistor should be equal to $Z_0/2$.

The Thevenin network shown in Fig. 9.14 uses two resistors, one connected to earth and the other connected to + $V_{CC}$. The midpoint d.c. voltage should not be allowed to be somewhere in between the high and the low input voltage levels since this would reduce the noise margin of the circuit. Typically, $R_1 = 5Z_0/3$ and $R_2 = 5Z_0/2$, when the effective terminating resistance is $Z_0$ and the midpoint d.c. voltage is + 3 V. This form of termination gives only a small increase in the propagation delay of the circuit but some power is dissipated.

The power dissipation problem can be reduced by the use of the

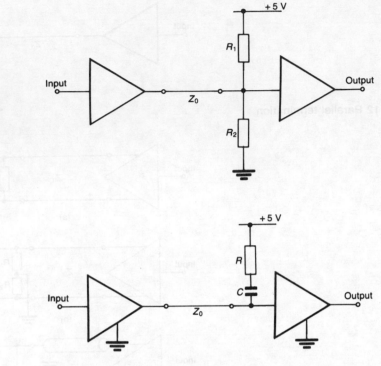

**Fig. 9.14** Thevenin termination

**Fig. 9.15** *RC* termination

*CR* termination shown in Fig. 9.15. The value of the resistor should be equal to $Z_0$ and the capacitor should be smaller than $1/(6Z_0 f_c)$, where $f_c$ is the clock frequency. The power dissipated in $R$ is then equal to $V_{CC}^2 f_c C$.

**Pulse Waveforms**

**Fig. 9.16** Digital signal

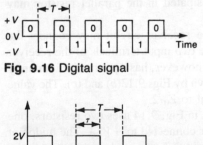

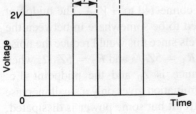

**Fig. 9.17** Pulse waveform

When a digital signal consists of alternate 1s and 0s it has a square waveform as shown by Fig. 9.16. The periodic time $T$ of the waveform is the time interval between the leading edges of consecutive pulses and it is equal to the reciprocal of the bit rate. The Fourier series representation of the waveform is given by equation (9.13).

$$v = (4V/\pi)[\sin \omega t + \sin (3\omega t)/3 + (\sin 5 \omega t)/5 + \ldots] \quad (9.13)$$

or

$$v = (4V/\pi)[\cos \omega t - (\cos 3\omega t)/3 + (\cos 5 \omega t)/5 -] \quad (9.14)$$

The series consists of a fundamental frequency component of amplitude $4V/\pi$, whose frequency is equal to one-half of the bit rate, plus components at the third, fifth, seventh, etc. odd harmonics. The amplitudes of the harmonics decrease in the same ratio as the order of the harmonic.

If the duty cycle is altered even harmonics will also appear in the waveform and the amplitude of each harmonic will change. A rectangular waveform has a d.c. component that is equal to the average value of the wave.

### Spectrum Diagram

A *spectrum diagram* shows the amplitudes of all the various frequency components plotted to a base of frequency. To simplify the arithmetic the digital waveform is moved in the positive direction so that its amplitude is changed from $\pm$ V to 2 V. The pulse waveform shown in Fig. 9.17 has an amplitude of 2 V, a pulse width (or duration) of $\tau$ and a periodic time of $T$. The expression for the instantaneous voltage of the waveform can be written as

$$v = V_{DC} + (4V/n\pi) \sin (n\pi\tau)/T \cos n\omega t \qquad (9.15)$$
$$= V_{DC} + (4V\tau/T)[\sin (n\pi\tau/T)/(n\pi\tau/T)] \cos n\omega t \qquad (9.16)$$

Equation (9.16) expresses the amplitude of the components of a rectangular wave in the form of $(\sin x)/x$, or sinc $x$, and tables of this function are readily available.

From equation (9.15) or (9.16) the amplitude of the fundamental frequency component, i.e. $n = 1$, is $(4V/\pi) \sin (\pi\tau/T)$, the amplitude of the second harmonic component, $n = 2$, is $(4V/2\pi) \sin (2\pi\tau/T)$, and so on.

For the square waveform, $\tau/T = 1/2$ and the amplitude, $v_f$, of the fundamental component is

$$v_f = (4V/\pi) \sin (\pi/2) = 4V/\pi$$

the amplitude of the third harmonic component is

$$v_3 = (4V/3\pi) \sin (3\pi/2) = 4V/3\pi$$

and so on as in equation (9.13).

The spectrum diagram for the waveform shown in Fig. 9.17 is given in Fig. 9.18. The spectrum diagram has a number of lobes spaced apart at frequency intervals equal to the reciprocal of the pulse width. Within each lobe are a number of spectral lines which represent the amplitudes of the component frequencies, these lines are spaced apart at frequency intervals equal to the reciprocal of the periodic time.

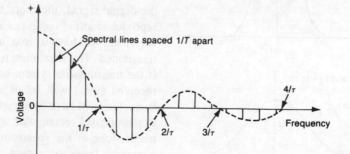

**Fig. 9.18** Spectrum diagram of waveform in Fig. 9.17

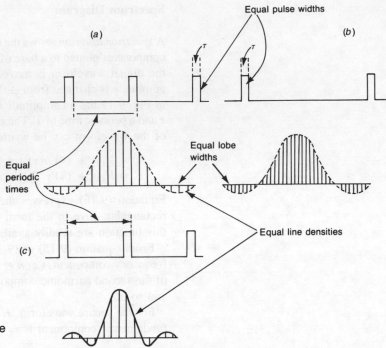

**Fig. 9.19** Effect on spectrum diagram of altering either the bit rate or the pulse duration

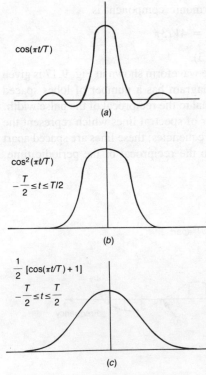

**Fig. 9.20** Pulse waveforms: (a) cosine; (b) cosine$^2$; (c) raised cosine

Figure 9.19 shows the effect upon the spectrum diagram of altering either the bit rate or the duration of each pulse. If the waveform in Fig. 9.19(a) has its periodic time increased with no change in the pulse duration as in Fig. 9.19(b) the width of each lobe will be unchanged but the spectral lines will become closer together. Conversely, increasing the pulse duration without altering the periodic time gives narrower lobes but the line density is unchanged (see Fig. 9.19(c).

A perfectly rectangular wave contains a very large number of harmonic components and it is not economically possible to transmit all of them. Some distortion of the waveform must therefore always be accepted. The wider the bandwidth that is made available to transmit the digital signal, the more faithfully will the digital waveform be reproduced at the receiver. The fundamental frequency of a digital waveform will have its maximum value when alternate 1s and 0s are transmitted. This maximum frequency is equal to one-half the bit rate. If the transmission system has a bandwidth equal to this figure the received signal will be of sinusoidal waveshape since all of the harmonics will have been suppressed. If a reasonable approximation to the original rectangular waveform is to be obtained the first, or major, lobe of the spectrum diagram must be transmitted since it contains a large percentage of the signal power. Often the shape of the transmitted pulses are modified before transmission in order to concentrate a greater percentage of the total signal power in the major lobe. Some possible waveforms are shown in Fig. 9.20.

**Attenuation**

The attenuation, or loss, of a transmission line is not a constant quantity but, instead, increases with increase in frequency in a non-linear manner. The attenuation of a line is generally quoted in decibels per metre and it will be negligibly small for all short lengths of line. The maximum attenuation $A_{(max)}$ that can be allowed in a longer line can be determined from equation (9.17).

$$A_{(max)} = 20 \log_{10}(V_D/V_R) \text{ decibels} \qquad (9.17)$$

where $V_D$ is the worst case, i.e. minimum, voltage the line driver will supply to the line, and $V_R$ the worst case voltage that must appear across the line receiver's input terminals. This latter figure includes a margin to give an adequate noise margin. Since the magnitude of a pulse waveform is mainly determined by its fundamental frequency component the effect of line attenuation can be approximately calculated by considering the attenuation of the line to a sinusoidal signal. For example, if a certain length of line attenuates a 10 MHz sinusoidal signal by 12 dB, then a digital waveform with the same fundamental frequency will also be attenuated by 12 dB.

*Example 9.6*

An RS 485 line driver/receiver system has $V_D = 1.5$ V and $V_R = 0.4$ V which includes 200 mV noise immunity. Calculate the maximum allowable line attenuation.

*Solution*
$A_{(max)} = 20 \log_{10}(1.5/0.4) = 11.5 \text{ dB}$     (*Ans.*)

*Example 9.7*

The digital waveform shown in Fig. 9.21 is transmitted over a line that has the attenuation/frequency characteristics given in Table 9.1. (*a*) Draw the spectrum diagram of the waveform. (*b*) Calculate the voltage of the fundamental, and the harmonic components at the end of the line.

*Solution*
(*a*)   $\tau = 0.25 \, \mu s$ and $T = 1.25 \mu s$. The spectrum diagram is shown in Fig. 9.22.
(*b*)   At the sending end of the line:

$$\text{Amplitude of fundamental} = (24/\pi) \sin (0.25\pi/1.25)$$
$$= 4.49 \text{ V}$$
$$\text{second harmonic} = (12/\pi) \sin (2 \times 0.25\pi/1.25)$$
$$= 3.57 \text{ V}$$
$$\text{third harmonic} = (8/\pi) \sin (3 \times 0.25\pi/1.25)$$
$$= 2.42 \text{ V}$$
$$\text{fourth harmonic} = (6/\pi) \sin (4 \times 0.25\pi/1.25)$$
$$= 1.12 \text{ V} \quad (\textit{Ans.}).$$

At 0.8 MHz loss = 12 dB = 3.98 voltage ratio;
$V_R = 4.49/3.98 = 1.13$ V    (*Ans.*)

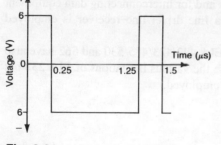

**Fig. 9.21**

**Table 9.1**

| Frequency (MHz) | 0.8 | 1.6 | 2.4 | 3.2 |
|---|---|---|---|---|
| Attenuation (dB) | 12 | 16 | 20 | 23 |

At 1.6 MHz voltage ratio $= 6.31$ so $V_R = 3.57/6.31$
$= 0.57$ V   (*Ans.*)

At 2.4 MHz voltage ratio $= 10$ so $V_R = 2.42/10$
$= 0.24$ V   (*Ans.*)

At 3.2 MHz voltage ratio $= 14.13$ so $V_R = 1.12/14.13$
$= 0.08$ V   (*Ans.*)

## Interface Standards and Line Drivers/Receivers

Standard interfaces are employed to link a computer to its various peripherals and to other equipment to ensure that there is compatibility between items bought from different manufacturers. The term 'data terminal equipment' (DTE) is used to refer both to computers and all other data terminals. The interface between a DTE and a modem — known as 'data communication equipment' (DCE) — must be able to (*a*) transfer data in both directions, (*b*) control the data flow, (*c*) pass clock signals and (*d*) select the bit rate at which the data are transmitted over the line. The Consultative Committee International Telephone and Telecommunication (CCITT) and the Electrical Industrial Association (EIA) have both produced specifications to standardize the interface for serial data transmission between a DTE and DCE.

The most commonly employed interfaces are the CCITT V24 and the EIA 232 (which until recently was known as the RS 232), which are functionally equivalent to one another. Both standards define a set of interface circuits over which the DTE can communicate with the DCE. The CCITT V28 recommendation defines the electrical characteristics of the V24 interface up to 20 kb/s. EIA 232 is essentially a combination of V24 and V28. Each circuit has a particular function and it is active when a positive voltage, representing binary 0 for a data circuit, or ON for a control circuit, is applied to it by either the DTE or the DCE. V24 is only employed for linking data terminals to modems but EIA 232 is also employed for linking peripherals to data terminals and for interconnecting data equipment over short distances when a line driver/line receiver is employed instead of a modem.

Newer standards such as EIA 422/423/485/530 and 562 have also been introduced to overcome the various limitations of EIA 232 but the standard is still widely employed.

### EIA 232

The RS 232 interface standard was first introduced in 1969 and was later superseded by the RS (EIA) 232D standard in 1987 and now by EIA 232E. The standard has now been renamed the EIA 232D. EIA 232 defines a single-ended interface for the connection of a DTE to a DCE for serial data transmission (see Fig. 9.22). Usually, the line is not terminated. The standard is also commonly employed for low-cost interfaces between a DTE and a peripheral equipment. The

**Fig. 9.22** EIA 232 connection

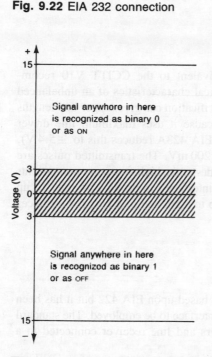

**Fig. 9.23** CCITT V24/EIA 232 voltage ranges

standard specifies that a positive voltage between 3 and 25 V should be used to indicate logic 0 for a data circuit, and ON for a control circuit; and a negative voltage of between −3 and −25 V should indicate logic 1 or OFF. The CCITT V24/EIA 232 voltage ranges are shown in Fig. 9.23. The latest 'E' version limits the voltages to ± 5 to ± 15 V. All the voltages are with respect to the signal earth line. The maximum speed of transmission is set by the relationship

$$\frac{\text{(time to change from } -3 \text{ to } +3 \text{ V})}{\text{(time duration of a bit)}} \tag{9.18}$$

and the EIA 232 specifies that for baud speeds up to 8000 this ratio should be 4%, and for higher baud speeds the risetime should be less than 5 μs.

Other parameters specified by EIA 232 are: (a) the receiver's sensitivity is ± 3 V; (b) its output resistance must be between 3000 and 7000 Ω; (c) the driver circuit must have a maximum slew rate of 30 V/μs.

As the length of the line is increased, or the voltages used are larger, the time it takes for the data at the end of a line to change from binary 1 to binary 0, or vice versa, increases also. The maximum bit rate is limited by both the capacitance of the line and the output current capability of the driving device. In the earlier 232C versions of the standard the maximum bit rate is specified as 20 kb/s for distances of up to 15 m, but the later 232D and E versions quote a maximum line capacitance of 2500 pF instead.

At these relatively low bit rates transmission line effects may be ignored and the line's resistance and capacitance may be regarded as lumped parameters. Another disadvantage of the EIA 232 standard is that since it is single-ended it has a high noise susceptibility. However, the performance of an EIA 232 interface is perfectly adequate for many applications and it is still widely employed.

### EIA 422

The EIA 422 standard, equivalent to the CCITT V11 recommendation, defines the electrical characteristics of a balanced or differential interface (see Fig. 9.24). It specifies that line drivers and line receivers are employed for bit rates of up to 10 Mb/s over distances of up to 12 m, or for bit rates of up to 100 kb/s over distances of up to 1200 m. The voltage ranges are shown by Fig. 9.25.

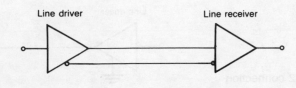

Line driver          Line receiver

**Fig. 9.24** EIA 422 connection

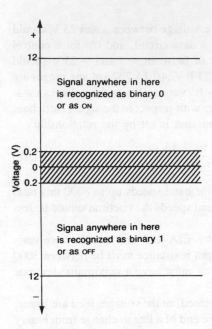

**Fig. 9.25** EIA 422 voltage ranges

### EIA 423

The EIA 423 standard is equivalent to the CCITT V10 recommendation. It defines the electrical characteristics of an unbalanced digital interface circuit. The specification can give longer line lengths than the EIA 232 standard because it uses maximum line driver voltages of only ± 6 V (latest EIA 423A reduces this to ±5.4 V), and a receiver sensitivity of ± 200 mV. The transmitted pulses are pre-shaped to reduce the amplitudes of the high-frequency components in the waveform. An EIA 423 interface can provide bit rates of up to 100 kb/s over distances of up to 10 m and of 1 kb/s for distances of up to 1200 m.

### EIA 485

The EIA 485 standard has been based upon EIA 422 but it has been modified to allow a multi-point interface to be employed. The standard can handle up to 32 line drivers and line receiver connected to a common line.

### EIA 530

The EIA 530 is a new standard which is intended to allow interfaces to be operated in the 20 kb/s to 2 Mb/s region. The standard is electrically similar to EIA 422.

### EIA 562

This is another recently introduced standard for single-ended interfaces operating at bit rates of up to 64 kb/s.

Table 9.2 shows the main parameters of the different EIA interface standards.

### Line Drivers and Receivers

The function of a line driver is to convert TTL or CMOS signals to the voltage levels that are specified by the interface standard that is

**Table 9.2**

|  | EIA 232D | EIA 232E | EIA 423A | EIA 422A | RS 485 | EIA 562 |
|---|---|---|---|---|---|---|
| Operation | Single | Single | Single | Differential | Differential | Single |
| Line length | 2500 m | 2500 pF | 1.22 km | 1.22 km | 1.22 k | 2500 pF |
| Data rate (b/s) | 20k | 20k | 100k | 10M | 10M | 64k |
| Voltage range | $\pm 3$ to $\pm 25$ | $\pm 5$ to $\pm 15$ | $\pm 3.6$ to $\pm 5.4$ | $\pm 2$ to $\pm 5$ | $\pm 1.5$ to $\pm 5$ | $\pm 3.7$ to $\pm 13.2$ |
| Slew rate (V/$\mu$s) | 30 | 30 | — | — | — | 30 |
| Receiver sensitivity | $\pm 3$ V | $\pm 0.2$ V | $\pm 0.2$ V | $\pm 0.2$ V | $\pm 3$ V | |

being used. An EIA 232 line driver, for example, will convert the + 5 V logic 1 level of a TTL system into the −3 to −25 V required by the EIA 232 specification. Line drivers are offered by the various manufacturers that operate to a particular specification, e.g. the SN 75150 and the $\mu$A 9636AC are EIA 232D line drivers, and the MC 3487 and $\mu$A 9638C are EIA 422 line drivers.

Conversely, the function of a line receiver is to convert received interface signals into the signals required by the kind of logic employed. The SN 75154 quad line receiver, for example, will convert an input negative voltage representing logic 1 into the + 5 V required by TTL circuitry.

# 10 System Design

An electronic system consists of a number of electronic circuits that are connected together to perform a particular function that is defined by the equipment specification. To simplify the design process the initial definition of the project is used to produce a block diagram of the system to be produced. Each of the blocks shown in the block diagram will represent a separate circuit, e.g. an amplifier, an oscillator, a counter or an analogue-to-digital converter. A functional block diagram is particularly useful for a digital system since the diagram can often specify the operation of the system. The operation of each block shown in such a diagram must be defined, often by a truth or a function table. If the circuit blocks each have a low output impedance and a high input impedance the blocks may be interconnected as required with little, if any, interaction between them. Wherever such impedances do not exist it will be necessary to consider the interface between two blocks since inter-stage loading will occur. Each of the circuit blocks is separately designed using, particularly with digital circuits, standard MSI/LSI devices wherever possible. As each circuit, or subsystem, is designed a prototype should be made and then tested to check that its operation is as expected; if it is not the circuit design will need to be modified and retested until it does work correctly. This method of design is not suitable for VSLI circuits because of their sheer complexity and the need to implement directly in silicon. Such circuits are therefore designed using a CAD package in which software tools are employed to model and evaluate designs and layouts.

The number of possible systems is so large, and the types of systems are so varied, that this chapter will only give a few representative block diagrams as examples.

## Specifications

A specification is a detailed description of the required characteristics of a material, component, circuit, system or equipment. The specification explains in detail what the item concerned should be able to do and under what environmental conditions this performance is to be guaranteed. The usual format for a specification is: (i) a description of what the equipment does and its intended applications; (ii) electrical data such as voltages and frequencies; (iii) power requirements, e.g. mains and/or battery operated; (iv) environmental data, e.g. the working temperature range; (v) mechanical data, e.g.

physical dimensions and weight. The specification ought to state clearly exactly what the system is required to do; it should concentrate on the functions to be performed rather than how the objects are accomplished.

Two examples of what might be included in a specification follow:

(a) an r.f. signal generator:
  (i) carrier frequency: 10 kHz to 510 MHz;
  (ii) frequency stability: $5 \times 10^{-6}$ per 20 minutes;
  (iii) r.f. output voltage: $0.2 \mu V$ to 300 mV e.m.f;
  (iv) output impedance: $50 \Omega$. VSWR (voltage standing wave ratio) better than 1.2 : 1;
  (v) amplitude modulation: depth 0–90% from 300 to 3000 Hz continuously variable;
  (vi) frequency modulation: deviation 0–100 kHz for modulation frequency 300 to 3000 Hz continuously variable.

(b) an audio power amplifier:
  (i) maximum output power into $8 \Omega = 3$ W;
  (ii) total harmonic distortion at 1 W output power < 1%;
  (iii) quiescent current = 15 mA;
  (iv) input sensitivity for 50 mW output power = 1.6 mV;
  (v) input impedance = 200 k$\Omega$;
  (vi) 3 dB bandwidth = 25 Hz to 16 kHz;
  (vii) output signal-to-noise ratio = 60 dB.

A typical specification for a digital system could be one for a drinks vending machine. The machine should provide a choice between eight different drinks some of which, such as tea and coffee, are to be made on demand. Hot drinks are to be delivered at a specified temperature after cold water has been heated in the machine. Any combinations of coins are to be accepted up to a total value that can be reset as, and when, it becomes necessary to increase prices. No change is to be given.

When the correct sum, or more, has been inserted into the machine all the drink-choice buttons should light up and become active. When a particular button is pressed, (i) all the inserted coins should fall into the machine, (ii) all the drink-choice buttons should become inactive, (iii) a cup should be made to fall into the correct position ready to receive the chosen drink, and (iv) the requested drink should be delivered in the quantity that very nearly fills the cup. After a short time delay the machine should be ready for the next customer.

A specification like this could be met in a number of different ways and it would demand both electronic and mechanical design. Factors such as cost and reliability are taken into account in the choice between competing technologies. Essentially, the choice is between SSI/MSI devices, PLDs or a microprocessor-based system. The use of SSI/MSI devices is relatively easy and it requires no programming equipment or expertise. The use of PLDs reduces the number of ICs used in the system, and hence increases the reliability, but some programming is necessary. A PLD design will be physically smaller but it may dissipate more power and it may not be as fast. Microprocessor control is usually best for any system that requires a function to be controlled via an input/output port, but again software is necessary. A microprocessor solution to a design problem will be slower than a hardware solution, because of the time taken to execute the instructions, and it will prove to be more expensive for a small system. On the other hand, microprocessor systems are more versatile, since any change in the system's requirements will only need a change in the programming. Microprocessor systems are outside the scope of this book.

The specification of a piece of equipment may be used by a potential buyer to evaluate and compare different equipments from different manufacturers in order to decide which one is best suited, on both technical and economic grounds, to his application. Sometimes an equipment specification is derived as the result of discussions between the manufacturer and the customer. The customer may request that certain materials/components are used and perhaps decree exactly what the equipment must be able to do. In many cases British Standards are quoted in parts of a specification.

A *test specification* is written by the system designer once the system has passed the prototype stage. It details the measurements and tests that should be carried out on a piece of equipment. It will give details of the test points to be used and the measured values that ought to be obtained at those points. A test specification is used by the manufacturer's test department to test each piece of equipment as it is made. It often also specifies the test instruments that ought to be used, and when this is so it is important that the specified instruments are used, otherwise different instrument loading may invalidate the test results. The testing instructions given in the handbook for the equipment, when produced, may be the same as the test specification or at least be based upon it.

The test specification should contain detailed instructions on how to set the system up for test, how to perform the tests, and the expected results if the system is working correctly. Often included is a check list so that items may be ticked off as each test is completed. It is best, and faster in the end, to test each block as it is designed/manufactured rather than wait until all the blocks have been assembled to produce the complete system.

Two short examples of test specifications follow.

## Power Supply Voltage Checks

Test equipment: digital multi-meter (DMM); make and model given.

*Method*

(a) With the amplifiers switched off check the resistance of each voltage rail using the DMM. Expected resistance values would then be given.

(b) Switch the power supply on and check the voltage at each rail. Expected voltages would then be given, e.g. $12 \pm 0.1$ V.

(c) Check that the voltage at the test point 1 is $6 \pm 0.1$ V. Adjust the variable resistor $R_{99}$ if necessary to obtain the correct voltage.

**Attenuator Test**

Test equipment: dB meter; make and model given.

*Method*

(a) Connect the dB meter to the output terminals of the attenuator.

(b) Set the attenuation of the attenuator to 0 dB and apply a 10 kHz tone at 0 dBm to the input terminal of the attenuator.

(c) Check that the dB meter reads 0 dBm.

(d) Increase the attenuation of the attenuator in 10 dB steps. At each step check that the dB meter reading has decreased by 10 dB.

**Design of an Electronic System**

The design of an electronic circuit or system is based upon the specification that has been obtained from a request by a customer, from market research or from a circuit designer. The designer must analyse the specification to determine whether it is possible to make a circuit/system that can perform the specified function(s). If the specified requirements seem to be reasonable then the various ways in which it may be possible to implement the circuit/system will need to be evaluated to select the one that appears to be the best.

The essential steps to be taken in the design of an electronic system are as follows:

(a) Draw a block diagram of a possible system that will meet the specification.

(b) For each block decide on which seems to be the most suitable type of technology and/or circuit to use. For example, in the design of a digital system should SSI/MSI ICs, or PLDs, or a microprocessor be used? Or some appropriate mixture of those approaches? If the system to be designed is analogue should op-amps be used, or some other linear ICs, or is a discrete component approach more appropriate in a particular case? When one of these options has been decided upon it will then be necessary to decide on the particular circuit(s) to be used and to select the components/active devices.

(c) Design the hardware circuitry and if a PLD or microprocessor is used also design or otherwise obtain the required software.

(d) The chosen circuit can then be designed and if possible breadboarded. The breadboarded circuit can then be tested and if it does not work correctly it can be checked to see if wiring error(s) exist. If not it may be necessary to redesign part, or all, of the circuit, or even, in some cases, scrap the circuit and try another approach.

(e) Once the breadboarded circuit works correctly a prototype can be built. When the prototype has been tested the test results should be evaluated against the design criteria. If necessary the design can be modified to improve the circuit's performance and then retested. The prototype may need to be tested and redesigned several times before the specification is met.

(f) When all the prototypes are satisfactory the various blocks can be connected together to form the complete electronic system. This must then be tested to check its performance against the specification.

(g) Design a suitable test procedure and specify suitable test points and test equipment. Then measure the voltages, etc. at each test point and record the values obtained.

An essential part of any designed circuit/system is documentation that tells other people how the circuit/system works, construction details, test procedures, etc. This information should be written down before it is lost and even the designer is no longer sure of how he arrived at the finished design. British Standard symbols and terms should be employed to ensure that any reader will understand the printed information. The documentation should include such items as the circuit/system specification, a schematic diagram, a component layout diagram, a parts list and the test procedure.

**Electronic Systems**

The basic block diagrams of a few electronic systems are now given.

*(a) Digital Rev Counter*

Figure 10.1 shows the block diagram of a digital rev counter for a four-cylinder engine. The range of revs to be counted must be specified within a ± tolerance. The counter should be updated more frequently than the human eye can detect, i.e. counter changes should occur more frequently than every 0.2 s.

$$\text{Revs/min} = (1 \text{ rev})/\text{min} \times (1 \text{ min})/(60 \text{ s}) \times (4 \text{ pulses})/\text{rev}$$
$$= 1 \text{ pulse}/15 \text{ s}.$$

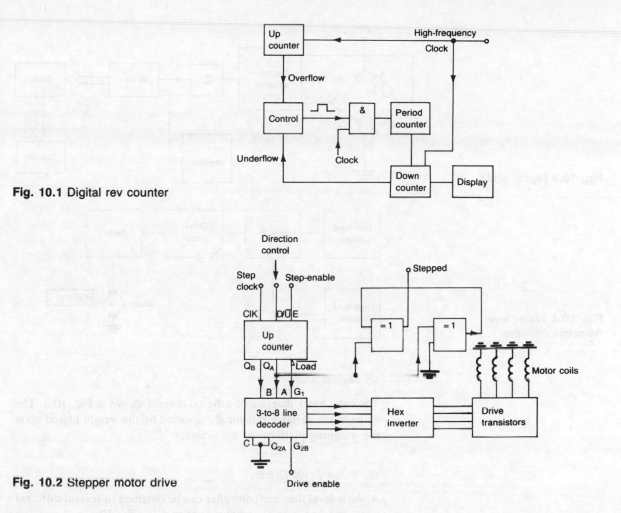

**Fig. 10.1** Digital rev counter

**Fig. 10.2** Stepper motor drive

Hence

$$\text{rpm} = 15 \times (\text{pulses/s}) = 15/(\text{s/pulse})$$

### (b) Stepper Motor Drive

Figure 10.2 shows the block diagram of a possible stepper motor drive. The four coils of the stepper motor are driven by the currents supplied by the four transistors. These transistors, in turn, are fed by the inverted outputs of the 3-to-8 line decoder. Disabling and enabling the stepper motor, and controlling its direction of rotation, are accomplished by applying the control signals shown in Fig. 10.2 to an up/down counter. When the counter is controlled to count up the motor will rotate in one direction and when it counts down rotation is in the reverse direction.

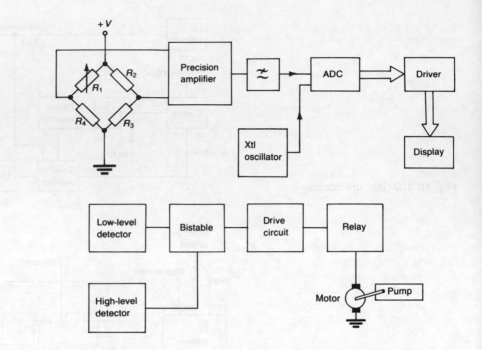

**Fig. 10.3** Digital scale

**Fig. 10.4** Water level detector/controller

### (c) Digital Scale

The basic block diagram of a digital scale is shown in Fig. 10.3. The value of the variable resistor $R_1$ is varied by the weight placed upon the weighing platform of the scales.

### (d) Water Level Detector

A water level detector/controller can be obtained in several different ways and one possible system is shown in Fig. 10.4.

# Exercises

## Exercises 1

**1.1** An ADC is to be selected for a particular application. List, and briefly discuss, the main factors that should be considered in the selection of a suitable device. What is integral non-linearity and why should it be good? An ADC has ideal differential non-linearity. How accurate is the translated output voltage?

**1.2**

(a)  Show that the input resistance of the circuit given in Fig. E.1 is $R_{in} = R_1(1 + R_3/R_2)$.

(b)  Calculate component values for the input resistance to be 100 MΩ.

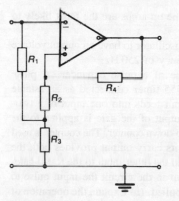

**Fig. E.1**

**1.3** Compare the relative merits of the various logic technologies. Explain why the HCMOS devices are increasingly employed in modern electronic circuitry. List the various families of HCMOS and compare their merits. What is meant by saying that a CMOS device is 'TTL' compatible?

**1.4** The data sheet of the Panasonic MN74HC138 3-to-8 line decoder includes:

| | | |
|---|---|---|
| Propagation time | $V_{CC} = 2$ V | 200 ns |
| (Enable $G_1$ to $Y$) | $V_{CC} = 4.5$ V | 40 ns |
| | $V_{CC} = 6$ V | 34 ns |

for an input transition time of no more than 6 ns and a load

capacitance of 50 pF. Determine the propagation delay if the device is to supply 10 HC loads at 4.5 V. (Each HC load has an input capacitance of 10 pF.)

**1.5** The maximum operating frequency of a digital IC is often listed in the data sheet. If not, it can be estimated as being equal to [1/(total worst case propagation time)]. If only one delay is listed then use twice that delay. For each of the devices whose data sheets are given in the book (see p. 178, 185, 191, 199 and 202) calculate the maximum operating frequency at 25 °C.

**1.6** Refer to the data sheet of the MC54/74HC573 on p. 48. At 25 °C (a) What is its maximum operating frequency? (b) What is the input capacitance? (c) The IC is used with a $V_{CC}$ of 5 V. Calculate its no-load dynamic power dissipation. (d) What is meant by saying that the latch is 'transparent'? (e) If $V_{CC} = 4.5$ V what are the minimum $V_{IH}$ and the maximum $V_{IL}$ values?

**1.7** A 500 mW dissipation op-amp has a slew rate of 5 V/μs and uses a + 15 V power supply. If the output voltage is to be 1 V calculate the full-power bandwidth. If the output capacitance is 2 nF calculate the output slew rate and hence the full-power bandwidth obtained. Assume that the maximum output current is 80% of the supply current. Also calculate the full-power bandwidth obtained when the supply voltage is reduced to ± 5 V.

**1.8** What is meant by the resolution of (a) an ADC and (b) a DAC? An ADC has a resolution of 16 bits. How many digital codes does it use? What is meant by the range of an ADC? What three factors determine the smallest change in input voltage that an ADC can detect? The ideal code width of an ADC is given by

$$\text{(voltage range)}/(\text{gain} \times 2^n)$$

where $n$ is the number of bits used. Calculate the ideal code width if the voltage range is 0–5 V, the gain is 500, and $n$ is (a) 12, and (b) 16.

**1.9** Explain the importance of the following parameters which are found in the data sheet of an op-amp: input offset voltage, input offset current, input bias current, CMRR, PSRR and slew rate. Explain how these parameters vary with (a) temperature and (b) frequency for the 741 op-amp.

## Exercises 2

**2.1** Derive an expression for the attenuation of the filter shown in Fig. 2.5.

**2.2** Show that the roll-off frequency of the filter shown in Fig. 2.6 is

$$f_c = [R_s + R_L + 2R_1]/[2\pi C(R_1 + R_s)(R_1 + R_L)]$$

Show that the maximum tolerable risetime (or falltime) is given by

$$t_{max} = 2.2C_1[(R_1 + R_s)(R_1 + R_L)]/(R_s + R_L + 2R_1)$$

**2.3** Design a loss-free matching network to match a load impedance of $250 + j125 \, \Omega$ at 40 MHz to a 50 $\Omega$ source.

**2.4** A rectangular waveform of peak-to-peak voltage 5 V is applied to an $RC$ high-pass filter that has a time constant of 1 ms. Determine the output waveform if the periodic time of the input waveform is (a) 10 ms and (b) 0.1 ms, and the mark–space ratio is $2:1$.

**2.5** The phase advance circuit shown in Fig. 2.9 is to give a phase shift of 17.8° at a frequency $f$. Calculate this frequency if $R_1 = R_2 = 36 \, k\Omega$, and $C_1 = 0.1 \, \mu F$.

**2.6** A $CR$ low-pass filter is to have a 3 dB point at 2000 Hz. Determine suitable values for its components. At what frequency will the loss of the filter be (a) 10 dB, and (b) 20 dB?

**2.7** Explain the principle of operation of a switched capacitor filter. When and why are they used? At what frequency must a switched capacitor filter with a capacitance of 100 pF switch to simulate a resistance of 1 k$\Omega$?

**2.8** Design a $\pi$ attenuator to have a voltage loss of 10 dB when connected between a source of 600 $\Omega$ and a 150 $\Omega$ load.

## Exercises 3

**3.1** Why is the platinum resistance thermometer a common method of measuring temperature?

In the circuit of Fig. E.2 the thermometer is represented

by resistor $R_T$. What is the necessary relationship between $R_1$, $R_2$, $R_3$ and $R_4$ for op-amps A and B to supply a constant current to the thermometer?

Design the circuit if the thermometer has a resistance of 120 $\Omega$ at 0 °C, 168 $\Omega$ at 100 °C and 196 $\Omega$ at 200 °C. The digital voltmeter is to indicate 0 V for a temperature of 0 °C and 1 V for a temperature of 100 °C.

**3.2** Draw the circuit of an op-amp voltage-to-current converter. Derive an expression for the ratio (load current)/(input voltage) and obtain the necessary relationship between the resistors in the circuit for the output current to be independent of the load impedance.

**3.3** Explain the basic operation of a Schmitt trigger circuit and give some typical uses for the circuit.

Design an inverting op-amp Schmitt trigger circuit that has output voltages suitable for interfacing with TTL logic, and input threshold voltages of 0.8 and 1.8 V.

A sinusoidal input voltage that varies between 0 and +3 V is applied to the circuit. Determine the duty factor of the output waveform.

**3.4** A square wave generator consists of an op-amp integrator followed by an op-amp Schmitt trigger circuit. Design the circuit to operate at 1 kHz with an input voltage of 1 V.

Which parameters of the op-amps are the most likely to cause errors?

**3.5** Design a sawtooth oscillator to have an output voltage of 12 V peak at a frequency of 250 Hz.

**3.6** Explain the purpose of a pulse expander. A pulse expander consists of a 555 timer connected as an astable multivibrator whose output feeds into one input of a two-input NAND gate. The output of the gate is applied to the clock input of a CMOS up-down counter. The counter is used as a binary counter and its carry output provides both the expanded pulse output and the other input to the NAND gate.

(a) State to which point in the circuit the input pulse to be expanded should be applied. (b) Explain the operation of the circuit.

**3.7** The circuit shown in Fig. E.3 oscillates at a frequency

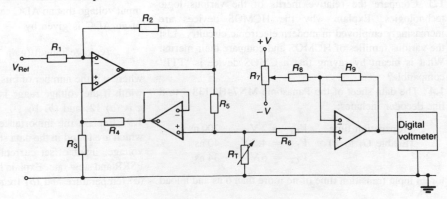

**Fig. E.2**

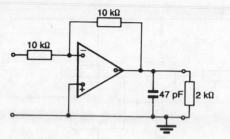

**Fig. E.3**

in the region of 5 MHz. The gain—frequency characteristic of the op-amp is shown in Fig. 1.6(a). Explain why. How can the circuit be made stable? If the input capacitance of the op-amp is 6 pF calculate the value of the necessary compensation capacitor.

How would the compensation be applied if the op-amp were connected in its non-inverting configuration?

**3.8** Figure E.4 shows the circuit of a thermometer that uses an $R^T$D as the temperature sensor. The dual op-amp used has the following parameters: $V_{OS} = 25\,\mu V$, $I_B = 2\,nA$, open-loop gain = 112 dB, CMRR = 114 dB, PSRR = 110 dB, gain band product = 500 kHz and slew rate = 0.15 V/$\mu s$.

Obtain the component values for the circuit to produce an output of 0 V at 0 °C and 10 V at 100 °C. The RTD has a resistance of 1000 Ω at 0 °C.

**3.9** Draw the circuit of a quadrature oscillator. Determine component values for the circuit to oscillate at 1020 Hz. Let all resistors have equal values and all capacitors have the same value.

**3.10** Show that if an op-amp summer amplifier has $n$ inputs each with a resistor of $R$ ohms and a single feedback resistor whose value is $R/n$ ohms, the output voltage will be equal to the average of the input voltages.

Design a three-input inverting averaging amplifier using a 741 op-amp.

### Exercises 4

**4.1** Describe the principle of operation of a successive approximation type of ADC.

Show how such a device can be interfaced to a microprocessor to form a data logging system.

**4.2** Draw the block diagram of a data acquisition system that stores data obtained from a transducer. If the transducer's output voltage has a maximum amplitude of 0.25 V, and a maximum frequency of 10 kHz, estimate the time for which data may be continuously collected if the microprocessor has 512 bytes of directly addressable memory. Assume a 12-bit ADC is employed.

**4.3** Draw the circuits of (a) a NAND gate and (b) a NOR gate in CMOS logic. Show how these circuits can be interfaced with TTL logic circuitry when the supply voltage is (i) 5 V and (ii) 9 V.

**4.4** Figure E.5 shows the circuit of a missing pulse detector. (a) Explain its operation. (b) Design the circuit.

**4.5** A transducer is connected to a DAS. (a) Why should the transducer signals be amplified before being processed? (b) To what extent should the signals be amplified? (c) Why is linearization required? (d) Why is an input filter employed?

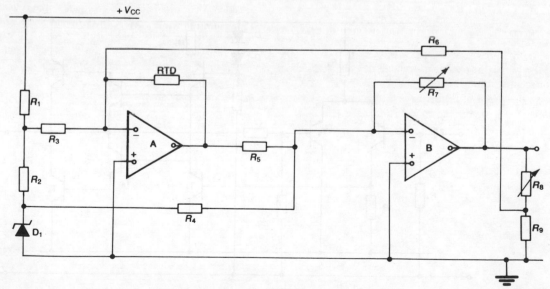

**Fig. E.4**

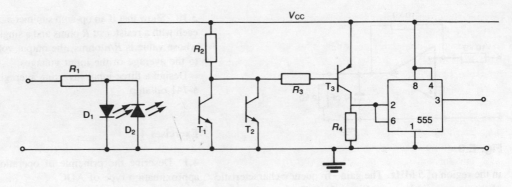

**Fig. E.5**

(e) A DAS uses a multiplexer that samples 10 input channels at 100k samples/s. Calculate the rate at which each channel is sampled.

**4.6** When is it desirable that the input channels of a DAS have different gains? Discuss the errors that may occur in the conversion of an analogue signal into the corresponding digital word. What determines the number of conversions that are made per second? What is the minimum sampling rate allowable? Why is it that temperature transducers do not require a high sampling rate? Why does noise reduce the resolution of a DAS?

**4.7** Which specifications of a DAC determine the quality of the analogue output signal? What differences in performance are required when the DAC is to provide (a) an audio signal output and (b) an output that will control a heater? What is meant by the output resolution of a DAC?

**4.8** A DAS/microprocessor system is used to process data from a transducer that has a maximum output voltage of 0.2 V and a frequency range of 15 Hz to 8 kHz. Describe the main

features of the system if the accuracy is to be about 0.1%.
**4.9** Explain, with the aid of diagrams, how (a) NMOS and (b) CMOS gates may be interfaced with TTL gates when the supply voltage to the NMOS/CMOS gates is (i) 5 V, and (ii) 10 V.

### Exercises 5

**5.1** Figure E.6 shows the basic circuit of an audio power amplifier. The circuit is to supply a maximum power of 40 W to an 8 Ω load, have a voltage gain in the region of 30, and an input resistance of at least 100 kΩ. The transistors $T_1$ through to $T_6$ all have a current gain $h_{fe}$ of 100. The bandwidth of the amplifier is restricted to 5 kHz by a capacitor connected between the base and collector pins of transistor $T_4$. Design the circuit adding any extra components that will improve its performance.
**5.2** The power supply for the amplifier in Exercise 5.1 is to employ a rectifier unit and a single reservoir capacitor.

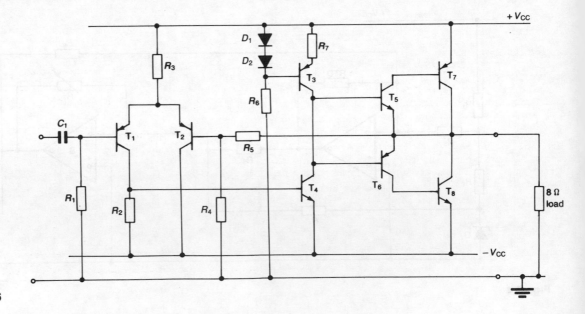

**Fig. E.6**

Design the circuit to have a peak-to-peak ripple voltage of not more than 5 V.

**5.3** Design a complementary pair Class B audio-power amplifier to deliver 64 W to a 8 Ω load. If the output transistors employed have to be able to dissipate up to 14 W what would be the advantage gained by using 20 W rated devices? Use the circuit given in Fig. 5.11.

**5.4** Figure E.7 shows the basic diagram of a audio power amplifier that is to supply 32 W to a 8 Ω load. (a) Estimate the required power supply voltages. (b) Calculate the voltage, current and power ratings of the output transistors. (c) Determine suitable values for the three resistors shown if the full power output is to be obtained for an input voltage of 0.5 V.

**5.5** Obtain the component values given for the a.f. power amplifier shown in Fig. E.8. The amplifier has a power output of 60 W into 8 Ω, an input sensitivity of 1 V and a total harmonic distortion (THD) of less than 0.2%. Leave the short-circuit protection part of the circuit. Some design equations are:

$$R_1 = V_{in}R_f/V_{out}$$

and

$$R_2 = (V_{CC} - V_{D_1})/[I_{bias}(T_1 \text{ and } T_2) + I_{D_1}]$$

**5.6**

(a)   What is the function of the circuit shown in Fig. E.9?

(b)   Explain its operation.

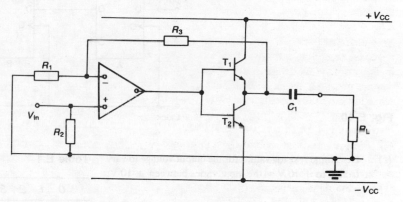

**Fig. E.7**

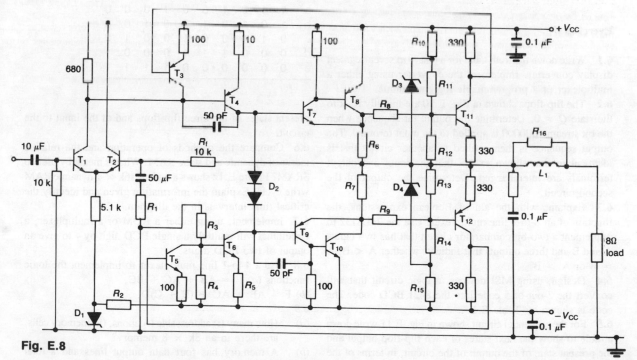

**Fig. E.8**

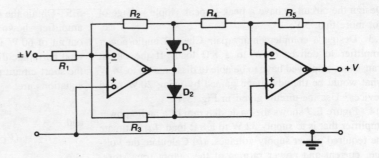

**Fig. E.9**

**Fig. E.10**

(c) Determine resistor values for the output voltage to vary from 0 to + 10 V as the input varies between ± 10 V.

**Table E.1**

|   | 0 | 1 | 2 | 3 | 4 | 5 | 6 | 7 | 8 | 9 |
|---|---|---|---|---|---|---|---|---|---|---|
| a | 1 | 1 | 0 | 1 | 0 | 0 | 1 | 0 | 0 | 0 |
| b | 1 | 0 | 1 | 0 | 1 | 0 | 0 | 1 | 0 | 0 |
| c | 0 | 1 | 1 | 0 | 0 | 1 | 0 | 0 | 1 | 0 |
| d | 0 | 0 | 0 | 1 | 1 | 1 | 0 | 0 | 0 | 1 |
| e | 0 | 0 | 0 | 0 | 0 | 0 | 1 | 1 | 1 | 1 |

**Exercises 6**

**6.1** Write down the truth table for a binary to seven-segment display converter. Implement the converter using either a multiplexer or a programmable logic technique.

**6.2** The flip-flops shown in Fig. E.10 are initially reset to the state $Q = 0$. Determine the output bit sequence when the bit stream 1000000 is applied to the input terminal. The output sequence is then applied to another circuit that is identical to that shown except that the input and output terminals are interchanged. Determine the output of the second circuit.

**6.3** Explain, with the aid of Boolean expressions, the function of an 8-to-1 line multiplexer. Use the 74LS152 to implement a two-bit comparator circuit that has two inputs A and B and three outputs that indicate whether A < B, A = B or A > B.

**6.4** Design, using MSI devices, a logic circuit that will convert the 2-out-of-5 code into the 8421 BCD code. The code is shown by Table E.1.

**6.5** For the sequential circuit shown in Fig. E.11 write down a table to show the next states of each flip-flop output and the present state of the output of the circuit, in terms of the present states of the three flip-flops and of the input to the circuit.

**6.6** Compare the methods of operation and the relative merits of dynamic and static RAM. What is meant by pseudo SRAM? Figure E.12 shows a typical 16k × 1 dynamic RAM write cycle. Explain the information given and identify the critical time intervals in the diagram.

**6.7** Implement, using either a ROM or a multiplexer, a circuit that will multiply a single BCD digit by 4 to give an output of two BCD digits.

**6.8** Use a 4-to-1 line multiplexer to implement the logic functions (a) $F = AB + \bar{A}C + B\bar{C}$,
(b) $F = \bar{A}B + A\bar{C} + (B + C)\bar{D}$.

**6.9**

(a) How many (i) addressable locations, (ii) memory cells, are there in an 8k × 8 memory?

(b) A memory has four data output lines and a total

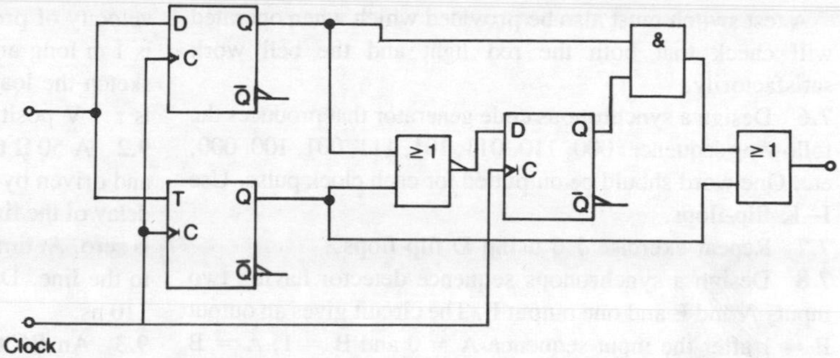
D
C
Q
T
Q
C
&
≥1
D
Q
C
Q
≥1
Clock

Fig. E.11

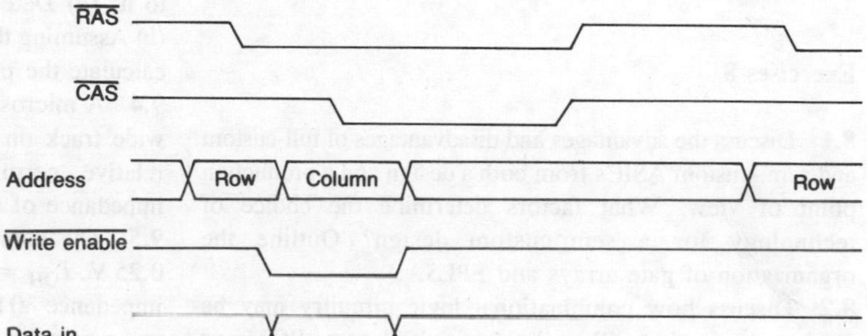
RAS
CAS
Address
Row
Column
Row
Write enable
Data in

Fig. E.12

capacity of 1 Mb. How many address pins will it have?

(*c*) A memory has 10 multiplexed address pins and 4 input/output lines. What is its organization?

**6.10** Design a 48k × 8 memory system that consists of 16k bytes of ROM and 32k bytes of RAM.

**6.11**

(*a*) Discuss the advantages to be gained by using a NOVRAM device instead of an EEPROM.

(*b*) Both EEPROMs and battery backed CMOS RAM are forms of non-volatile memory. Discuss their relative merits.

## Exercises 7

**7.1** Design a circuit using flip-flops and gates that follows the sequence listed in Table E.2 to generate a three-phase output voltage.

**Table E.2**

| | *0* | *1* | *2* | *3* | *4* | *5* |
|---|---|---|---|---|---|---|
| A | 0 | 1 | 1 | 1 | 0 | 0 |
| B | 0 | 0 | 1 | 1 | 1 | 0 |
| C | 0 | 0 | 0 | 1 | 1 | 1 |

**7.2** Design a synchronous logic circuit that will detect the sequence 1101 in an input serial data stream. Detection of the sequence is to be signalled by the output going HIGH. The output should then stay HIGH until a 0 bit is received.

**7.3** Discuss the differences and similarities between synchronous and non-synchronous logic design. Draw the state diagram for a digital circuit that has two inputs A and B and one output F. The output F goes high whenever input A is held high and input B follows the sequence 101. The output then stays high until input A goes low whatever the value of B.

**7.4** A synchronous sequential circuit has a single input x and a single output F and uses four D flip-flops. The inputs to the flip-flops are:

$$D_A = A(BC + \bar{C}\bar{x}) + B(\bar{A}x + AC)$$
$$D_B = \bar{A}(Bx + \bar{C})$$
$$D_C = \bar{A}\bar{C}(A + x) + A\bar{B}\bar{C}(x + \bar{C}) + C(A\bar{x} + Bx)$$
$$D_D = B(A\bar{B}x + \bar{A}B\bar{x}).$$

Deduce the state table and the state diagram for the circuit.

**7.5** Design a circuit that will ring a bell and light a red alarm light whenever a fault occurs in a piece of equipment. Correct operation should be indicated by a green light. A switch should be provided that when operated will turn off the bell but not the red light while the fault is being put right. A reset button should be provided to reset the circuit once the fault has been cleared.

A test switch must also be provided which when operated will check that both the red light and the bell work satisfactorily.

**7.6** Design a synchronous code generator that produces the following sequence: 000, 110, 011, 101, 111, 001, 100, 000, etc. One word should be outputted for each clock pulse. Use J–K flip-flops.

**7.7** Repeat exercise 7.6 using D flip-flops.

**7.8** Design a synchronous sequence detector having two inputs A and B and one output F. The circuit gives an output F = 1 after the input sequence A = 0 and B = 1, A = B = 1, A = 1 and B = 0. Design the circuit using D flip-flops.

## Exercises 8

**8.1** Discuss the advantages and disadvantages of full-custom and semi-custom ASICs from both a design and a production point of view. What factors determine the choice of technology for a semi-custom design? Outline the organization of gate arrays and FPLS.

**8.2** Discuss how combinational logic circuitry may be implemented using SSI packages, multiplexers, ROMs or PLDs. What are the relative merits of PLAs and PALs?

**8.3** Program the TICPAL 16R4 to have the same logic function as the 153 dual 4-to-1 data selector.

**8.4** Programme the TICPAL 16R4 to have the same logic function as the 485 4-bit magnitude comparator.

**8.5** Design a PLA to implement a circuit that will (a) convert BCD numbers into XS3 form, and (b) convert XS3 numbers into BCD form.

Use a PLA that has at least eight inputs and eight outputs.

**8.6** Use (a) a PLA, and (b) a PAL to implement the equations

$$F_0 = A\bar{B}\bar{C} + \bar{A}B\bar{C} + \bar{A}\bar{B}C + A\bar{B}C$$

and

$$F_1 = ABCD + A\bar{B}\bar{C}D + A\bar{C}\bar{D} + B\bar{D}.$$

**8.7** Show how an active high version of the TICPAL 16L8 could be used to provide (a) a two-input AND gate, and (b) a two-input NOR gate.

**8.8** Compare the relative merits of the programmable devices (a) PAL, (b) PLA, (c) GAL, (d) FPGA, (e) ELPD and (f) PEEL. A PLD is quoted by the manufacturer as having the dimensions 6 × 8 × 3. How many inputs and outputs has it?

## Exercises 9

**9.1** The output of an ECL circuit is connected by a transmission line of characteristic impedance 50 Ω and velocity of propagation $2 \times 10^8$ m/s to its load. If the line is 1 m long and the load impedance is 100 Ω calculate and sketch the load voltage when the signal applied to the line is a 1 V positive step.

**9.2** A 50 Ω transmission line is open-circuited at one end and driven by a digital circuit at the other. The propagation delay of the line is 100 ns. In the steady state the line voltage is zero. At time $t = 0$ a voltage transition of + 2 V is applied to the line. Determine the voltage at the open-circuit after 110 ns.

**9.3** An $RC$ termination for a digital line of characteristic impedance 80 Ω has a clock signal at 12 MHz applied to it. (a) Determine a suitable value for the capacitance. (b) Assuming the receiving device is in the TTL logic family calculate the power dissipated in the termination.

**9.4** A microstrip line is formed by a 1 mm thick, 0.5 cm wide track on a PCB board that is 0.75 cm thick and of relative permittivity 4.9. Calculate the characteristic impedance of the line.

**9.5** The connection between two gates having $V_{OH} = 0.25$ V, $I_{OH} = 0.45$ mA is made by a line of characteristic impedance 80 Ω. The steady-state voltage on the line is 2 V and a negative 2 V transition is applied to the sending end. Draw the lattice diagram of the system.

**9.6** Explain, with the aid of sketches, the various ways in which a digital line may be terminated. List, and discuss, the relative merits of the methods shown.

**9.7** A train of rectangular pulses has an amplitude of 1.5 V, a pulse width of 100 $\mu$s and a pulse repetition frequency of 1 kHz. Draw the spectrum diagram of the waveform. What is the approximate bandwidth needed to transmit this signal and retain a more or less rectangular waveshape?

## Exercises 10

**10.1** Figure E.13 shows a digital pulse length measurement system that is used to measure the pulse widths to within ± 5 $\mu$s using a 200 kHz clock. Design the logic for the system. The discrimination unit is to provide clock pulses of full length for the duration of the pulse input, to provide a transfer pulse after the pulse input has ended, and to provide a reset pulse after the transfer pulse, as shown. Use MSI/LSI devices.

**10.2** The sketch of a hall with four doors is shown in Fig. E.14(a). People are allowed to enter or leave the hall at any of the doors but the doors are only wide enough to allow one person to pass at a time. Each time a person goes through a door a light beam is broken, generating the pulses shown in Fig. E.14(b). Assuming that the minimum time separation between two people at any door is 500 ms design the logic that will maintain a count of the number of people in the hall.

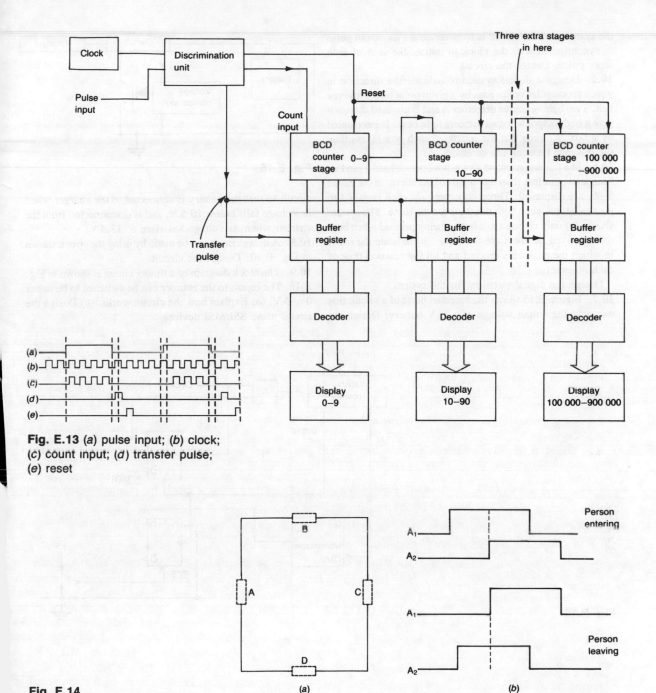

**Fig. E.13** (a) pulse input; (b) clock;
(c) count input; (d) transfer pulse;
(e) reset

**Fig. E.14**

(a)

(b)

**10.3** Four spring-loaded push-buttons A, B, C and D
operate four similarly labelled lights. Initially light A only
is ON. It is required that the lights are operated in the
sequence A, B, C, D, C, B, A, etc. with only one light ON
at any time. Whenever a particular light is lit it should not
be possible to turn ON any light other than the next one in
the given sequence. Design the digital circuitry that will meet
this requirement.

**10.4** A digital system uses 1 MHz clock pulses. Non-
synchronous input pulses of 1 ms duration are inputted to

the system and the circuit is to generate a 1 μs output pulse in synchronism with the clock to initiate the start of each input pulse. Design the circuit.

**10.5** Design a digital system to indicate the direction in which rectangular boxes pass by a given point on a conveyor belt. Two light-sensitive detectors A and B are used that each give a high output when an incident light beam is interrupted by a box. Assume that the length of each box is less than the distance between the detectors.

**10.6** In a training exercise a switch is operated and a random time later, but less than 3 s, a pair of numbers, in the range 1–6, are displayed. Three observers are each required to press a button when both numbers are the same. The system should (a) only respond to a button being pressed when both the displayed numbers are the same, (b) indicate the order in which the observers reacted and (c) the reaction time of each observer.

Design the digital circuitry for the system.

**10.7** Figure E.15 shows the essential parts of a circuit that monitors the output voltage of a 12 V battery. Design the

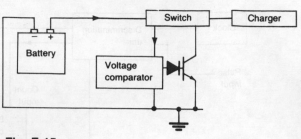

**Fig. E.15**

circuit so that the battery is connected to the charger when its voltage falls below 10.5 V, and is disconnected from the charger when its voltage has risen to 13.5 V.

**10.8** A delay circuit can be made by using the circuit shown in Fig. E.10. Design the circuit.

**10.9** The block diagram of a timing circuit is shown in Fig. E.16. The inputs to the counter can be switched to be either 0 or 5 V. (a) Explain how the circuit works. (b) Design the circuit using SSI/MSI devices.

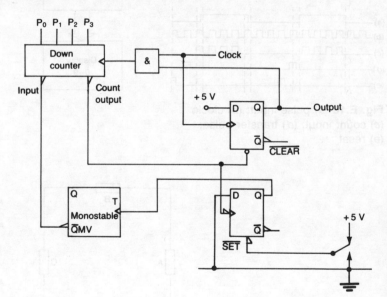

**Fig. E.16**

# Answers to Numerical Exercises

**1.1** Within ± + 0.5 LSB

**1.2** $R_1 = 10\,M\Omega$, $R_2 = 1\,k\Omega$, $R_3 = 10\,k\Omega$

**1.4** 80 ns

**1.5** 573, 4.17 MHz; 192/3, 32 MHz; 150/1/2, 13.16 MHz; 138, 12.2 MHz; 283, 23.8 MHz

**1.6** (a) 4.17 MHz, (b) 10 pF (max), (c) 375 mW, (e) 3.15 V, 0.9 V

**1.7** 795.8 kHz, 2.12 MHz, 6.37 MHz

**1.8** 65536, (a) 2.44 $\mu$V, (b) 153 nV

**2.3** $C_1 = 231\,pF$, $C_2 = 31.8\,pF$, $L_1 = 497\,nH$

**2.4** (a) −5.52 V, 4.48 V, 2.3 V, −7.7 V.
(b) −6.7 V, 3.3 V, 3 V, −7 V

**2.5** 9.75 Hz

**2.6** $C = 0.1\,\mu F$, $R = 787\,\Omega$, 6 kHz, 19.9 kHz

**2.7** 10 MHz

**2.8** $R_1 = 1488\,\Omega$, $R_2 = 925\,\Omega$, $R_3 = 169\,\Omega$

**3.1** $R_1/R_2 = R_3/R_4$

**3.3** 0.31

**3.6** Via a diode and resistor to discharge pin of the 555

**3.7** > 6 pF

**4.2** 10.8 ms

**4.5** 10k samples/s per channel

**5.4** $V_{CC} = 50\,V$, $I_{C(max)} = 3\,A$, $P_{D(max)} = 12\,W$, $R_1 = R_2 = 1\,k\Omega$, $R_3 = 91\,k\Omega$

**5.6** All resistors 10 kΩ

**6.9** (a) (i) 8192, (ii) 65536, (b) 9, (c) 1M × 4

**9.2** 4 V

**9.3** 174 pF, 45 mW

**9.4** 52.65 Ω

**9.7** 10 kHz

# Index